TURKISH GRAMMAR

G. L. LEWIS

*Fellow of the British Academy,
Emeritus Professor of Turkish,
and Emeritus Fellow of
St. Antony's College
in the University of Oxford*

OXFORD · NEW YORK
OXFORD UNIVERSITY PRESS

Oxford University Press, Walton Street, Oxford OX2 6DP
Oxford New York
Athens Auckland Bangkok Bombay
Calcutta Cape Town Dar es Salaam Delhi
Florence Hong Kong Istanbul Karachi
Kuala Lumpur Madras Madrid Melbourne
Mexico City Nairobi Paris Singapore
Taipei Tokyo Toronto
and associated companies in
Berlin Ibadan

Oxford is a trade mark of Oxford University Press

Published in the United States by
Oxford University Press Inc., New York

© Oxford University Press 1967

All rights reserved. No part of this publication may be reproduced,
stored in a retrieval system, or transmitted, in any form or by any means,
without the prior permission in writing of Oxford University Press.
Within the UK, exceptions are allowed in respect of any fair dealing for the
purpose of research or private study, or criticism or review, as permitted
under the Copyright, Designs and Patents Act, 1988, or in the case of
reprographic reproduction in accordance with the terms of the licences
issued by the Copyright Licensing Agency. Enquiries concerning
reproduction outside these terms and in other countries should be
sent to the Rights Department, Oxford University Press,
at the address above

This book is sold subject to the condition that it shall not, by way
of trade or otherwise, be lent, re-sold, hired out or otherwise circulated
without the publisher's prior consent in any form of binding or cover
other than that in which it is published and without a similar condition
including this condition being imposed on the subsequent purchaser

British Library Cataloguing in Publication Data
Data available

Library of Congress Cataloging in Publication Data
Lewis, Geoffrey L.
Turkish grammar.
English and Turkish.
Bibliography: p.
Includes index.
1. Turkish language—Grammar. I. Title.
PL123.L4 1985   494'.355   85-15517
ISBN 0-19-815838-6 (Pbk)

7 9 10 8 6

Printed in Hong Kong

TO MY WIFE

CONTENTS

INTRODUCTION	*Page*	xix
BIBLIOGRAPHY		xxiii
ABBREVIATIONS		xxiv
I. ORTHOGRAPHY AND PHONOLOGY		1
1. The alphabet		1
2. The apostrophe		2
3. The circumflex accent		2
4. Consonants: general observations		2
5. b, p		3
6. ç		3
7. d, t, n		3
8. f, v		3
9. ģ, k		3
10. ğ		4
11. h		5
12. l		6
13. r		7
14. y		7
15. The glottal stop		7
16. Doubled consonants		8
17. Consonant-clusters and epenthetic vowels		9
18. Foreign diphthongs		10
19. Alternation of consonants		10
20. Consonant assimilation in suffixes		12
21. Vowels: general observations		12
22. a		13
23. ı		13
24. o		13
25. u		14
26. e		14

CONTENTS

27. i	14
28. ö	14
29. ü	14
30. Vowel length	14
31. Vowel harmony	15
32. Exceptions to the rules of vowel harmony	17
33. Vowel harmony in foreign borrowings	17
34. Vowel harmony of suffixes	17
35. Vowel harmony of suffixes with foreign borrowings	19
36. Alternation of vowels	20
37. Accentuation: general observations	20
38. Word-accent	21
39. Exceptions	21
40. Enclitic suffixes	23
41. Enclitic words	24
42. Group-accent	24
43. Intonation	24

II. THE NOUN — 25

1. Gender	25
2. Number: the Turkish plural	25
3. Arabic plurals	26
4. Other plurals	28
5. The Arabic dual	28
6. The cases	28
7. Summary of case-endings	34
8. Uses of the cases	35
9. The absolute form	35
10. The accusative case	35
11. The genitive case	36
12. The dative case	36
13. The locative case	37
14. The ablative case	37
15. Personal suffixes	38
16. Personal suffixes followed by case-suffixes	40
17. The izafet group	41
18. Words indicating nationality	44

CONTENTS

19. The izafet chain	45
20. Place-names consisting in an izafet group	47
21. Culinary terms without izafet	48
22. Third-person suffix with substantivizing and defining force	48
23. The Janus construction	48
24. Suffixes with izafet groups	49
25. The vocative use of the third-person suffix	50
26. Persian izafet	50

III. THE ADJECTIVE — 53
1. General observations — 53
2. Attributive adjectives — 53
3. The indefinite article — 53
4. Comparison of adjectives — 54
5. Arabic and Persian comparatives — 55
6. Intensive adjectives — 55

IV. NOUN AND ADJECTIVE SUFFIXES — 57
1. Diminutives — 57
2. Diminutives of personal names — 58
3. -(i)msi, -(i)mtrak, -si — 58
4. -ci — 59
5. -li — 60
6. ... -li ... -li — 61
7. -siz — 62
8. -lik — 62
9. -daş — 64
10. -gil — 65
11. -(s)el — 65
12. -varî — 66
13. -cil — 66
14. -hane — 66
15. -ane — 66

V. PRONOUNS — 67
1. Personal pronouns — 67
2. Uses of the personal pronouns — 68
3. -ki — 69

CONTENTS

    4. kendi   70
    5. Demonstratives   71
    6. Interrogatives   72
    7. Indefinite, determinative, and negative   74
    8. bazı, kimi   74
    9. birtakım   75
    10. her   75
    11. hep   75
    12. çok   75
    13. az   75
    14. birkaç   76
    15. bütün   76
    16. başka, diğer   76
    17. öbür   76
    18. birbir   76
    19. aynı   76
    20. şey   77
    21. falan   77
    22. insan   77
    23. hiç   77
    24. kimse   78

VI. NUMERALS   79

    1. Cardinals   79
    2. Classifiers   80
    3. Fractions   81
    4. Ordinals   82
    5. Distributives   83
    6. Collectives   84
    7. Dice numbers   84

VII. POSTPOSITIONS   85

    1. General observations   85
    2. Primary postpositions with absolute case   85
    3. Primary postpositions with absolute or genitive case   85
    4. Primary postpositions with dative case   87
    5. Primary postpositions with ablative case   89

6.	Secondary postpositions: I	89
7.	Secondary postpositions: II	92
8.	Secondary postpositions: III	94
9.	**leh, aleyh**	94
10.	The preposition **ilâ**	95

VIII. THE VERB — 96

1.	The stem	96
2.	The verb 'to be'	96
3.	The present tense of 'to be'	96
4.	Uses of **-dir**	97
5.	Examples of the present tense of 'to be'	98
6.	Forms based on **i-**	99
7.	The past tense of 'to be'	99
8.	The present conditional of 'to be'	99
9.	The past conditional of 'to be'	100
10.	The inferential	101
11.	The inferential conditional	102
12.	The negative of 'to be'	103
13.	Interrogative	105
14.	Negative-interrogative	106
15.	The regular verb	106
16.	Present I	108
17.	Uses of the present	109
18.	Paradigms of the present	109
19.	Present II	111
20.	Future I	112
21.	Uses of the future	112
22.	Paradigms of the future	113
23.	Future II	115
24.	Aorist	115
25.	Uses of the aorist	117
26.	Paradigms of the aorist	118
27.	**miş**-past	122
28.	Pluperfect	123
29.	Other paradigms of the **miş**-past	123
30.	Necessitative	125

CONTENTS

31. **di**-past	127
32. Uses of the **di**-past	128
33. Other paradigms of the **di**-past	128
34. Conditional	130
35. Subjunctive	132
36. Uses of the subjunctive	133
37. Other paradigms of the subjunctive	134
38. Synopsis of the verb	135
39. Imperative	137
40. **-sindi**	138
41. **-sin için**	139
42. **-dir** suffixed to finite verbs	139
43. **-dir** with a following verb	141
44. Summary of the forms of 'to be'	141
45. **var, yok**	142
46. Extended stems	143
47. The reciprocal or co-operative verb	143
48. The causative verb	144
49. Doubly causative verbs	146
50. Syntax of the causative	146
51. The repetitive verb	148
52. The reflexive verb	149
53. The passive verb	149
54. Uses of the passive	150
55. The potential verb	151
56. The order of extensions	152
57. Auxiliary verbs	154
IX. PARTICIPLES	**158**
1. Present	158
2. Future I	158
3. Future II	161
4. Aorist	161
5. **miş**-past	162
6. **di**-past	162
7. The personal participles	163
8. **-eceği gel-**	165

CONTENTS xiii

9. -esi gel- 165
10. -eceği tut- 166
11. Periphrastic tenses and moods 166

X. VERBAL NOUNS 167
1. Introductory 167
2. -mek 167
3. The infinitive with subject 169
4. -mekli 169
5. Common nouns in -mek 169
6. -meklik 170
7. -me 170
8. Common nouns in -me 171
9. -meli 172
10. -masyon 172
11. -iş 172
12. -mezlik, -memezlik 173

XI. GERUNDS 174
1. Introductory 174
2. -e 174
3. -erek 176
4. olarak 177
5. -ip 177
6. -ince 179
7. -inceye kadar, -inceyedek, -inceye değin 180
8. -ene kadar, -enedek, -ene değin 180
9. -esiye 180
10. -eli, -eli beri, -eliden beri, -dim -eli 181
11. oldum olası(ya) 181
12. -meden, -mezden 181
13. -r -mez 182
14. -dikçe 183
15. -diğince 183
16. -dikte 183
17. -dikten sonra 183
18. -dikten başka 184

CONTENTS

19. -diğinden başka	184
20. Gerund-equivalents	184
21. -diği müddetçe	185
22. -diği halde	186
23. -diği takdirde	186
24. -diği için, -diğinden	186
25. -diği nispette	186
26. -diği kadar	186
27. -diği gibi	187
28. -eceğine, -ecek yerde	187
29. -mekle	187
30. -mektense, -mekten ise, -medense	188
31. -meksizin	188
32. Equivalents of 'as if'	188
33. -mecesine	189
34. iken	190
35. Compound verbs	191

XII. ADVERBS — 193

1. General observations	193
2. -ce	194
3. Nouns used adverbially	195
4. Foreign adverbs	196
5. Comparison of adverbs	196
6. bir	197
7. bir türlü	197
8. . . . bile, hattâ . . .	197
9. âdeta	198
10. Adverbs of place	198
11. aşırı	199
12. -re	199
13. Adverbs of time	200
14. Telling the time	202
15. ertesi	202
16. evvelsi, evvelki	202
17. evvel, sonra	203
18. şimdi	203

19.	artık, bundan böyle, gayrı	203
20.	daha	203
21.	hemen	204
22.	gene, yine	204
23.	The verb 'to be' in temporal expressions	204
24.	derken	204

XIII. CONJUNCTIONS AND PARTICLES — 206

1.	ve	206
2.	de	206
3.	ne . . . ne (de) . . .	207
4.	gerek . . . gerek(se) . . .	209
5.	hem . . . hem (de) . . .	209
6.	ha . . . ha . . .	209
7.	ister . . . ister . . .	209
8.	. . . olsun . . . olsun	209
9.	ya . . . ya . . . veya . . ., yahut, veyahut	210
10.	ama, fakat, lâkin, ne var ki	210
11.	ancak, yalnız	211
12.	mamafih, bununla beraber, bununla birlikte	211
13.	madem(ki), değil mi (ki)	211
14.	meğer(se)	211
15.	ki	211
16.	meğerki	214
17.	nitekim, netekim	214
18.	halbuki, oysa(ki)	215
19.	çünkü, zira	215
20.	demek	215
21.	diğer taraftan, öte yandan	216
22.	gerçi	216
23.	gûya	216
24.	hani (ya)	216
25.	hele	217
26.	herhalde	217
27.	ise	217
28.	işte	217
29.	sakın	218

CONTENTS

30. sanki — 218
31. şöyle dursun — 218
32. ya — 218
33. yok — 219
34. yoksa — 219

XIV. WORD-FORMATION — 220

1. Deverbal substantives — 220
2. -ici — 220
3. -men — 220
4. -ik — 221
5. -i — 221
6. -ti, -inti — 222
7. -gi — 222
8. -ç — 222
9. -ek, -k — 223
10. -gen, -eğen — 223
11. -gin — 223
12. -it, -t — 224
13. -im, -m — 224
14. -in — 225
15. -geç, -giç — 225
16. -tay — 226
17. -ev, -v — 226
18. -ey, -y — 226
19. Denominal verbs — 227
20. -e- — 227
21. -le- — 227
22. -len- — 228
23. -let- — 228
24. -leş- — 228
25. -el-, -l- — 229
26. -er- — 229
27. -se- — 230
28. -imse- — 230
29. -de- — 231
30. Compound nouns and adjectives — 231

CONTENTS   xvii

 31. Abbreviated nouns   232
 32. Izafet groups'   232
 33. Frozen izafet groups   233
 34. Proper names consisting in izafet groups   233
 35. Adjective+noun   233
 36. Noun+noun+-li   234
 37. Adjective+noun+-li   234
 38. Noun+adjective   234
 39. Noun+third-person suffix+adjective   234
 40. Noun+verb   234
 41. Onomatopoeic word+verb   235
 42. Verb+verb   235
 43. Hyphenated compounds   235
 44. Repetitions   235
 45. Doublets   236
 46. **m**-doublets   237

XV. THE ORDER OF ELEMENTS IN THE SENTENCE   239
 1. Nominal sentences and verbal sentences   239
 2. The principles of word-order   239
 3. The inverted sentence; **devrik cümle**   241
 4. The sentence-plus   244

XVI. NUMBER, CASE, AND APPOSITION   246
 1. Concordance of subject and verb   246
 2. Singular and plural in izafet groups   247
 3. Idiomatic uses of the plural   247
 4. The accusative with **bir**   248
 5. Two idiomatic uses of the dative case   248
 6. The genitive as logical subject   250
 7. Apposition   252

XVII. THE NOUN CLAUSE AND THE SUBSTANTIVAL SENTENCE   254
 1. The verbal noun in **-me** and the personal participles 254
 2. The substantival sentence   256
 3. The substantival sentence as adjectival qualifier   256

CONTENTS

    4. The substantival sentence as qualifier in izafet — 257
    5. The sentence with case-endings — 258

XVIII. ADJECTIVAL PHRASES AND PARTICIPIAL QUALIFIERS — 259
    1. The başıbozuk construction — 259
    2. Translation of English relative clauses — 260
    3. Two variant types of participial qualifier — 262

XIX. THE SUBJUNCTIVE AND Kİ — 264
    1. Clauses of purpose — 264
    2. The subjunctive after a negative main verb — 264
    3. The subjunctive in noun clauses — 265

XX. CONDITIONAL SENTENCES — 267
    1. Open conditions — 267
    2. Alternatives to the conditional verb — 267
    3. Remote and unfulfilled conditions — 268
    4. Apodosis to an unexpressed protasis — 268
    5. Alternative protases — 269
    6. Concessive clauses — 269
    7. 'Whatever, whenever, whoever, wherever' — 269
    8. eğer, şayet — 270
    9. Conditional sense of the future personal participle — 270
    10. The conditional base — 271
    11. -sene, -senize — 271
    12. -se beğenirsin? — 272
    13. olsa olsa — 272
    14. olsa gerek — 273

XXI. ASYNDETIC SUBORDINATION — 274

XXII. PUNCTUATION — 276

XXIII. SENTENCE-ANALYSIS — 278

XXIV. FURTHER EXAMPLES — 283

INDEX — 291

INTRODUCTION

The subject of this book is the official and literary language of the Republic of Turkey. Until 1922 the language of Turkey was known as **Osmanlıca** or Ottoman Turkish. The downfall of the Ottoman dynasty in that year, after six centuries of sovereignty, made it necessary to find a new name to distinguish this language from all the other members of the same linguistic family, whose dominions extend from the Mediterranean to China. The Turks themselves distinguish it informally as **Türkçemiz**, 'our Turkish', formally as **Türkiye Türkçesi**, 'Turkey-Turkish'; this latter locution is followed by French and German scholars: 'Turc de Turquie', 'Türkei-Türkisch'. The growing practice among English-speaking scholars, which is followed in this book, is to refer to the language of Turkey simply as 'Turkish', while calling the linguistic family to which it belongs 'Turkic'.

Turkish is a member of the south-western or Oghuz group of the Turkic languages, the other members being: the Turkic dialects of the Balkans; Azeri or Azerbaijani, spoken in northwest Persia and Soviet Azerbaijan; the Qashqai of south Persia; the Turkmen or Turcoman of Soviet Turkmenistan.

The problem of the classification of the Turkic languages and dialects is a complicated one. The migrations of the Turkic peoples in the course of history, and their consequent intermingling with one another and with peoples of non-Turkic speech, have created a linguistic situation of vast complexity, which has not yet been investigated sufficiently to permit the last word to be said. An indication of the provisional nature of the various solutions so far offered is that the editors of the first volume of *Philologiae Turcicae Fundamenta* (Wiesbaden, 1959) invited two scholars to write independent contributions to the chapter on classification. Johannes Benzing's scheme shows five main divisions comprising eight groups, while Karl Menges distinguishes six divisions comprising twelve groups. The map at the end of the volume shows six groups.

It is still being debated whether or not the Turkic family is itself a branch of a larger 'Altaic' family, including Mongol,

Tunguz, and possibly Korean. The nineteenth-century concept of a 'Ural–Altaic' family, embracing Finnish and Hungarian as well as the 'Altaic' languages, no longer commands support. It was based chiefly on the fact that these languages share three features: agglutination, vowel harmony, and lack of grammatical gender.

An introductory word must be said about agglutination, as it is this feature which English-speakers find most alien, although it does occur in English to a limited extent in such a word as *carelessness*. But in Turkish the process of adding suffix to suffix can result in huge words which may be the equivalent of a whole English phrase, clause, or sentence: **sokaktakiler**, 'the people in the street'; **gelirlerken**, 'while they are coming'; **avrupalılaştırılamıyabilenlerdenmişsiniz**, 'You seem to be one of those who may be incapable of being Europeanized'. Our English sentences are like drystone walls, with one chunk of meaning dropped into place after another. The Turk's ideas are laid in place like bricks, each cemented to the next. Unwieldy though we may find his massive **çalıştırılmamalıymış**, we must remember that he finds equally unwieldy our fragmented and monosyllabic 'they say that she ought not to be made to work'.

A brief explanation is also necessary for the references that will be found throughout the book to the language-reform movement.[1] The Turks had begun to convert to Islam and to adopt the Arabo-Persian alphabet from the tenth century onward, in the course of their migration into western Asia. In the eleventh century, when under the leadership of the Seljuk dynasty they overran Persia, Persian became the language of their administration and literary culture. Persian had by this time borrowed a great many words from Arabic. These, together with a host of Persian words, were now at the disposal of educated Turks, who felt free to use any they wished as part of their vocabulary. The bulk of these Arabic and Persian borrowings were never assimilated to Turkish phonetic patterns. More, with the foreign words came foreign grammatical conventions. To offer an English analogy, it was as if we said not 'for obvious reasons' but 'for rationes obviae', or 'what is the conditio of your progenitor reverendus?' instead of 'how's your father?'

This hybrid language became the official language of the Otto-

---

[1] See Uriel Heyd, *Language Reform in Modern Turkey* (Israel Oriental Society, 1954); Geoffrey Lewis, 'The Present State of the Turkish Language', in *Proceedings of the British Academy*, LXXI (1985).

man dynasty, who at the end of the thirteenth century entered upon the inheritance of the Seljuks. It attained an extraordinary degree of flexibility, expressiveness, and grandeur, but it was caviar to the general; the speech of the majority of Turks was dismissed by the speakers of Ottoman as **kaba Türkçe**, 'crude Turkish'.

The rise of journalism in the nineteenth century led to a movement in favour of a simplification and 'purification' of the literary language, but this movement did not become truly effective until the establishment of the Republic. A requirement of the populism which was one of the cardinal principles of Atatürk's Republican People's Party was to reduce and eventually to eliminate the gap between the language of the administration and that of the people. Moreover, Atatürk wanted his people to turn their backs on their Asian past, which is why in 1928 he introduced the Latin alphabet in place of the Arabo-Persian. The Turkish Language Society (**Türk Dil Kurumu**) churned out lists of 'pure Turkish' (**öztürkçe**) replacements for Arabic and Persian words. Some were old words that survived in spoken Turkish; some were obsolete words resurrected, some were borrowed from other Turkic languages, some were deliberate inventions. The Society was a private body, but thanks to Atatürk's patronage it was able to channel its neologisms to the Press, the schools and the general public. It is this aspect of the reform that has attracted most attention, but at least as effective was the official encouragement given to the elimination of non-Turkish grammatical constructions.

Despite certain excesses and absurdities, the success of the movement was such that even its critics found it hard to express themselves without using some of the neologisms, at least if they wanted to appeal to a mass audience. In August 1983, however, a half-century of conservative hatred of the Language Society culminated in its being brought under State control. The era of sustained—though never systematic—tampering with the language appeared to have ended, though it was generally agreed that most of the changes already wrought were irreversible. But danger to the purity of Turkish had long threatened from the opposite quarter, in the form of a steady flow of English and French words. Ask for your bill (**hesap**) in a **restoran**, and the odds are that what the **garson** brings you will be headed **Adisiyon**.

The borrowing of French words began in the nineteenth century, and there seems to be still a preference for the French forms of words common to French and English; thus 'detergent' appears as **deterjan**, and **kilosikl** is far commoner than **kilosaykl**.

Although this book is, in principle, concerned with written Turkish, it will be found to contain a good many references to the colloquial, for two reasons. The first is that the gap between written and spoken language has been considerably narrowed in recent years, so that it is not always possible to draw the line. The second reason is reminiscent of the problem that vexed the Islamic theologians concerning the utterances of Satan that are quoted in the Koran: as Satan's lies are part of Holy Writ, must they not therefore be true? Now if a novel or a newspaper happens to report conversations between speakers of sub-standard Turkish, it can scarcely be argued that the sub-standard is thereby rendered literary; nevertheless the student is entitled to expect some guidance on how to translate it, if he has bought what claims to be a Turkish grammar. For a similar reason, some obsolescent and even obsolete features of the language are discussed, as they may occur in quotations or in Ottoman texts transcribed into the modern alphabet. The aim in fact has been to present every form and construction that the reader may meet in print, including some which certain of my Turkish friends, and even I myself, may dislike.

Although our familiar Latin-based grammatical terms do not exactly fit the facts of Turkish, they have been used as far as possible. The Turkish adjective, for example, although it does not behave in all respects like the English adjective, resembles it closely enough to be permitted to share its name. Occasionally I have used the word 'substantive' to include nouns and adjectives and sometimes pronouns. Any other less familiar terms are explained at their first occurrence.

In translating the examples I have made extensive use of word-for-word renderings; for the peculiarity of the resulting English I must ask the reader's indulgence.

It is a pleasure to record my indebtedness to Professor Fahir İz, Dr. Ercüment Atabay, and Mr. Berent Enç for their patience in answering my questions. A few of the grammatical examples I owe to the late Mr. Philip Lechmere Stallard, who shared my love for this most fascinating language.

BIBLIOGRAPHY

1. *Grammatical works and dictionaries*. The list below is limited to the books and the one article which have been constantly by my side during the actual writing of this grammar. To show the full extent of my indebtedness would involve listing the complete works of their authors, as well as of Ömer Asım Aksoy, Saadet Çağatay, Vecihe Hatiboğlu (*née* Kılıçoğlu), Mecdut Mansuroğlu, Zeynep Korkmaz, and Talât Tekin.

The surnames or initials preceding some of the items of this list are used as abbreviations for them in the present work.

Banguoğlu	Tahsin Banguoğlu, *Türk Grameri* i (TDK, 1959).
Deny	Jean Deny, *Grammaire de la langue turque* (Paris, 1921).
	A. Dilâçar, *Türk Diline Genel Bir Bakış* (TDK, 1964).
Ediskun	Haydar Ediskun, *Yeni Türk Dilbilgisi* (Istanbul, 1963).
Elöve	Ali Ulvi Elöve, *Türk Dili Grameri* (a Turkish translation of Deny with some additional notes; Istanbul, 1941).
	Ahmet Cevat Emre, *Türk Dilbilgisi* (TDK, 1945).
Ergin	Muharrem Ergin, *Türk Dil Bilgisi* (Istanbul, 1962).
Gabain	Annemarie von Gabain, *Alttürkische Grammatik* (Leipzig, 1950).
Mundy	C. S. Mundy, 'Turkish Syntax as a System of Qualification' (article in *BSOAS* (1955), xvii/2, pp. 279–305).
OTD	H. C. Hony and Fahir İz, *A Turkish-English Dictionary*, second edition (Oxford, 1957).
TS	*Türkçe Sözlük*, third edition (TDK, 1959).
YİK	*Yeni İmlâ Kılavuzu* (TDK, 1965; revised second impression, 1966).

2. *Other works*. The examples given to illustrate the rules of grammar have been drawn, over many years, from Turkish publications of all kinds. I have indicated the sources of relatively few, partly because most could have been written by any literate Turk and exhibit no special individuality of style, but partly too, I must confess, because I was not consistent about recording the authors of the extracts with which I filled so many notebooks.

ABBREVIATIONS

(A)	of Arabic origin
abl.	ablative
abs.	absolute
acc.	accusative
BSOAS	*Bulletin of the School of Oriental and African Studies*, University of London
coll.	colloquial
dat.	dative
gen.	genitive
intr.	intransitive
lit.	literally
loc.	locative
OTD	see Bibliography
(P)	of Persian origin
pl.	plural
sing.	singular
TDAYB	*Türk Dili Araştırmaları Yıllığı Belleten*
TDK	Türk Dil Kurumu, Ankara
tr.	transitive
TS	see Bibliography
YİK	see Bibliography
<	derived from
>	becoming
*	marks a postulated form never actually found
'	marks an accented syllable
ŋ	represents the sound of ng as in *singer*, a sound formerly occurring in some Turkish words now written with **n**
[ ]	in literal translations, enclose words not required in normal English
⟨ ⟩	in literal translations, enclose words which must be added in order to make intelligible English
§ 3	A reference of this form is to another section within the chapter in which it occurs
II, 3	A reference of this form is to a section of a chapter other than the one in which it occurs

I

ORTHOGRAPHY AND PHONOLOGY

1. The alphabet

Form		Name	Value
A	a	a	French a in *avoir*
B	b	be	b
C	c	ce	j in *jam*
Ç	ç	çe	ch in *church*
D	d	de	d
E	e	e	French ê in *être*
F	f	fe	f
G	ğ	ğe	g in *gate* or in *angular*
Ğ	ğ	yumuşak ğe	lengthens preceding vowel
H	h	he	h in *have*
I	ı	ı	i in *cousin*
İ	i	i	French i in *si*
J	j	je	French j
K	k	ke	c in *cat* or in *cure*
L	l	le	l in *list* or in *wool*
M	m	me	m
N	n	ne	n
O	o	o	French o in *note*
Ö	ö	ö	German ö
P	p	pe	p
R	r	re	r
S	s	se	s in *sit*
Ş	ş	şe	sh in *shape*
T	t	te	t
U	u	u	u in *put*
Ü	ü	ü	German ü
V	v	ve	v
Y	y	ye	y in *yet*
Z	z	ze	z

Yumuşak ğe ('soft g') cannot begin a word. Note that the capital form of the dotted **i** is also dotted. The letter **k** is often called **ka** instead of **ke**; less often, **h** is called **ha**.

2. *The apostrophe*. In addition to these twenty-nine letters, two orthographic signs are used in the writing of Turkish. The apostrophe ('), known as **kesme işareti**, is used:

(*a*) To mark the glottal stop in Arabic borrowings.

(*b*) To separate proper nouns, or words specially emphasized, from grammatical endings: **Atatürk'ten** 'from Atatürk'; **Ankara'da** 'in Ankara'; **vecizemiz, halka hizmet'tir** 'our slogan is "service to the people"'. It is thus regularly used before the case-suffixes of the third-person pronoun **o** when this is written with a capital letter as a mark of respect, the normal practice when writing of Atatürk and other great men (though not usually of Allah): **O'nun** 'His'; **O'na** 'to Him'.

(*c*) To distinguish between homonyms: **karın** 'stomach', **kar'ın** 'of snow', **karı'n** 'your wife'; **halk oyunu** 'folk-dance', **halk oyu'nu** 'referendum' (acc.). It occurs in some surnames compounded of two words: **O'kan** 'that blood', which might otherwise be read as **ok+an** 'arrow-intellect'; **İş'er** 'work-man', which without the apostrophe could be mistaken for the aorist participle of **işemek** 'to urinate'.

(*d*) To mark the omission of a letter, as in **n'olacak** for **ne olacak** 'what will happen?'

3. The circumflex accent (ˆ), known as **düzeltme işareti**, is used primarily to indicate the palatalizing of a preceding **ğ, k**, or **l** and secondarily to mark a long vowel in Arabic borrowings, especially where ambiguity might otherwise arise: **nar** < Persian *nār* 'pomegranate' but **nâr** < Arabic *nār* 'hell-fire'; **adil** 'justice' but **âdil** 'just'; **tarihi** 'history' (acc.) but **tarihî** 'historical'. This rule is neglected in masculine names ending in the Arabic adjectival suffix *-ī*, because the final vowel is nowadays pronounced short: **Bedri, Rahmi, Ruhi**. The original vowel length and consequently the spelling with the circumflex are retained in pen-names of classical authors: **Nef'î, Fuzulî**.

4. *Consonants: general observations*. Native words do not, as a rule, begin with **c, f, j, l, m, n, r**, or **z**. The only notable exceptions, apart from such onomatopoeic words as **civciv** 'chick' and

lololo 'nonsensical jabber', are the interrogative particle **mi**, and **ne** 'what'? See also XIV, 46. **j** occurs only in foreign words and is often replaced by **c** in popular speech. A vowel is often inserted before **l**, **r**, and **n** when they occur initially in foreign words: **ilimon** for **limon** 'lemon', **irahmet** or **ırahmet** for **rahmet** 'divine mercy, rain', **inefes** for **nefes** 'breath' (especially used of breathing on someone for magical purposes). Some such pronunciations have become part of the written language: **orospu** 'harlot' < Persian *rūspī*, **oruç** 'fasting' < Persian *roza*.

The consonants **b**, **c**, and **ğ** do not occur finally in native words.

5. b, p. The voiced labial is pronounced as in English, but **p** is less heavily aspirated than English p.

6. ç. In rapid speech the first of two adjacent **ç**'s is often heard as **t**: **kaç çocuk** 'how many children?' pronounced as if written **kat çocuk**.

7. d, t, n. In English these letters are pronounced with the tip of the tongue touching the gums above the top teeth. In Turkish they are true dentals, the tongue touching the top teeth.

Modern **n** may represent an older **ŋ**, e.g. in **yeni** 'new' and **sonra** 'after'. The **n** of the latter word is frequently dropped in speech and sometimes in writing.

8. f, v. The pronunciation is lighter than that of the corresponding English consonants, particularly in the case of intervocalic **v**, which is heard as a weak w: **tavuk** 'hen' is pronounced tawuk and popularly misspelt **tauk**. The personal and local name **Mustafa Bey** is generally pronounced Mustābey or Mıstābey.

9. ğ, k. In conjunction with any of the back vowels **a**, **ı**, **o**, and **u**, these are pronounced as in *gate* and *kale* respectively: **kızğın** 'excited', **karğa** 'crow'. With the front vowels **e**, **i**, **ö**, and **ü** they are palatalized like English g in *angular* and c in *cure* respectively: **ğerçek** 'true' pronounced g[y]erchek[y]; **kesik** 'cut' pronounced k[y]esik[y]; **köşk** 'palace' pronounced k[y]öshk[y]. The palatalization of the initial **k** is responsible for the i in the English form of this last example, *kiosk*.

In some Arabic and Persian borrowings, however, **ğ** and **k** are also palatalized in conjunction with **a** and **u**. It is in such cases that the circumflex is used: **gâvur** 'infidel' pronounced g^yawur (hence the **i** in the old-established English spelling *giaour*; cf. *Kiazim*, the usual English transcription of the name **Kâzım**); **mahkûm** 'condemned', **kâbus** 'nightmare', pronounced mahk^yum, k^yābūs. The circumflex in these words is solely to indicate palatalization and has nothing to do with vowel length.

Some inconsistency arises from the fact that **k** may stand for both the Arabo-Persian ك and ق (respectively *k* and *q* in English transliteration). The second of these letters represents a k articulated at the uvula, the nearest English counterpart being the sound of c in *cough*. Further, initial **ğ** may represent Persian *g* or Arabic *gh*. In the combinations **ğa** < *ghā* and **ka** < *qā*, the circumflex cannot be used to show that the vowel is long because it would be taken rather as showing—falsely—that the **ğ** or **k** is palatalized. To avoid ambiguity in such cases, the length of the **a** is shown by writing it twice: **katil** 'murder' but **kaatil** 'murderer'. This device is regularly employed only in **kaatil** but may be met with also in **ğaamız** 'obscure', **ğaasıp** 'usurper', **kaabız** 'astringent', **kaabile** 'midwife', **kaadir** 'mighty', and **kaani** 'convinced', all of which are more usually spelt with a single **a**.

Another complication arises from the fact that it is no more natural for Turks than for English-speakers to pronounce a back consonant with a front vowel; e.g. if one tries to pronounce *king* with the initial consonant of *cough* the resulting sound is as much un-Turkish as it is un-English. Consequently Arabic *qi* is transcribed as **kı**, while *qī* (written *qiy* in Arabic letters) ought to be transcribed as **kıy**. In fact, however, although Arabic *qīmat*- 'value' appears as **kıymet**, in other Arabic borrowings in which *q* is followed by long *i*, such as *ḥaqīqat*- 'truth' and *taḥqīr* 'contempt', the convention is to use dotted **i**: **hakikat, tahkir**. The phonetic spelling **hakıykat**, occasionally seen in the early years of the new alphabet, is no longer in general use.

In most Anatolian dialects initial **k** is pronounced as g, medial and final **k** as the velar fricative kh, the sound heard in German *ach!* So **korkma** 'do not fear' may be heard as gorkhma, **çok** 'much' as chokh. In standard Turkish the **ğ** of **ğaliba** 'presumably' is often pronounced as **k**.

10. ğ. Yumuşak **ğe** is a concession to the traditional spelling of Turkish in the Arabo-Persian alphabet. It represents two separate letters of that alphabet, ك g and غ gh. The latter represents the voiced velar fricative, the gargling sound of the Parisian[1] or Tynesider's r, the 'Northumbrian burr'. Arabic initial *gh* becomes **g**: *ghāzī* 'warrior for the Faith' > **gazi**; *ghāfil* 'heedless' > **gafil**. Medial or final *gh* becomes **ğ**: *maghfūr* 'forgiven' > **mağfur**; *tablīgh* 'communication' > **tebliğ**. This **ğ**, whether in borrowings or in native words, though audible as a 'Northumbrian burr' of varying intensity in dialect, serves in standard Turkish to lengthen the preceding vowel, a following vowel being swallowed up. Thus **gideceğim** 'I shall go' is pronounced as gidejēm; **alacağız** 'we shall take' as alajāz; **ağir** 'heavy' as ār; **ağız** 'mouth' as āz; **ağa** 'master, landowner' as ā. Note particularly **ağabey** 'elder brother', pronounced ābī; **Boğaziçi** 'Bosphorus' pronounced Boazichi or Bāzichi. Between o and a or o and u, it may be heard as a weak v or w: **soğan** 'onion', **soğuk** 'cold'. The verbs **koğmak** 'to chase away', **oğmak** 'to rub', **oğalamak** 'to crumble', are pronounced and sometimes spelled **kovmak, ovmak, ovalamak** (also **uvalamak**). The same phenomenon occurs after ö in: **öğmek** 'to praise', **döğmek** 'to beat', **söğmek** 'to curse', **göğermek** 'to become blue', **öğün** 'portion', **öğür** 'accustomed', also spelled **övmek, dövmek, sövmek, gövermek, övün, övür**. Otherwise, ğ in conjunction with front vowels is heard as a weak y: **öğle** 'noon', **değer** 'worth'. In two common Persian borrowings, **diğer** 'other' and **eğer** 'if', the original hard g is sometimes heard instead of y, but never in **meğer** 'apparently'. **değil** 'not' is pronounced deyil, deil, or, rather preciously, dīl.

Intervocalic **k** regularly becomes **ğ**: **ayak** 'foot'+-ım 'my' > **ayağım** 'my foot'.

11. h. In conjunction with any of the narrow vowels ı, i, u, ü, particularly when it ends a syllable, **h** is sometimes pronounced more heavily than otherwise, like the Arabic pharyngal unvoiced fricative ح *ḥ*:[2] **mıh** 'nail', **ıhlamur** 'lime-tree', **hıyar** 'cucumber', **ihtiyar** 'aged', **ruh** 'soul'. This is not due, as some Arabists suppose, to a memory of the spelling of such words in the

---

[1] Ragıp Özdem, *Tarihsel Bakımdan Öztürkçe ve Yabancı Sözlerin Fonetik Ayraçları*, 2. Fasikül (Istanbul, 1939), p. 15, having indicated that ğ no longer has this sound in standard Turkish, employs it to show the pronunciation of French *programme* as **pğoğğam** and *carte postale* as **kağt postâl**.

[2] This sound may be achieved by uttering in a stage-whisper any word beginning with h. A useful sentence for practice is *Has Harry heard?*

Arabo-Persian alphabet; it is a popular and not a learned pronunciation.

The **h** in the common masculine name **Mehmet** is silent in standard Turkish, there being a compensatory lengthening of the first vowel. The spelling and pronunciation **Muhammed** are reserved for the name of the Prophet, while the intermediate stage **Mehemmed** is used in scholarly works for sultans of the name. In some regional dialects the **h** is pronounced in **Mehmet** but is silent in **Ahmet**, with a compensatory lengthening of the first vowel. The **h** of the name **Ethem** is also liable to be lost in pronunciation. The final **h** of **sahih** 'correct' is dropped in writing as well as speech when it is used as an adverb meaning 'really'; in speech the first **h** is sometimes dropped too. In slipshod speech intervocalic **h** is sometimes dropped together with its preceding vowel, so **muhafaza** 'protection' may be heard as māfaza. **Allah aşkına** 'for the love of God!' is pronounced as one word without the **ah**. The expletive **Allahını seversen** 'if you love your God, for Heaven's sake' is even further contracted, to allāsen. The **h** of **hanım** 'lady' regularly disappears, together with the preceding vowel, when following a name ending in **e** or **a**: **Ulviye hanım**, **Fatma hanım** are pronounced ulviyānım, fatmānım. So too in Persian borrowings compounded with *khāne* 'house': **postahane** 'post office', **hastahane** 'hospital', **eczahane** 'chemist's shop' are nowadays spelt and pronounced **postane, hastane, eczane**, all with long **a** in the middle syllable.

English-speaking students must take care not to mispronounce the letter-combinations **ph**, **sh**, and **th**: e.g. **ph** in **kütüphane** 'library' is pronounced as in *uphill*; **sh** in **İshak** 'Isaac' as in *mishap*; **th** in **methetmek** 'to praise' as in *nuthatch*.

12. 1. As in English, this letter represents two totally different sounds, the 'clear l' of *list* and the 'dark l' of *wool*. Clear l is formed towards the front of the mouth and is naturally produced in conjunction with the front vowels, while dark l, formed in the hollow of the palate, comes naturally with back vowels. Thus we find clear l in **yel** 'wind' and **köle** 'slave', dark l in **yıl** 'year' and **yol** 'way'. In foreign borrowings, however, a complication arises, such as we have met in considering **ğ** and **k**. In Arabic, l is clear except in the name of God, *Allāh*.[1] In French it is always clear.

[1] That is why the Arabic name *'Abdullāh* 'slave of Allah' is spelt **Abdullah**

In borrowings from these languages, l should be pronounced clear even when in conjunction with back vowels and, as with **g** and **k**, the circumflex is used as a reminder of this. Thus the l of **lâzım** 'necessary' and **plân** 'plan' is pronounced as in *list* not as in *lad*, and the best way to learn to pronounce it accurately is to insert a faint y after it. The y-sound is not so marked as in the English pronunciation of *lurid*, but is quite audible; in Turkish spelling, the English and American pronunciations of this word would be shown as lûrid and lurid respectively. It cannot be overemphasized that the primary function of the circumflex is to indicate palatalization and not vowel length; e.g. in **mütalâa** 'observation' the first and not the second **a** is long.

The latest impression of *YİK* recommends that the circumflex should be written over an **a** following an l in Arabic and Persian borrowings but not in western borrowings, and then only when the vowel was long in Arabic or Persian; thus **mütalaa** and **plan**, but **lâzım** 'necessary', **alâyiş** 'showiness'.

13. r. Turkish **r** is an alveolar, produced by the vibration of the tip of the tongue against the gums just above the top teeth. In the Rumelian dialects it is trilled, a practice to avoid. Finally it may be heard as a fricative, accompanied by a heavy aspiration, not unlike the sound of Welsh *rh*; this pronunciation is most commonly observable in **var** 'there is'.[1]

14. y. Following a front vowel and preceding a consonant, **y** is barely audible but lengthens the preceding vowel: **teyze** 'maternal aunt', **öyle** 'thus'.

15. The glottal stop. This is not native to Turkish but occurs in Arabic borrowings. It is the sound which replaces the t in the Cockney and Glasgow pronunciations of, for example, *Saturday* and which occurs in standard English between a final and an initial vowel; the glottal stop is what makes the difference in

---

in Turkish, whereas masculine names incorporating other divine names have **ü**: **Abdülkadir, Abdülaziz, Abdüllatif**, the clear l fronting the original Arabic *u* into **ü**.

[1] Jozef Blaskovics, in '"R" Sessizinin Söylenişi', *X. Türk Dil Kurultayında Okunan Bilimsel Bildiriler* (Ankara, 1964), pp. 5–10, likens it to the Czech *ř*. He describes it as particularly frequent after front vowels, which does not square with the present author's observations.

pronunciation between *siesta* and *see Esther*. In Turkish it may be primary, standing for an original Arabic glottal stop (*hamza*), or secondary, standing for ʻ*ayn*. The latter is a voiced pharyngal gulp; to produce it, students of Arabic are sometimes told to sing as far down the scale as they can and then one note lower. It is as difficult for Turks as for other non-Arabs. The glottal stop, both primary and secondary, is preserved in spelling:

(*a*) To avoid ambiguity: **telin** 'of the wire' (gen. of **tel**) but **tel'in** 'denunciation' (Arabic *talʻīn*); **kura** 'villages' (Arabic *qurā*) but **kur'a** 'conscription by lot' (Arabic *qurʻa*).

(*b*) In high style, out of respect for the traditional Arabic spelling particularly of religious terms: **şer'î** 'pertaining to the sacred law' (*sharʻī*), **Kur'an** 'the Koran' (*Qurʼān*).

It is preserved in pronunciation but not in writing when intervocalic, as in **müdafaa** 'defence' (*mudāfaʻa*), **teessüf** 'regret' (*taʼassuf*). Otherwise, the modern practice is to omit the apostrophe in writing and to neglect the glottal stop in speaking: **sanat** 'art' (*sanʻat-*), **memur** 'official' (*maʼmūr*), **mesele** 'problem' (*masʼalat-*).[1] If the glottal stop is heard it is because the speaker is elderly, pedantic, or speaking slowly and deliberately.

16. Doubled consonants. These are not pronounced separately, but their enunciation is spread over a longer time than that of a single consonant: **batı** 'west' but **battı** 'it sank'; **eli** 'his hand' but **elli** 'fifty'; **ĝitti mi** 'did he go?' but **ĝittim mi** 'did I go?'

The final consonant of the accented syllable of interjections is sometimes doubled: **yazık** or **yazzık** 'a pity!', **bravo** or **bravvo** 'well done!' Similarly **o saat** 'straight away', the **o** being accented, is regularly pronounced with a doubled s, though not so spelt.

Doubled final consonants in Arabic borrowings are simplified into a single consonant, except when the addition of a suffix

[1] The ending *-at-* of these examples represents the Arabic *tā' marbūṭa* 'linked *t*', a feminine ending pronounced as *a* except if followed by a vowel, when it is pronounced *at*; cf. the French *il a* but *a-t-il*. The Arabic short *a* appears in Turkish as *a* or *e*, according to the quality of the neighbouring consonants. As some Arabic words in *-at-* were taken into Turkish with the *t*, others without it, this ending appears in four forms: **a, e, at, et**. There are a few instances of differentiation of meaning by the retention or non-retention of the *t*: e.g. **izafe** 'attribution' but **izafet** 'nominal annexation'; **maliye** 'finance department' but **maliyet** 'cost'.

consisting of or beginning with a vowel makes the doubled consonant pronounceable: *ḥaqq* 'right' > **hak**, acc. **hakkı**; *ḥadd* 'limit' > **had**, acc. **haddi**; *kull* 'totality' > **kül**, acc. **küllü**. *ʿafw* 'pardon' > **af**, acc. **affı**, with the original *w* assimilated to the *f*. The doubled *s* is lost in *tamāss* 'contact' > **temas**, acc. usually **temasi**, pedantically **temasi**.

17. Consonant-clusters and epenthetic vowels. Two consonants never occur together at the beginning of native words, if we except the colloquial **brakmak** for **bırakmak** 'to leave'. Within a word, it is rare to find more than two consonants adjoining. When consonant-clusters occur in foreign borrowings they are simplified by the addition of a vowel (*a*) before or (*b*) within an initial cluster, or (*c*) within a final cluster:

(*a*) French *station* > **istasyon**; *statistique* > **istatistik**; *splendide* > **ispilândit** (name of an apartment-building in Istanbul). Italian *scala* > **iskele** 'quay'; *sgombro* > **uskumru** 'mackerel'; *spirito* > **ispirto** 'alcohol'. English *screw* > **uskur**; *steam* > **istim** or **islim**.

(*b*) German *Schlepp* > **şilep** 'cargo-boat'; *Groschen* > **kuruş** 'piastre'. English *train* > **tiren**. French *sport* > **sıpor**; *club* > **kulüp**; *classeur* > **kılâsör** 'file'. The tendency among the educated is towards dispensing with such epenthetic vowels in initial clusters. The time-honoured **kuruş** and **şilep** have no alternative forms, but **tren** is used side by side with **tiren**, **klüp** with **kulüp**, **klâsör** with **kılâsör**.

(*c*) Numerous borrowed nouns end in two consonants, which Turks have difficulty in pronouncing unless the first is **l** or **r** or unless a vowel is suffixed: Arabic *ism* 'name' > **isim**, acc. **ismi**; *ʿadl* 'justice' > **adil**, acc. **adli**; *qism* 'part' > **kısım**, acc. **kısmı**; *ʿaql* 'intelligence' > **akıl**, acc. **aklı**; *matn* 'text' > **metin**, acc. **metni**; *ʿumr* 'life' > **ömür**, acc. **ömrü**; *fikr* 'thought' > **fikir**, acc. **fikri**. Persian *shahr* 'city' > **şehir**, acc. **şehri**.

Some borrowings of this shape, however, retain the epenthetic vowel even when a vowel is suffixed: Arabic *ṣinf* 'class' > **sınıf**, acc. **sınıfı**; *saṭr* 'line' > **satır**, acc. **satırı**; *shiʿr* 'poetry' > **şiir**, acc. **şiiri**. Persian *zahr* 'poison' > **zehir**, acc. **zehiri**; *tukhm* 'seed' > **tohum**, acc. **tohumu**. Presumably the retention of the vowel was originally a vulgarism. There are signs that the number of such words is on the increase; particularly frequent in newspapers is **şehire** instead of **şehre** for the dative of **şehir**.

Conversely, some native words ending in consonant+vowel+consonant drop the original vowel when a vowel is suffixed: **oğul** 'son', acc. **oğlu**; **beyin** 'brain', acc. **beyni**.

18. Foreign diphthongs. The Arabic ai diphthong, written *ay*, is treated in Turkish as consisting of vowel+consonant. It may appear as **ay** or **ey**; in either event an epenthetic vowel appears before a following consonant unless that consonant is followed by a vowel: *khayr* 'good' > **hayır**, acc. **hayrı**; *Ḥusayn* (masculine name) > **Hüseyin**, acc. **Hüseyni**; *meyl* 'tendency' > **meyil**, acc. **meyli**. Exception: the Arabic dual ending *-ayn* > **-eyn**, with no epenthetic vowel.

The Arabic au diphthong, written *aw*, similarly appears as **av** or **ev**, the former generally taking an epenthetic vowel: *qawm* 'people' > **kavim**; *qawl* 'word' > **kavil** 'agreement'; *qaws* 'bow' > **kavis**; *ḥawḍ* 'pool' > **havuz**; the accusatives being **kavmi, kavli, kavsi**, but **havuzu**. Two English sporting terms, *foul* and *round*, appear as **favl** and **ravnt**, with no epenthetic vowel, the **v** being pronounced as a semivowel. The combination **ev**+consonant generally has no epenthetic vowel: *mawt* 'death' > **mevt**; *sawq* 'driving' > **sevk**; *shawq* 'desire' > **şevk**; *dhawq* 'taste' > **zevk**. Exceptions: *jawr* 'tyranny' > **cevir**, acc. **cevri**; *jawz* 'walnut' > **ceviz**, acc. **cevizi**; *nawʿ* 'sort' > **nevi** (for **neviʾ**), acc. **nevʾi**.

19. Alternation of consonants. k/ğ. Final postvocalic **k** in polysyllabic substantives becomes **ğ** when a vowel is added: **ayak** 'foot', acc. **ayağı**; **ekmek** 'bread', acc. **ekmeği**; **sokak** 'street' (< Arabic *zuqāq*), acc. **sokağı**; **elektrik** 'electricity', acc. **elektriği**; **trafik** 'traffic, traffic-police', acc. **trafiği**. A few monosyllables exhibit the same phenomenon: **çok** 'much', acc. **çoğu**; **gök** 'sky', acc. **göğü** as well as **gökü**. Most, however, follow the pattern of **kök** 'root', acc. **kökü**, and **dok** 'dock, warehouse' (English through French), acc. **doku**.

ğ/k. Final postconsonantal *g* in foreign borrowings (cf. § 4, end) becomes **k** but reappears if a vowel is added: Persian *rang* 'colour' > **renk**, acc. **rengi**; *āhang* 'harmony' > **ahenk**, acc.

ahengi. In borrowings from western languages the **g** is usually retained in writing but pronounced as **k** except by the learned: **kliring** 'clearing' (a financial term), **miting** 'political meeting'.

ğ/ğ. Final postvocalic **ğ**, vulgarly pronounced **k**, changes to **ğ** before an added vowel: **katalog** 'catalogue', acc. **kataloğu**; **jeolog** 'geologist', acc. **jeoloğu** or **jeolog'u**, depending on the extent to which the user of the word regards it as a naturalized part of his vocabulary. In the latter example, as in the next, the apostrophe does not mark a glottal stop but preserves the original **ğ** from the usual intervocalic change to **ğ**. **lig** 'football league', acc. **ligi** or **lig'i**; the colloquial form, however, is **lik**, acc. **liki**.

b/p, c/ç, d/t. An original **b**, **c** (i.e. Arabic or Persian *j*), or **d** becomes **p**, **ç**, or **t** at the end of most polysyllabic borrowings and some monosyllabic, but reappears before a vowel: Arabic *kitāb* 'book' > **kitap**, acc. **kitabı**; *iḥtiyāj* 'need' > **ihtiyaç**, acc. **ihtiyacı**; Persian *tāj* 'crown' > **taç**, acc. **tacı**; *dāmād* 'son-in-law' > **damat**, acc. **damadı**; French *sérénade* > **serenat**, acc. **serenadı**.

Final **p**, **ç**, and **t** are voiced before vowels into **b**, **c**, and **d**, regularly in polysyllables, occasionally in monosyllables: **dip** 'bottom', acc. **dibi**; **ağaç** 'tree', acc. **ağacı**; Arabic *sharīṭ* 'tape' > **şerit**, acc. **şeridi**; French *groupe* > **grup**, acc. **grubu**; *principe* 'principle' > **prensip**, acc. **prensibi** (the accusatives **grupu**, **prensipi** are pedantic). Care must be taken not to confuse **at** 'horse', acc. **atı**, with **ad** 'name', acc. **adı**, or **ot** 'grass', acc. **otu**, with the archaic **od** 'fire', acc. **odu**.

A few verb-stems change final **t** to **d** when a vowel is added, e.g. **et-** 'to do', aorist **eder**; **git-** 'to go', aorist **gider**; **güt-** 'to pasture', aorist **güder**.

Final **p** and **t** may arise from a doubled final *b* and *d* in Arabic borrowings. When a vowel is added, the original voiced double consonant reappears: *muḥibb* 'friend' > **muhip**, acc. **muhibbi**; *radd* 'rejection' > **ret**, acc. **reddi**. There is no current example of the change from *jj* to **ç**; *ḥajj* 'pilgrimage' becomes **hac**, acc. **haccı**, thus avoiding confusion with **haç** 'crucifix', acc. **haçı** (< Armenian *khach*). The original voiced consonant is similarly preserved in **had** 'limit' (§ 16, last paragraph), which is thus distinguished from **hat** 'line', acc. **hattı** (< Arabic *khaṭṭ*). The normal unvoicing occurs, however, in **serhat** 'frontier' (< Persian *sar* 'head, chief'+Arabic *ḥadd* 'limit'), acc. **serhaddi**.

In foreign words which have become part of popular speech the original voiced consonant does not reappear before a vowel: Arabic *ḥabb* 'pill' > **hap**, acc. **hapı**; French *tube* > **tüp**, acc. **tüpü**. **set** 'parapet' (< Arabic *sadd*) has alternative learned and popular forms: acc. **seddi** and **seti** respectively.

There was a short-lived fashion in the nineteen-fifties for consistently spelling (though not pronouncing) with **b**, **c**, and **d** the absolute forms of all nouns subject to the alternations **b/p**, **c/ç**, **d/t**: **kitab, muhib, ihtiyac, tac, şerid, red**, etc. This fashion is reflected in the second edition of *OTD* (1957) and the third edition of *TS* (1959), but not in the fifth edition of *İmlâ Kılavuzu* (1959) or *YİK* (1965), which print **kitap, muhip, ihtiyaç, taç, şerit, ret**, etc. Survivals of the practice may be regarded as idiosyncratic.

20. Consonant assimilation in suffixes. When a suffix beginning with **c**, **d**, or **ġ** is added to a word ending in one of the unvoiced consonants **ç, f, h, k, p, s, ş, t**, the initial consonant of the suffix is unvoiced to **ç, t**, or **k**. The suffix **-ci** denotes occupation: **eski** 'old', **eski-ci** 'old-clothes dealer', but **elektrik-çi** 'electrician'.[1] The suffix of the locative case is **-de**: **İzmir'de** 'in Izmir' but **Paris'te** 'in Paris'. The locative case of **iş** 'work' is sometimes spelled **işde** to avoid confusion with **işte** 'behold'; similarly the locative of **üs** 'base' (naval, etc.) is spelt **üsde** to avoid confusion with the dative case **üste** of **üst** 'top'. The adjectival suffix **-ġan**: **atıl-ġan** 'reckless' but **unut-kan** 'forgetful'.

Those writers who follow the fashion mentioned at the end of § 19, i.e. who do not, in writing, recognize the unvoicing of final **b**, **c**, and **d**, do not unvoice the initial voiced consonant of suffixes either; e.g. **kitapçı** 'bookseller' they write as **kitabcı** and **sevinçte** 'in joy' as **sevincde**.

21. Vowels: general observations. Turkish vowels are normally short but may be long in three situations; see § 30. The difference between short and long vowels is of quantity not quality: the positions of the speech organs is the same; the change is in the length of time during which the breath flows. That is why long **a** may be written **aa**; see § 9, third paragraph and, § 30 (*c*).

---

[1] Suffixes are not hyphenated except in grammars, to illustrate the processes of word-formation.

Vowels are classified according to three criteria:

(*a*) 'Front' or 'back', according to whether it is the front or back of the tongue which interrupts the flow of breath.

(*b*) 'Open' or 'close', according to the amount of space left between tongue and palate; alternative terms are 'high' and 'low'.

(*c*) 'Rounded' or 'unrounded', according to the position of the lips.

The eight vowels of standard Turkish are tabulated thus:

	Unrounded		Rounded	
	Open	*Close*	*Open*	*Close*
Back	a	ı	o	u
Front	e	i	ö	ü

22. a. A back, open, unrounded vowel, like the a of French *avoir* or northern English *man*. Short, **baba** 'father'; long, **dağa** 'to the mountain'.

There is also a front sound of **a**, verging on that of **e**, which can be heard in careful speakers' pronunciation of some Arabic borrowings and in the Istanbul word **anne** 'mother' (elsewhere **ana**). Short, **dikkat** 'attention'; long, **cahil** 'ignorant'.

23. ı. A back, close, unrounded vowel. It is very close to the sound of i in *cousin*, but a closer approximation can be achieved by spreading the lips as if to say *easy* but saying *cushion* instead; the result will be the Turkish **kışın** 'in winter'. Short, **dış** 'exterior'; long, **yaptığım** 'which I did'. The corresponding capital letter is **I**, whereas the capital form of **i** is **İ**.

24. o. A back, open, rounded vowel, like French o in *note*: **çok** 'much', **yok** 'non-existent'. Long, as in **doğdu** 'he was born', it is much like the vowel of English *daw* without the final u-glide. A word of warning is necessary here. Some English-speakers, aware that Turkish **o** is not the same as English o in *hot*, go to the other extreme and pronounce **çok** and **yok** exactly like *choke* and *yoke*, thus providing their Turkish friends with a good deal of amusement. Turkish **o** is in fact closer to the vowel of *hot* than to that of *choke*.

25. u. A back, close, rounded vowel, between the vowels of English *put* and *pool*. Short, **burun** 'nose'; long, **uğur** 'luck'.

26. e. A front, open, unrounded vowel, like French e in *être*. Short, **sever** 'he loves'; long, **tesir** 'effect'. It also has a closer pronunciation, verging on the sound of **i**, which is sometimes heard especially in the first syllables of **vermek** 'to give' and **gece** 'night'. These two sounds of **e** are not separate phonemes in standard Turkish. In dialect, however, **el** 'hand' and **el** 'people', homophones in the standard language, are pronounced differently, with open and close **e** respectively.

27. i. A front, close, unrounded vowel, as in French *si*, closer than in English *pin*. Short, **diş** 'tooth'; long, **iğne** 'needle'.

28. ö. A front, open, rounded vowel, as in German; the French eu in *peur*. Short, **görmek** 'to see'; long, **öğrenmek** 'to learn'.

29. ü. A front, close, rounded vowel, as in German; the French u in *mur*. Short, **üzüm** 'grapes'; long, **düğme** 'button'.

30. Vowel length. The three situations in which long vowels occur are:

(*a*) In foreign borrowings: *ghāzī* (A) 'warrior for the Faith' > **gazi** (pronounced gāzī); *barābar* (P) 'together' > **beraber** (long a). Many originally long vowels, however, are shortened: *kabāb* (A) 'roast' > **kebap**; *baqqāl* 'greengrocer' > **bakkal** 'grocer'. This tendency is still in progress, as may be seen in so recent a borrowing as *jeep* > **cip**, with short **i**.

A long vowel in Arabic borrowings may represent an original short vowel+*hamza* or short vowel+*'ayn*: *ta'thīr* 'effect' > **tesir** (e long); *ma'lūm* 'known' > **malûm** (both vowels long).

A large number of Arabic borrowings retain an original long vowel in the last syllable when a vowel is added, but shorten it otherwise; **tesir** is one such, acc. **tesiri** pronounced tēsīri. Others in common use include **hayat** 'life', acc. **hayatı**; **zaman** 'time', acc. **zamanı**; **mal** 'property', acc. **mali**; **cevap** 'answer', acc. **cevabı**, all the accusatives with **a** long. All such words are indicated in the *OTD*.

Rarely one sees an idiosyncratic use of **iy** to denote long **i**, particularly in the pietistic spelling **iyman** for **iman** 'faith'. Yakup Kadri Karaosmanoğlu, one of Turkey's greatest writers, habitually spells, for example, **ilân** 'notice' and **itibar** 'regard' as **iylân** and **iytibar**; he also uses the spelling **kıy** for the Arabic *qī* (see § 9, penultimate paragraph). The spelling **liynet** for Arabic *līnat*- 'looseness of the bowels', however, is conventional.

(*b*) Any vowel followed by **ğ**+consonant (except when **ğ** is interchangeable with **v**; see § 10), any front vowel followed by **y**+consonant, or any back vowel followed by **ğ** is pronounced long: **değdi** 'he touched'; **yağmur** 'rain'; **meyva** 'fruit'; **dağ** 'mountain'; **çığ** 'avalanche'.

(*c*) When it is desired to emphasize a word, one vowel may be dwelled on, i.e. lengthened. This lengthening may be shown in writing by repeating the vowel-letter, often to an extent that would surprise an English printer: **asla** 'never', **aslaaa** 'never!'; **çok** 'much', **çoook** 'much too much'; **çoktan** 'for a long time', **çoktaaan** 'for ages and ages'; **fakat** 'but', **fakaaaaaat** '*but*'. **vay anam**, literally 'oh, my mother', an expression of distress, may be spelled **vay anaam**.

Doubled vowels originally separated in Arabic by *hamza* or *'ayn*, as well as doubled vowels arising from compounding words, are pronounced separately: **şâşaa** 'glitter' < *sha'sha'a* (A); **karaağaç** 'elm' < **kara** 'black'+**ağaç** 'tree'.

31. Vowel harmony. The principle of vowel harmony, which permeates Turkish word-formation and suffixation, is due to the natural human tendency towards economy of muscular effort. It is expressed in three rules:

(*a*) If the first vowel of a word is a back vowel, any subsequent vowel is also a back vowel; if the first is a front vowel, any subsequent vowel is also a front vowel.

(*b*) If the first vowel is unrounded, so too are subsequent vowels.

(*c*) If the first vowel is rounded, subsequent vowels are either rounded and close or unrounded and open.

The first rule is based on the phenomenon of palatal assimilation: that part of the tongue which interrupts the breath-flow over the palate in the production of the first vowel of a word remains in use for the production of the subsequent vowels of the word.

The second and third rules are based on labial assimilation: if the lips are rounded for the first vowel they may stay rounded for subsequent vowels, whereas if they are unrounded for the first vowel the speaker does not make the effort to round them subsequently. There is a partial exception to the second rule: the special case of words whose first vowel is **a** followed by **b, m, p,** or **v**, as the lip-position for the production of these labial consonants is not far removed from the rounded position necessary for the production of **u**.

The practical effect of these rules may be set out thus:

    a may be followed by **a** or **ı**.

It may also be followed by **u**, if **b, m, p,** or **v** intervenes.

    ı may be followed by **a** or **ı**
    o ,, ,, ,, ,, **u** ,, **a**
    u ,, ,, ,, ,, **u** ,, **a**
    e ,, ,, ,, ,, **e** ,, **i**
    i ,, ,, ,, ,, **e** ,, **i**
    ö ,, ,, ,, ,, **ü** ,, **e**
    ü ,, ,, ,, ,, **ü** ,, **e**

If the vowel of the first syllable is, say, **e**, that of the second will be **e** or **i**, so, for example, **ġelen** and **ġelin** are possible[1] words but gelan and gelön are not. To find the possible third vowels of a word beginning **ġelin**, we look at **i** in the first column and see that it can be followed by **e** or **i**; thus **ġeline** and **ġelini** are possible but gelinö and gelinı are not. It will be observed that, as a rule, **o** and **ö** occur only in first syllables.

Vowel harmony is a process of progressive assimilation, the first vowel affecting the second, and so on. There are some instances of regressive assimilation; for example, in **o bir** 'the other' the **o** is fronted to **ö** by regressive assimilation to the **i**, which it in turn rounds to **ü**, giving the form **öbür**. See also **bu ġün > büġün** in § 32 (*b*) and **imparator, menecer, madalya, apolet,** and **ütüv** in § 33.

---

[1] By 'possible' is meant possible as a standard modern word of native origin. In dialect, exceptions to the second rule are not uncommon, e.g. **hanuk** 'stubborn', **karuk** 'stalk'. In the older language, **u** and **ü** regularly followed unrounded vowels: **içün** 'for', now **için**; **yazup** 'having written', now **yazıp**.

32. Exceptions to the rules of vowel harmony. These are of four classes:

(*a*) Native words, simple. The following words contain both back and front vowels: **dahi** 'also', **elâ** 'light brown', **elma** 'apple', **hangi** 'which?', **hani** 'where?', **haydi** 'come on!', **inanmak** 'to believe', **kardeş** 'brother' (see IV, 9), **katmer** 'the state of being folded', **şişman** 'fat'. Note also **anne** (§ 22).

(*b*) Compound words, e.g. **bu** 'this'+**gün** 'day' > **bugün** 'today', sometimes heard as **bügün** and even **büğün**; **baş** 'head'+**müfettiş** (A) 'inspector' > **başmüfettiş** 'chief inspector'.

(*c*) Invariable suffixes: **-daş, -yor, -ken, -leyin, -imtrak, -ki,** and **-gil**. **-ki** and the rare suffixed form of **için** 'for' sometimes exhibit an exceptional variation **i/ü**, appearing as **-ki** and **-çin** after unrounded vowels, **-kü** and **-çün** after rounded vowels.

(*d*) Foreign words, e.g. **beyan** (A) 'declaration', **ferman** (P) 'command', **mikrop** (French) 'microbe', **feribot** (English) 'car-or train-ferry', **piskopos** (Greek) 'bishop'.

33. Vowel harmony in foreign borrowings. The effect of vowel harmony extends to non-Turkish words too, bringing as many vowels as possible of a foreign borrowing into one class, or pressing a foreign borrowing whose vowels happen to be all of one class still further into Turkish form. Thus Serbo-Croat *imperator* 'emperor' > **imparator**. English *manager*, in the special sense of manager of a football team, appears as **menecer**, while a 'penalty' in football is **penaltı**. Italian *medaglia* 'medal' > **madalya**. French *épaulette* > **apolet**; *étuve* 'sterilizer' > **etüv** > **ütüv**. Arabic *mumkin* 'possible' > **mümkin** > **mümkün**; *mudīr* 'administrator' > **müdir** > **müdür**; *muftī* 'jurisconsult' > **müfti** > **müftü**; *qibṭī* 'Copt' > **kıpti** > **kıptı** 'gipsy'.

34. Vowel harmony of suffixes. Apart from the exceptions noted in § 32 (*c*), all suffixes are subject to the rules of vowel harmony, the quality of the last syllable of the word determining the quality of the vowel of the suffix. Some suffixes are twofold, their vowel appearing as **e** after front-vowel words, as **a** after back-vowel words. Others are fourfold, their vowel being **i** or **ü** after unrounded and rounded front vowels respectively, **ı** or **u** after unrounded and rounded back vowels respectively. The suffix of the dative case, for example, is twofold: **-e** with front-vowel words, **-a** with back-vowel words. The suffix of the genitive is fourfold: **-in** after **e** or **i**, **-ün** after **ö** or **ü**, **-ın** after **a** or **ı**, **-un** after **o** or **u**. As for words with more than one suffix, the two tables below show the possible vowels (*a*) of a twofold suffix after a fourfold suffix, and (*b*) vice versa. It will be seen that these tables contain no new information but are based entirely on § 31.

(*a*)	Last vowel of word	Vowel of first suffix: fourfold	Vowel of second suffix: twofold
	e or i ö or ü	i ü	e
	a or ı o or u	ı u	a

If the vowel of the second suffix is also fourfold it will be as shown for the first suffix.

(*b*)	Last vowel of word	Vowel of first suffix: twofold	Vowel of second suffix: fourfold
	e, i, ö, ü	e	i
	a, ı, o, u	a	ı

If the vowel of the second suffix is also twofold it will be as shown for the first suffix.

The tables also hold good for suffixes of two syllables; e.g. reference to table (*a*) shows that the suffix **-ince** will appear as **-unca** after a word whose last vowel is **u**, while reference to (*b*) shows that **-esi** will appear as **-ası** after a word whose last vowel is **ı**.

The convention followed in this book is to refer to suffixes in their **e** or **i** forms; thus we shall speak of the plural suffix, which is **-ler** after front vowels and **-lar** after back vowels, as **-ler**. Similarly the genitive suffix will be referred to as **-in**, which must be read as short for 'the appropriate close vowel+**n**, i.e. **-in**, **-ün**, **-ın**, or **-un**, according to the nature of the preceding vowel'. Some grammars indicate whether a suffix undergoes the twofold or the fourfold mutation by the use of indices; e.g. the plural suffix may be shown as **-ler**[2], the genitive suffix as **-in**[4], but there is no need for this once the principle has been grasped.

The consonant-alternations described in § 20 add to the protean nature of the Turkish suffix. The suffix conventionally referred to as -ci, for example, has eight possible forms, illustrated in these eight words: **kahveci** 'coffee-maker', **tütüncü** 'tobacconist', **kapıcı** 'janitor', **sporcu** 'sportsman', **bekçi** 'watchman', **sütçü** 'milkman', **telgrafçı** 'telegraphist', **topçu** 'artilleryman'.

35. Vowel harmony of suffixes with foreign borrowings. Some foreign words with a back vowel in the last syllable nevertheless take front-vowel suffixes. These may be grouped as follows:

(a) Arabic or French words ending in *l* (§ 12): **mahsul** 'produce', acc. **mahsulü**; **rol** 'role', acc. **rolü**. The fact that **gol** 'goal' is similarly treated—acc. **golü**—shows that it is borrowed from French and not directly from English.

(b) Arabic words ending in *k*: **idrak** 'perception', acc. **idraki**; **iştirak** 'participation', acc. **iştiraki**. But Arabic words ending in *q* take back-vowel suffixes: *sharq* 'east' > **şark**, acc. **şarkı**; see the penultimate paragraph of this section.

(c) Arabic words ending in *t* or *-at-* (p. 8, footnote): **dikkat** 'attention', acc. **dikkati**; **saat** 'hour, clock', acc. **saati**. But Arabic feminine plurals in *-āt* take back-vowel suffixes: **riyaziyat** 'mathematics', acc. **riyaziyatı**; **ruhiyat** 'psychology', acc. **ruhiyatı**. So too do words ending in **t** derived from the unvoicing of Arabic final *d*: *iqtiṣād* 'economy' > **iktisat**, acc. **iktisadı**; *maqṣad* 'purpose' > **maksat**, acc. **maksadı**.

(d) Arabic monosyllables with an *a* followed by two consonants, the second of which is a front consonant: *ḥarb* 'war' > **harp**, acc. **harbi**; *ḥarf* 'letter of the alphabet' > **harf**, acc. **harfi**; *gharb* 'west' > **garp**, acc. **garbi**. When the two consonants in such words represent a sound-combination unpronounceable by Turks, the epenthetic vowel is a front vowel, because of the final front consonant; cf. **kavil, kavim**, and **kavis** in § 18, and note also: *waqt* 'time' > **vakit**, *qabr* 'tomb' > **kabir**, *baḥth* 'discussion, topic' > **bahis**; accusatives **vakti, kabri, bahsi**.

(e) The Persian **yâr** 'friend, beloved' (confined to poetry) has front-vowel suffixes: acc. **yâri**, gen. **yârin**, dat. **yâre**, and **yârim** 'my friend'. The explanation that this is to avoid confusion with the native words **yara** 'wound', **yarı, yarım** 'half', and **yarın** 'tomorrow' is too facile; it could never have happened if **yâr** had not ended in a front consonant.

Some such foreign words, however, have become completely naturalized, because they conform to Turkish phonetic patterns, and they therefore take back-vowel suffixes: **sanat** 'art' (§ 15) could perfectly well be a native word, like **kanat** 'wing', whereas **saat**, with its double **a**, could not. Similarly **kanal** (French *canale*) looks as Turkish as **kartal** 'eagle' and its accusative is **kanalı**. So too **asıl** 'origin' (Arabic *aṣl*), acc. **aslı**. Partly by analogy with this word, partly through its frequent use as a musical term, *faṣl* (A) 'division' > **fasıl**, acc. **faslı**, although its initial **f** marks it as non-Turkish. Likewise **rahat** 'ease, comfort', though marked as non-Turkish by its initial **r**, belongs to the back-vowel class because of its everyday use.

Arabic words ending in *q*, even if they have a front vowel in the last syllable, take back-vowel suffixes: *shawq* 'desire' > **şevk**, acc. **şevkı**; *sawq* 'drive' > **sevk**, acc. **sevkı**; *fawq* 'top' > **fevk**, acc. **fevkı**.

There is a tendency to eliminate more and more of these exceptional vowel harmonies. Some elderly people still give **sanat**, or rather **san'at**, front-vowel suffixes; for a young person to do so would be regarded as affectation, and it is a fairly safe prediction that **rolü, idraki, harbi**, and so on will one day yield to **rolu, idrakı, harbı**, first in vulgar speech, then in educated speech, and finally in writing.

36. Alternation of vowels. If a suffix beginning with **y** is added to a verb-stem ending in **e** or **a**, the **y** narrows the **e** or **a** into **i** or **ı** respectively, unless (*a*) the vowel after the **y** is **i** or **ı**, when the **e** or **a** remains unchanged, e.g. in **konuşma+yıverdi**, or (*b* both the vowel before the **e** or **a** and the vowel after the **y** are rounded, when the **e** or **a** becomes **ü** or **u**. Thus **bekle+yecek > bekliyecek; anla+yacak > anlıyacak; türe+yor > türüyor; kolla+yor > kolluyor**. Some writers disregard these changes, writing **bekleyecek, anlayacak, türeyor, kollayor**. *Yeni İmlâ Kılavuzu* recommends disregarding them except before **-yor**.

37. Accentuation: general observations. There is little unanimity about accentuation among writers on Turkish grammar. As one listens to Turkish being spoken one notices that some syllables are more marked than others. The problem is to identify the way they are marked; is it by stress or a change in musical pitch?[1] In the present work 'accent' means a rise in the pitch of the voice. But apart from the nature of the accent, there is some disagreement, even among native authorities, about which syllable in a given word is accented. The reason why such disagreement is possible is, firstly, that word-accent in Turkish is not so powerful as in English, where the accented syllable often swamps the unaccented ('Extr'órd'n'ry!'), or as in Russian, grammars of which have to give rules for the pronunciation of unaccented syllables. Secondly, group-accent and sentence-accent (i.e. intonation) both override word-accent so completely that some authorities have denied the existence of word-accent altogether. An English parallel will make this clearer. If one were asked to mark where the word-accent comes in *machine*, one would naturally put it on the second syllable: *machíne*. But if the word is used as the second element of a compound noun its accent is lost and the group-accent prevails: *séwing-machine*. If a manufacturer of sewing-machines tells his wife that he has bought one for her, her reply may well be an incredulous 'You've *bought* a sewing-machine?' with both word- and group-accent lost and the sentence-accent on 'bought' prevailing.

38. Word-accent. With the exceptions stated below, Turkish words are oxytone, i.e. accented on the last syllable; when an oxytone word is extended by suffixes the accent is on the last syllable of the word thus formed: **çocúk** 'child',[2] **çocuklár** 'children', **çocuklarımíz** 'our children', **çocuklarımızín** 'of our children'; **odá** 'room', **odadá** 'in the room', **odadakí** 'that which is in the room', **odadakilér** 'those who are in the room', **odadakilerdén** 'from those who are in the room'. Non-oxytones keep the accent on the original syllable: **téyze** 'aunt', **téyzeniz** 'your aunt', **téyzenize** 'to your aunt'; **Ánkara'da** 'in Ankara'.

39. Exceptions:

(*a*) Place-names are not oxytone: **Anádolu** 'Anatolia', **İstánbul**. Most are accented on their first syllable: **Páris, Zónguldak**. This rule is particularly to be noted with regard to place-names

[1] For a summary account of recent views see Robert B. Lees, *The Phonology of Modern Standard Turkish* (The Hague, 1961), pp. 71–75.
[2] The acute accent here used to indicate the accented syllable is never written in Turkish.

which are spelt the same as common nouns: mısír 'maize',
Mísır 'Egypt'; sirkecí 'vinegar-seller', Sírkeci, a district of
Istanbul; bebék 'doll, baby', Bébek, a village on the Bosphorus;
karamán 'fat-tailed sheep', Káraman, a city of southern
Anatolia; ordú 'army', Órdu, a city on the Black Sea.

Polysyllabic place-names of non-Turkish origin generally retain
their original accentuation: İnġiltére 'England', İspánya 'Spain',
Antálya, Malátya. There is, however, a tendency for the accent
to go back to the beginning of the place-name; residents of
Malatya, for example, generally accent the name of their city on
the first syllable.

(*b*) Foreign nouns usually retain their original accentuation:
lokánta 'restaurant' (Italian *locanda*); ólta 'fishing-line' (Greek
βόλτα); rádyo 'radio, wireless'; táksi 'taxi'; kókteyl 'cocktail,
cocktail-party'; ġazéte 'newspaper' (Italian *gazzetta*).

(*c*) A number of nouns denoting relatives and living creatures:
ánne 'mother', ábla 'elder sister', ġörűmce 'husband's sister',
yénġe 'brother's wife', hála 'paternal aunt', téyze 'maternal
aunt', ámca 'paternal uncle', çekírġe 'grass-hopper', karínca
'ant', kokárca 'pole-cat'.[1]

(*d*) Adverbs are usually accented on the first syllable: şímdi
'now', sónra 'after', évvelâ 'firstly', ánsız or ánsızın 'suddenly',
áncak 'only'. This applies also to adverbs formed from nouns
with the addition of a case-suffix: ġerçektén 'from the truth' but
ġérçekten 'truly'. In several adverbs the suffix of the old instrumental case appears: kíşın 'in winter', yázın 'in summer' (the
genitives, 'of winter, of summer', are spelt identically but are
accented on the last syllable). The nouns of place (XII, 12) are
accented on the first syllable: búra 'this place', óra 'that place',
búrada 'here', óraya 'thither'. So are nouns used adverbially
without suffixes: nihayét 'end' but níhayet 'finally'; çoklúk
'multitude' but çókluk 'often'; artík 'residue' but ártık 'henceforth'. So too are some, but not all, adjectives used as adverbs:
yalníz 'alone' but yálnız 'only'; sahíh 'correct' but sáhi 'really'
(§ 11). On the other hand, iyi 'good', fena 'bad', and çabuk
'quick' remain oxytone even when used to mean 'well, badly,

[1] Banguoğlu (p. 184) gives fourteen examples of names of living creatures
which are not oxytone, while recognizing that they may also be heard as oxytone.
It is a measure of the elusiveness of the word-accent that *OTD* agrees with him
on only one of the fourteen and *TS* only on three, both dictionaries tacitly
showing the remainder as oxytone.

quickly'. The third-person singular of the aorist of **olmak** 'to become', **olur**, means not only 'becomes' but also 'all right, O.K.'. In this sense it may be accented on the first syllable as an adverb.

(*e*) In compound words the accent usually falls on the end of the first element: **çıplák** 'naked', **çırílçıplak** 'stark naked'; **baş** 'head'+**bakán** 'minister' > **báṣbakan** 'prime minister'; **bir** 'a'+**takím** 'set' > **bírtakım** 'several'.

(*f*) Diminutives in **-cik** are accented on the first syllable: **úfacık** 'tiny', **évcik** 'little house'.

(*g*) Polysyllabic suffixes, except **-leyin** and **-cesine** (§ 40) are accented on the first syllable: **ġid-ínce** 'having gone'; **yap-árak** 'by doing'.

So too are interjections and vocatives: **háydi** 'come on!' **áferin** 'bravo!' **ġarsón** 'waiter' but **ġárson** 'waiter!'

40. Enclitic suffixes. The following suffixes are enclitic; i.e. they themselves are never accented but throw the accent on to the preceding syllable:

(*a*) **-le** 'with': **memnuniyétle** 'with pleasure'; **onúnla** 'with him'.

(*b*) **-ken** 'while': **yazárken** 'while writing'.

(*c*) The adverbial suffix **-ce** and its extension **-cesine**: **iyíce** 'well', **hayváncasına** 'bestially'.

(*d*) The adverbial suffix **-leyin** (compounded with the instrumental **-in**): **ġecéleyin** 'by night', **akşámleyin** 'at evening'.

(*e*) The negative suffix **-me**: **ġel** 'come', **ġélme** 'do not come'; **anladí** 'he understood', **anlámadı** 'he did not understand'.

(*f*) The suffix **-yor** of the present tense: **ġelíyor** 'he is coming'.

(*g*) The suffixed forms of the verb 'to be'; see § 41 (*a*).

(*h*) The vowel of the Persian izafet; see II, 26.

Part of the controversy about Turkish accentuation is over the question whether these suffixes are properly described as enclitic or atonal, i.e. without accent. The former seems correct, as we see if we take a non-oxytone such as **başbakan** (§ 39 (*e*)) and add **-le**: **báṣbakánla**, with the accent before **-la** at least as noticeable as that on **baş**. Those who take the opposing view adduce, e.g., **sádece** 'simply' from **sadé** 'simple' and **áyrıca** 'separately' from **ayrí** 'separate', in which the syllables immediately before the suffix are not accented; these words, however, simply have the accentuation normal in adverbs.

41. Enclitic words. The following words are enclitic:

(a) Those parts of the verb 'to be' which are not formed from the stem **ol-**; they are enclitic both when independent words and when suffixed: **arkadaşím idi** or **arkadaşímdı** 'he was my friend'; **evlí ise** or **evlíyse** 'if she is married'; **kímse** 'person' (V, 24); **memnúnum** 'I am glad'.

(b) The interrogative particle **mi**. The rise in pitch before this particle is most noticeable, predominating over the word-accent: **anlámadı** 'he did not understand' but **anlamadí mi?** 'did he not understand?' When the present tense, however, is followed by this particle it retains the accent on the penultimate: **gelíyor** 'he is coming'; **gelíyor mu?** 'is he coming?'

(c) The postpositions: **sizín gibi** 'like you'; **bením için** 'for me'.

(d) The conjunction **ki**: **diyorlár ki ólmıyacak** 'they are saying that it will not happen' (note also the accent before the negative suffix in the last word).

(e) The adverb **de**: **bíz de** 'we too'.

If more than one of these words come together, the accent is on the word preceding them: **bíz de mi gidelim?** 'shall we go too?'

42. Group-accent. The two most obvious features of this are:

(a) That in izafet-groups (II, 17) the accent is normally on the first element, on the syllable which carries the accent when the word is spoken in isolation: **Túrkiye Cumhuriyeti** 'Republic of Turkey'; **yemék odası** 'dining-room'.

(b) Demonstratives are accented when they qualify nouns: **bú çocuk** 'this child' but **çocúk bu** 'it's only a child' (V, 5).

43. Intonation. Sentence-accent or intonation is partly emotional, depending on the feelings and emphasis which the speaker wishes to convey, and partly syntactical and automatic. The general rule is that a rise in pitch denotes that the thought is not yet complete, whereas a fall in pitch marks its end. Thus the subject is uttered with a rising intonation, the predicate with a falling. The protasis of a conditional sentence likewise has a rising intonation, the apodosis a falling. Questions and exclamations have a rising intonation.

II

THE NOUN

1. **Gender.** Turkish is devoid of grammatical gender, nor does the sex of persons affect the forms of words. The third-person pronoun **o** does duty for 'he', 'she', and 'it'; **gelir** means 'he/she/it comes'. There are totally distinct names for the male and female of most domestic animals: **aygır** 'stallion', **kısrak** 'mare'; **boğa** 'bull', **inek** 'cow'. The gender of other animals is indicated if necessary by the use of **dişi** 'female' or **erkek** 'male': **dişi kedi** 'female cat'; **erkek ayı** 'male bear'.

With nouns which may denote persons of either sex, femininity may be shown by using **kız** 'girl' or **kadın** 'woman' in apposition: **kız talebe** 'girl student'; **kadın garson** 'waitress', **kadın kahraman** 'heroine'; **kardeş** 'brother' or 'sister', **kızkardeş** (written as one word) 'sister'.

Advantage is also taken of the facilities possessed by French and Arabic for showing gender: 'actress' is **aktris**; 'female dancer' is **rakkase (A)** or **dansöz**; 'female clerk' is **kâtibe (A)**.

The Serbo-Croat feminine suffix *-ica* appears in three old borrowings: **kıraliçe** 'queen', **imparatoriçe** 'empress', **çariçe** 'tsarina' (< *kraljica, imperatorica, carica*). Modelled on these is the neologism **tanrıça** 'goddess', from the name of the old Turkish sky-god **Tanrı**.

The word **hanım** 'lady', originally 'wife of a Khān (**han**)', is held by some to contain an ancient Turkish feminine ending; cf. the Eastern Turkish **begüm** 'begum', originally 'wife of a Beg or Bey', though the evidence is slight. The ending is more probably the possessive suffix of the first-person singular.[1]

2. **Number: the Turkish plural.** The plural is formed by adding **-ler** to the singular: **talebeler** 'students', **kızlar** 'girls'. The

---

[1] Saadet Çağatay in her article 'Türkçede "Kadın" İçin Kullanılan Sözler', *TDAYB*, 1962, pp. 13–49, suggests that **hanım** may derive from Persian *khānumān* 'household' and **begüm** from **begnüŋ** 'of the Beg', with an ellipsis of a word meaning 'wife'.

'singular' form of the noun, however, is numerically neutral (hence its use after numbers), denoting a category or an individual member of that category: **polis** 'the police' or 'the policeman', **bir polis** 'a policeman', **polisler** 'the policemen'; **şiir yazar** 'he writes poetry', **bir şiir yazar** 'he writes a poem', **şiirler yazar** 'he writes poems'; **casusluk** 'espionage', **bir casusluk** 'a case of espionage'; **kahramanlık** 'heroism', **kahramanlıklar** 'deeds of heroism'; **iyilik** 'goodness', **iyilikler** 'benefactions'. Thus **padişahın biri**, lit. 'one of sultan', means 'one of the category "sultan", a certain sultan' and may well begin a fairy-tale. But **padişahların biri** means one out of all the historical individuals who have held the title, 'one of the Sultans'.

A singular verb is commonly used with an inanimate plural subject, the plural verb being used for individuals. The use of a plural verb with a singular subject, second or third person, is a mark of respect. See further XVI, 1 and 3 (*b*).

Personal names may be used in the plural like our 'the Joneses' to refer to a family; in Turkish the plural may be of a given name as well as of a surname: **Mehmetler** 'Mehmet and his family'; cf. § 15, end, and IV, 10.

The plural word **erenler** was used as a form of address among dervishes, even when speaking to a single person. Popularly supposed to be the plural of the present participle of **ermek** 'to attain', it is in fact the plural of **eren**, itself an ancient irregular plural of **er** 'man'.

Numerals are usually followed by a singular noun: **kırk harami** 'forty thieves', **üç silâhşor** 'three musketeers', **on iki ada** 'twelve islands'. The use of a plural noun after a numeral shows that the persons or things in question form a particularly well-known and distinct entity: **kırk haramiler** 'The Forty Thieves', **üç silâhşorlar** 'The Three Musketeers', **on iki adalar** 'the Dodecanese'.[1] The plural suffix may be added directly to the numeral: 'fourteen' is **on dört** and the fourteen officers dropped from the Committee of National Unity in November 1960 were referred to in the press and in conversation as **ondörtler** 'The Fourteen'. So **kırklar** 'The Forty ⟨Saints⟩'.

3. *Arabic plurals.* Arabic has two types of plural.

(*a*) The external or sound plural masculine is made by adding -*ūn* for the nominative, -*īn* for the accusative and genitive. Turkish

[1] Now that the Dodecanese are not often in the news, **On iki Ada** is more usual.

borrowed only the latter form, using it as a nominative (as in colloquial Arabic): **memur** 'official', pl. **memurin**; this ending is obsolete in Turkish, except that this particular example is occasionally used.

The external or sound plural feminine is formed by adding *-āt*. Arabic uses it as Greek and Latin use the neuter plural; the modern Turkish vocabulary still retains a number of words such as **varidat** 'revenues', **mülâhazat** 'observations', **iktisadiyat** 'economics', **haşarat** 'vermin'. There are two jocular formations with this suffix added to Turkish words: **ğidişat** 'goings-on' < **ğidiş** 'manner of going'; **saçmaviyat** 'stuff and nonsense' < **saçma** 'nonsense'. As **saçma** ends with a vowel, the latter formation is analogous to that of, e.g., **küreviyat** 'spherical trigonometry' < **küre** 'sphere'. Both **ğidişat** and **saçmaviyat** have something of the quasi-learned flavour of English *shambolical* < *shambles*.

(*b*) The internal or broken plural is made not by suffixation but by wresting the singular out of shape: *waqf*, pl. *awqāf*, 'pious foundation'; *sulṭān*, pl. *salāṭīn*, 'sultan'; *'ālim*, pl. *'ulamā'*, 'scholar'. Many broken plurals survive in Turkish, most being treated as Turkish singulars: *ṭalaba* 'students', pl. of *ṭālib*, appears as **talebe** 'student'; *'amala* 'workmen', pl. of *'āmil*, as **amele** 'workman'; *khadama* 'servants', pl. of *khādim*, as **hademe** 'manservant'; *tujjār* 'merchants', pl. of *tājir*, as **tüccar** 'merchant'. The reason is that the Arabic broken plural, unlike the sound plural but just like the Turkish singular, can denote a class (in Arabic it can be construed with a singular verb) and therefore it fitted naturally into place in Turkish as a singular. The Arabic sound plural, on the other hand, denotes a number of individuals and was therefore felt by the Turks to be a true plural, not requiring the Turkish plural suffix.[1]

Some Arabic broken plurals are used in Turkish with a sense different from that of their singulars: *juz'* 'part' and its plural *ajzā'* appear respectively as **cüz** 'fascicle' and **ecza** 'the unbound fascicles of a book' or 'chemicals, drugs'; *shay'* 'thing' and its plural *ashyā'* appear as **şey** 'thing' and **eşya** 'luggage, belongings', the latter usually with the plural suffix, **eşyalar**.

[1] One ancient exception, **raşidinler** 'the Rightly-Guided' (*rāshidīn* being an Arabic plural), may be explained as an honorific plural, referring as it does to the first four Caliphs; cf. the last paragraph of § 2. A modern parallel—**hâşa huzurdan**—is **Beatles'ler** 'the Beatles'.

4. Other plurals. In Ottoman, the Persian plural suffix *-ān* was frequently attached to Arabic singulars: **mebusan** 'Deputies to the Ottoman Parliament' < Arabic *mab'ūth*+Persian *-ān*. Still in occasional use is **zabitan** 'officers' < Arabic *ḍābiṭ*+Persian *-ān*.

domates 'tomatoes' and **patates** 'potatoes' are both direct borrowings from modern Greek and both are treated as Turkish singulars.

One quasi-Mongoloid plural, **erat**, a product of the language-reform movement, has replaced, in official parlance, the Ottoman **efrat** 'private soldiers and N.C.O.s' < Arabic *afrād*. It is apparently a cross between **efrat** and the Turkish **er** 'man'.

5. The Arabic dual. Arabic has a dual number, indicated by *-ān* in the nominative, *-ayn* in the accusative and genitive: *ṭaraf* 'side'; *ṭarafān*, *ṭarafayn* 'two sides'. As with the sound plural masculine, only the accusative–genitive form was taken into Ottoman; it survives in three obsolescent words: **tarafeyn** 'the two parties', and **valideyn** and **ebeveyn**, both meaning 'the two parents'.

6. The cases. There are six. The simplest form of a noun, with no suffixes, is termed the absolute case; it is used not only for the nominative and vocative but also for the indefinite accusative. The term accusative will be applied to what should strictly be called the defined accusative. The other cases are: the genitive denoting possession; the dative denoting the indirect object and the end of motion; the locative denoting place where; the ablative denoting point of departure. Their commonest functions are illustrated below; for a more detailed examination see §§ 9–14 and XVI, 5, 6.

absolute	**ev satıldı**	the house has been sold
	bir ev arıyoruz	we are seeking a house
accusative	**evi aldık**	we have bought the-house
genitive	**evin bahçesi**	the garden of-the-house
dative	**eve geldim**	I came to-the-house
locative	**evde kaldı**	he has stayed in-the-house
ablative	**evden uzak**	far from-the-house

As will be seen from these examples, the suffix of the accusative is **-i**, genitive **-in**, dative **-e**, locative **-de**, and ablative **-den**. The first two are subject to the fourfold, the others to the twofold harmony. The case-suffixes follow the -ler of the plural.

To prevent those suffixes which consist in or begin with a vowel from being lost when added to a word ending with a vowel, a separator or 'buffer-letter' is used, **n** for the genitive, **y** for the accusative and dative. The sole exception is **su** 'water', which in the ancient language was **suw** and now has **y** before the suffix of the genitive as well as of the accusative and dative. Another relic of the original final **w** is the **v** in the verb **suvarmak** 'to water an animal'.

Examples will now be given to illustrate the changes wrought by vowel harmony and the other alternations described in Chapter I.

(*a*) Front-vowel class, consonant-stems; **el** 'hand', **köy** 'village':

Singular

abs.	el	köy
acc.	eli	köyü
gen.	elin	köyün
dat.	ele	köye
loc.	elde	köyde
abl.	elden	köyden

Plural

abs.	eller	köyler
acc.	elleri	köyleri
gen.	ellerin	köylerin
dat.	ellere	köylere
loc.	ellerde	köylerde
abl.	ellerden	köylerden

Reference to the tables in I, 31 or 34 will show that suffixes after **i** have the same forms as after **e**, so that the declension of **el** serves as a model for consonant-stems whose last or only vowel is **i**, such as **deniz** 'sea', **diş** 'tooth'. Similarly, the declension of **köy** serves as a model for consonant-stems whose last or only vowel is **ü**, such as **tütün** 'tobacco', **gün** 'day'.

(*b*) Front-vowel class, vowel-stems; **gece** 'night', **ölçü** 'measure':

Singular

abs.	gece	ölçü
acc.	geceyi	ölçüyü

gen.	gecenin	ölçünün
dat.	geceye	ölçüye
loc.	gecede	ölçüde
abl.	geceden	ölçüden

Plural

abs.	geceler	ölçüler
acc.	geceleri	ölçüleri
gen.	gecelerin	ölçülerin
dat.	gecelere	ölçülere
loc.	gecelerde	ölçülerde
abl.	gecelerden	ölçülerden

Like **gece** are declined vowel-stems in **i**, such as **gemi** 'ship', **sergi** 'exhibition'. Like **ölçü** are declined vowel-stems in **ö**, of which the only examples in common use are two French borrowings: **mösyö** 'Monsieur' and **banliyö** 'suburbs'.

(c) Back-vowel class, consonant-stems; **akşam** 'evening', **son** 'end':

Singular

abs.	akşam	son
acc.	akşamı	sonu
gen.	akşamın	sonun
dat.	akşama	sona
loc.	akşamda	sonda
abl.	akşamdan	sondan

Plural

abs.	akşamlar	sonlar
acc.	akşamları	sonları
gen.	akşamların	sonların
dat.	akşamlara	sonlara
loc.	akşamlarda	sonlarda
abl.	akşamlardan	sonlardan

Like **akşam** are declined consonant-stems whose last or only vowel is **ı**: **kadın** 'woman', **çığ** 'avalanche'. Like **son** are declined consonant-stems whose last or only vowel is **u**: **sabun** 'soap', **kuş** 'bird'.

(d) Back-vowel class, vowel-stems; **tarla** 'field', **korku** 'fear':

Singular

abs.	tarla	korku
acc.	tarlayı	korkuyu
gen.	tarlanın	korkunun
dat.	tarlaya	korkuya
loc.	tarlada	korkuda
abl.	tarladan	korkudan

Plural

abs.	tarlalar	korkular
acc.	tarlaları	korkuları

etc., as type (c).

Like **tarla** are declined vowel-stems in ı: **kapı** 'door', **darı** 'millet'. Like **korku** are declined vowel-stems in o: **palto** 'overcoat', **kadro** 'staff, cadre'.

(e) Nouns exhibiting alternation of consonants; **kitap** 'book', **ağaç** 'tree', **damat** 'son-in-law', **renk** 'colour', **ekmek** 'bread' (I, 19):

	b/p	c/ç	d/t	ğ/k	k/ğ
Singular					
abs.	kitap	ağaç	damat	renk	ekmek
acc.	kitabı	ağacı	damadı	rengi	ekmeği
gen.	kitabın	ağacın	damadın	rengin	ekmeğin
dat.	kitaba	ağaca	damada	renge	ekmeğe
loc.	kitapta	ağaçta	damatta	renkte	ekmekte
abl.	kitaptan	ağaçtan	damattan	renkten	ekmekten
Plural					
abs.	kitaplar	ağaçlar	damatlar	renkler	ekmekler
acc.	kitapları	ağaçları	damatları	renkleri	ekmekleri
	etc.	etc.	etc.	etc.	etc.

(f) Foreign borrowings with exceptional vowel-harmony; **hal** 'case', **rol** 'role', **saat** 'hour', **harp** 'war' (I, 35):

Singular

abs.	hal	rol	saat	harp
acc.	hali	rolü	saati	harbi
gen.	halin	rolün	saatin	harbin

dat.	hale	role	saate	harbe
loc.	halde	rolde	saatte	harpte
abl.	halden	rolden	saatten	harpten

Plural

abs.	haller	roller	saatler	harpler
acc.	halleri	rolleri	saatleri	harpleri
	etc.	etc.	etc.	etc.

(*g*) Nouns which add or drop a vowel in the last syllable; **isim** 'name' (I, 17 (*c*)), **ağız** 'mouth', **koyun** 'bosom', **oğul** 'son', **alın** 'forehead' (I, 17, end):

Singular

abs.	isim	ağız	koyun	oğul	alın
acc.	ismi	ağzı	koynu	oğlu	alnı
gen.	ismin	ağzın	koynun	oğulun	alnın
dat.	isme	ağza	koyna	oğula	alına
loc.	isimde	ağızda	koyunda	oğulda	alında
abl.	isimden	ağızdan	koyundan	oğuldan	alından

Plural

abs.	isimler	ağızlar	koyunlar	oğullar	alınlar
acc.	isimleri	ağızları	koyunları	oğulları	alınları
	etc.	etc.	etc.	etc.	etc.

Most native nouns of this sort are names of parts of the body. Of the examples, **ağız** and **koyun** are typical, whereas **oğul** and **alın** retain the vowel of the second syllable in the dative, as does **burun** 'nose'; **oğul** retains it in the genitive too.[1] As well as 'son', **oğul** can mean 'swarm of bees'; as well as 'bosom', **koyun** can also mean 'sheep'. In these latter senses, both retain the **u** in all cases. There is yet another word **koyun**, the genitive of **koy** 'bay'.

(*h*) Nouns originally ending in ʻ*ayn* (I, 15) are increasingly treated as vowel-stems except by the learned. Examples: **mevzu** 'topic, subject' < Arabic *mawḍūʻ*; **cami** 'mosque' < Arabic *iāmiʻ*.

---

[1] The word **oğul** 'son' is seldom used except with a personal suffix, and the genitive **oğulun** happens to be of particularly rare occurrence; cf. § 23.

THE NOUN

	Learned	Popular	Learned	Popular
Singular				
abs.	mevzu		cami	
acc.	mevzuu	mevzuyu	camii	camiyi
gen.	mevzuun	mevzunun	camiin	caminin
dat.	mevzua	mevzuya	camie	camiye
loc.		mevzuda		camide
abl.		mevzudan		camiden
Plural				
abs.		mevzular		camiler
acc.		mevzuları		camileri
		etc.		etc.

(*i*) Nouns originally ending in a doubled final consonant (I, 16, 19); **hak** 'right', **muhip** 'friend', **hat** 'line', **serhat** 'frontier'. Note that the dative and locative of **hat** are identical in shape; **hatta** can represent **hatt+a** or **hat+ta**.

Singular

abs.	hak	muhip	hat	serhat
acc.	hakkı	muhibbi	hattı	serhaddi
gen.	hakkın	muhibbin	hattın	serhaddin
dat.	hakka	muhibbe	hatta	serhadde
loc.	hakta	muhipte	hatta	serhatte
abl.	haktan	muhipten	hattan	serhatten

Plural

abs.	haklar	muhipler	hatlar	serhatler
acc.	hakları	muhipleri	hatları	serhatleri
	etc.	etc.	etc.	etc.

(*k*) Nouns combining the characteristics of types (*f*) and (*g*); in this type are included nouns which in Arabic have medial or final ʿ*ayn* or medial *hamza*. Examples: **vakit** 'time' < Arabic *waqt*, **nevi** 'sort' < Arabic *nawʿ*, **yeis** 'despair' < Arabic *yaʾs*, **kaır** 'profundity' < Arabic *qaʿr* (I, 17 (*c*), 18).

Singular

abs.	vakit	nevi	yeis	kaır
acc.	vakti	nev'i	ye'si	ka'rı

THE NOUN

gen.	vaktin	nev'in	ye'sin	ka'rın
dat.	vakte	nev'e	ye'se	ka'ra
loc.	vakitte	nevide	yeiste	kaırda
abl.	vakitten	neviden	yeisten	kaırdan

Plural

abs.	vakitler	neviler	yeisler	kaırlar
acc.	vakitleri	nevileri	yeisleri	kaırları
	etc.	etc.	etc.	etc.

To this type belong **şeri** 'Muslim religious law', acc. **şer'i**; **defi** 'repulsion', acc. **def'i**; **refi** 'elevation', acc. **ref'i**; **beis** 'harm', acc. **be'si**.

There are two nouns in use which in the original Arabic have *hamza* as their final consonant: **şey** 'thing' < *shay*'; **cüz** 'part' < *juz*'. These are declined as type (*a*), except that **cüz** is usually spelled with an apostrophe before vowel endings: acc. **cüz'ü**, gen. **cüz'ün**, dat. **cüz'e**.

7. Summary of case-endings. The letters in brackets appear after vowel-stems. The table is valid except for nouns of type (*f*) and for **su** 'water', which declines: sing. **su, suyu, suyun, suya, suda, sudan**; pl. **sular**, etc.

Last vowel of abs. sing.	e or i	ö or ü	a or ı	o or u
Singular				
acc.	-(y)i	-(y)ü	-(y)ı	-(y)u
gen.	-(n)in	-(n)ün	-(n)ın	-(n)un
dat.	-(y)e		-(y)a	
loc.	-de/te		-da/ta	
abl.	-den/ten		-dan/tan	
Plural				
acc.	-leri		-ları	
gen.	-lerin		-ların	
dat.	-lere		-lara	
loc.	-lerde		-larda	
abl.	-lerden		-lardan	

8. Uses of the cases. A case-ending is attached only to the final element in a nominal group; in this respect the Turkish case-endings behave like English prepositions and not like the case-endings of inflected languages such as Latin: 'good citizens', **iyi vatandaşlar**, *boni cives*; 'of good citizens', **iyi vatandaşlar-ın**, *bon-orum civ-ium*. **dört kere dokuz** 'four times nine'; **dört kere dokuz-un kare kökü** 'the square root of four times nine'.

9. The absolute form. This has five functions:

(*a*) Nominative, as subject of a sentence or as complement of a verb meaning 'to be, to become' or the like: **kapı açıldı** 'the door was opened'; **ben Başbakan olmıyacağım** 'I shall not become Prime Minister'. See also § 10, end.

(*b*) Vocative: **Ahmet! Taksi!**

(*c*) Indefinite accusative, i.e. as the undefined object of a verb: **gazete çıkarmak zor bir iş** 'to publish newspapers is a hard job'; **bilet satıyorlar** 'they are selling tickets'; **sigara içmez** 'he does not smoke cigarettes'; **öküz aldı** 'he bought oxen'; **bir öküz aldı** 'he bought an ox'.

(*d*) It may stand for any case in suspended affixation, i.e. when one grammatical ending serves two or more parallel words: **sıhhat ve afiyet-te** 'in health and well-being'. One can put the case-ending on both words—**sıhhat-te ve afiyet-te**—just as one can repeat the preposition in English—'in health and in well-being'— but this is less usual.

(*e*) Many adverbs of time are originally nouns in the absolute form, e.g. **bugün** 'today'.

10. The accusative case. It marks the definite object of a verb, i.e. an object defined:

(*a*) By a demonstrative adjective: **bu gazete-yi çıkarmak zor bir iş** 'to publish this newspaper is a hard job'.

(*b*) By a personal pronoun, suffixed or independent: **ev-imiz-i** or **bizim evi kiraladı** 'he has rented our house'.

(*c*) By its nature, e.g. as a place-name, a personal name or title, a personal or demonstrative pronoun: **Adana'yı gezdik** 'we toured Adana'; **Hasan'ı hemen tanıdım** 'I recognized Hasan immediately'; **Profesör-ü selâmladı** 'he greeted the Professor'; **siz-i ilgilendirmez** 'it does not concern you'; **bu-nu niçin yaptın** 'why have you done this?'

(d) By having been mentioned previously, i.e. in situations where English uses the definite article: **öküz-ü aldı** 'he bought the ox'; **kitab-ı okumadım** 'I have not read the book'.

(e) By being otherwise adequately defined, e.g. by a participle. The use of **bir**, the 'indefinite article', in such circumstances does not necessarily make the object indefinite; see XVI, 4.

A descriptive adjective is not in itself sufficient to make an object definite; compare **bir mavi kumaş istiyor** 'she wants a blue material' with **mavi kumaş-ı seçti** 'she chose the blue material'.

The second object of a factitive verb, i.e. a complementary object, remains in the absolute form: **onu Vali tayin ettiler** 'they appointed him Governor'; **İstanbul'u İstanbul yapan budur** 'what makes Istanbul Istanbul is this'; **seni arkadaş sanırdım** 'I used to think you a friend'.

11. The genitive case. The genitive suffix shows that the substantive to which it is attached stands in a possessive or qualifying relationship to another substantive; see § 17.

The substantive in the genitive case can also stand predicatively: **hâkimiyet millet-in-dir** 'sovereignty belongs to ("is of") the nation'; **bütün suç siz-in** 'all the guilt is yours' ('is of-you').

Certain postpositions, originally nouns, are construed with the genitive of personal pronouns; see VII, 3.

12. The dative case. This expresses:

(a) The indirect object of a verb: **mektubu Ali'ye gösterdim** 'I showed the letter to Ali'; **hizmetçi-ye bir palto vereceğiz** 'we are going to give the servant a coat'. It may translate the English 'for' as in **hizmetçi-ye bir palto alacağız** 'we are going to buy a coat for the servant'.

(b) Place whither: **Türkiye'ye döndüler** 'they returned to Turkey'; **yer-e düştü** 'it fell to the ground'; **şişeyi masa-ya koydu** 'he put the bottle on the table'; **borc-a batmıyalım** 'let us not plunge into debt'; **sandalye-ye oturdum** 'I seated myself on the chair' (but the locative is used in **sandalye-de oturuyordum** 'I was sitting on the chair'); **bir orman-a gizlendiler** 'they hid in a forest'.

(c) Purpose: **kız, çiçek dermeğ-e çıkıyor** 'the girl is going out to pick flowers'; **talebe, imtihan-a hazırlanıyor** 'the student is preparing for the examination'.

(*d*) Price: **bunu kaç-a aldın?** 'for how much did you buy this?'; **göz-e göz, diş-e diş** 'eye for eye, tooth for tooth'.

Turkish idiom requires a dative with a number of verbs whose English equivalents take a direct object,[1] among the commonest being: **başlamak** 'to begin', **değmek** and **dokunmak** 'to touch', **benzemek** 'to resemble', **devam etmek** 'to continue', **ermek** and **varmak** 'to reach', **girmek** 'to enter', **yardım etmek** 'to help'.

For postpositions with the dative, i.e. postpositions modifying or narrowing down the meaning of the dative, see VII, 4.

13. The locative case. This expresses location, which may be:

(*a*) In place: **tiyatro-da** 'at the theatre'; **su-da** 'in the water'; **yer-de** 'on the ground'; **ben-de para yok** 'I have no money on me'; **radyo-da bir vazo var** 'there is a vase on the wireless'.

(*b*) In time: **Ramazan-da** 'in Ramadan' (the month of fasting); **beş eylûl-de** 'on 5 September'.

(*c*) In an abstract: **radyo-da bir konuşma var** 'there is a talk on the wireless'; **ihtiyarlık-ta** 'in old age'; **sıhhat-te** 'in health'; **gitmek-te** 'in ⟨the act of⟩ going'. The locative is used with expressions denoting shape, size, colour, and age, where English idiom varies between 'of' and 'in': **yumurta şeklin-de bir taş** 'a stone in the shape of an egg'; **on metre uzunluğun-da bir ip** 'a cord of (lit. "in") ten metres' length'; **kahve rengin-de bir şapka** 'a hat of coffee-colour'; **yirmi yaşında** 'twenty years old' ('in the age of twenty'); **bu fikir-de değilim** 'I am not of this opinion'.

14. The ablative case. This case expresses point of departure:

(*a*) Place from which: **şehir-den ayrıldı** 'he departed from the city'; **rağbet-ten düştü** 'it fell from esteem, ceased to be in vogue'; **bu gidiş onu yerin-den edecek** 'this behaviour will cost him his job' ('will make him ⟨away⟩ from his position').

(*b*) Place through which: **pencere-den girdi** 'he entered by the window'; **hangi yol-dan gidilir?** 'by which road does one go?'; **sizi telefon-dan arıyorlar** 'you are wanted on the

---

[1] The most reliable guide to the case required with any given verb is *TS*.

telephone' ('they are seeking you through the telephone'); **haber radyo-dan yayıldı** 'the news was broadcast' ('was spread through the radio'); **hırsızı kolun-dan tuttum** 'I caught the thief by his arm'; **kitabı bir yerin-den daha açtım** 'I opened the book at another page' ('through one place more'). In such uses as **o kız kafa-dan sakattır** 'that girl is weak in the head' and **ihtiyar böbreklerin-den rahatsızdır** 'the old man has kidney-trouble' ('is ill through his kidneys'), the ablative is to be explained as like that in the two previous examples, i.e. as indicating the point through which someone or something is affected, rather than as causal.

(c) The causal use is very frequent: **muvaffakıyet-ten sarhoş** 'drunk from success'; **açlık-tan bitkin** 'exhausted from hunger'; **ne-den?** 'why?' ('from what?'); **on-dan** 'for that reason' ('from that'). Hence the use of the ablative with verbs such as **korkmak** 'to fear', **şüphelenmek** 'to suspect', **nefret etmek** 'to loathe', **hoşlanmak** 'to like'; what in English would be the object of the emotion is in Turkish its source.

(d) The second member of a comparison is put in the ablative: **Türkiye Lübnan'dan büyüktür** 'Turkey is bigger than Lebanon', i.e. Turkey is big if we take Lebanon as our point of reference.

(e) The ablative denotes the material from which something is made: **naylon-dan yapılmış bir balık ağı** 'a fishing-net made of nylon'; **söz gümüş-ten, sükût altın-dan** 'speech is silver, silence is gold'; **ateş-ten gömlek** 'shirt of fire' (a proverbial expression; cf. 'shirt of Nessus').

(f) The partitive use: **komşular-dan biri** 'one of the neighbours'; **üyeler-den birkaçı** 'several of the members'. Under this heading belongs **hafif-ten almak** 'to take lightly', lit. 'to take from the light', i.e. to take as belonging to the light.

(g) The ablative expresses price, but not synonymously with the dative: **bu elmaları kaç-tan aldın?** 'at what price did you buy these apples?' i.e. at what price each or per kilo. With the substitution of the dative **kaç-a** the meaning would be 'what was the total amount you paid for these apples?'

For postpositions with the ablative see VII, 5.

15. Personal suffixes. The suffixed personal pronouns, indicating possession, are:

THE NOUN

	After consonants	After vowels
Singular		
1	-im	-m
2	-in	-n
3	-i	-si
Plural		
1	-imiz	-miz
2	-iniz	-niz
3	-leri	

Thus a singular noun with the third-person plural suffix, e.g. **el-leri, çocuk-ları**, is identical in form with the plural of the noun with the third-person singular suffix (**eller-i, çocuklar-ı**) and with the accusative plural. Consonant stems with the third-singular suffix have the same form as the accusative singular, while with the second-singular suffix they have the same form as the genitive.

Consonant-stems:

	el	akşam	köy	çocuk
	hand	evening	village	child
my	elim	akşamım	köyüm	çocuğum
your (*sing.*)	elin	akşamın	köyün	çocuğun
his, her, its	eli	akşamı	köyü	çocuğu
our	elimiz	akşamımız	köyümüz	çocuğumuz
your (*pl.*)	eliniz	akşamınız	köyünüz	çocuğunuz
their	elleri	akşamları	köyleri	çocukları

Vowel-stems:

	anne	kapı	ölçü	korku
	mother	door	measure	fear
my	annem	kapım	ölçüm	korkum
your (*sing.*)	annen	kapın	ölçün	korkun
his, her, its	annesi	kapısı	ölçüsü	korkusu
our	annemiz	kapımız	ölçümüz	korkumuz
your (*pl.*)	anneniz	kapınız	ölçünüz	korkunuz
their	anneleri	kapıları	ölçüleri	korkuları

Two anomalies: **su** 'water' is treated as a consonant-stem (cf. § 7), while **ağabey** 'elder brother' (pronounced ābī, with the accent on the ā) behaves like a vowel-stem, though in the spelling this is acknowledged only with the suffix of the third-person singular:

			Pronounced
my	**suyum**	**ağabeyim**	ābim
your (*sing.*)	**suyun**	**ağabeyin**	ābin
his, her, its	**suyu**	**ağabeysi**	ābīsi
our	**suyumuz**	**ağabeyimiz**	ābīmiz
your (*pl.*)	**suyunuz**	**ağabeyiniz**	ābīniz
their	**suları**	**ağabeyleri**	ābīleri

The personal suffixes follow the suffix of the plural, except that two **-ler**s never occur together, so that **-i** and not **-leri** is used for the third-person plural suffix after plural nouns:

	eller	**çocuklar**
	hands	children
my	**ellerim**	**çocuklarım**
your (*sing.*)	**ellerin**	**çocukların**
his, her, its	**elleri**	**çocukları**
our	**ellerimiz**	**çocuklarımız**
your (*pl.*)	**elleriniz**	**çocuklarınız**
their	**elleri**	**çocukları**

Thus **çocukları** can mean 'his/her children', 'their children', or 'their child', as well as 'the children' (acc.), while **çocukların** can mean 'your children' or 'of the children'.

In the colloquial, **kardeşimler** means 'my brother and his family', **teyzemler** 'my aunt and her family' (cf. § 2, third paragraph), whereas **kardeşlerim** is 'my brothers' and **teyzelerim** 'my aunts'.

16. Personal suffixes followed by case-suffixes. An **n** appears between the suffix of the third person and any case-suffix, the result, with the singular of consonant-stems and all plurals, being identical in shape with the second-singular suffix plus the case-suffix. Thus the locative of **el-i** 'his hand' is **el-i-n-de** and of **el-in** 'your hand' **el-in-de**, while the dative of **eller-i** 'his hands' is **eller-i-n-e** and of **eller-in** 'your hands' **eller-in-e**. This ambiguity does not arise with the singular of vowel-stems: 'from his mother' is **anne-si-n-den** but 'from your mother' is **anne-n-den**; 'at his door' is **kapı-sı-n-da** but 'at your door' is **kapı-n-da**.

As late as the eighteenth century, the third-person suffix with the suffix of the accusative could be **-in** as well as **-ini**.

Some examples are given of the declension of nouns with the third-person suffix. There is no need to set out the declensions with the other personal suffixes, since, for example, **ellerimiz** 'our hands', **köyünüz** 'your village', **çocuklarım** 'my children' decline exactly like unsuffixed consonant-stems. Cf. the declensions of **el**, **köy**, and **akşam** respectively in § 6 (*a*) and (*c*).

	el-i his hand	anne-si his mother	köy-ü his village
acc.	elini	annesini	köyünü
gen.	elinin	annesinin	köyünün
dat.	eline	annesine	köyüne
loc.	elinde	annesinde	köyünde
abl.	elinden	annesinden	köyünden
	ad-ı his name	karı-sı his wife	çocuğ-u his child
acc.	adını	karısını	çocuğunu
gen.	adının	karısının	çocuğunun
dat.	adına	karısına	çocuğuna
loc.	adında	karısında	çocuğunda
abl.	adından	karısından	çocuğundan

The principle of suspended affixation (§ 9 (*d*)) must be borne in mind: **tebrik ve teşekkürlerimi sunarım** 'I offer my congratulations and thanks', the **-ler-im-i** applying to both nouns.

The suffix of the first-person singular added to **güzel** 'beautiful' and **can** 'soul' makes **güzelim** and **canım**, used as adjectives of endearment even with nouns with suffixes of other persons: **güzelim piyano-su** 'her lovely piano'; **canım Türkçemiz** 'our beloved Turkish'.

17. **The izafet group.** The commonest function of the suffix of the third person is to link one noun to another in a relationship most conveniently described by the Turkish term **izafet** 'annexation'. In English one noun may qualify another in two ways. In

the first, the qualifying noun is put into the genitive: Land's End, St. Antony's College, soldiers of the Queen. In the second, no grammatical mechanism but simple juxtaposition is involved: Lane End, Oxford University, Palace guard. The two types of izafet correspond fairly closely to these two English patterns, with the difference that in both Turkish types the qualified noun takes the third-person suffix.

The classical Turkish grammarians recognize a third type of izafet in which neither noun has a suffix, namely, when the first is a noun of material: **altın bilezik** 'gold bracelet', **demir perde** 'iron curtain'.[1] If we are concerned only with the facts of modern Turkish, however, it is more practical to regard names of materials as being indifferently used as nouns or adjectives, as in English. Leaving the 'izafet of material' aside, therefore, the two types of izafet are the definite or possessive and the indefinite.

The definite izafet is employed when the first element is a definite person or thing to which or within which the second belongs.[2] The first noun has the genitive suffix, the second has the suffix of the third person: **uzman-ın rapor-u** 'the expert's report' ('of-the-expert his-report'), **hafta-nın günler-i** 'the days of the week', **uzman-ın kendi-si** 'the expert himself' ('of-the-expert his-self'), **İstanbul'un kendi-si** 'Istanbul itself'.

The indefinite izafet is used when the relationship between the two elements is merely qualificatory and not so intimate or possessive as that indicated by the definite izafet. The second noun has the suffix of the third person, but the first noun remains in the absolute form. As a working rule, an indefinite izafet group can be turned into intelligible (though not necessarily normal) English by the use of a hyphen: **Ankara şehr-i** 'Ankara-city'; **seçim kurul-u** 'election-committee'; **Türkiye Cumhuriyet-i** 'the Turkey-Republic'.

The distinction between the two types is seen in the following pairs of examples: **Üniversite-nin profesörler-i** 'the professors

---

[1] This view is shared by S. S. Mayzel in his exhaustive monograph *Izafet v turetskom yazyke* (Akademia Nauk S.S.S.R., 1957). He states (pp. 98–99) that the suffixless izafet is used only with nouns denoting worked materials, but his example of **mercan terlik** 'coral slippers' as opposed to **mercan kıyılar-ı** 'coral shores' is irrelevant in that these slippers are not made of coral but derive their name from the **Mercan** quarter of Istanbul.

[2] The term 'belongs' here implies grammatical possession; e.g. in **ev-in sahib-i** 'the house's owner' the first element, though legally and logically the property of the second, is grammatically its possessor.

of the University'; **üniversite profesörler-i** 'university professors'. **Orhan'ın ism-i** 'Orhan's name'; **Orhan ism-i** 'the name "Orhan"'. **kimse-nin cevab-ı** 'nobody's answer'; **kimse cevab-ı** 'the answer "nobody"'. **Sultan Ahmed'in türbe-si** 'Sultan Ahmet's tomb'; **Sultan Ahmet cami-i** 'the Sultan Ahmet Mosque'. **Atatürk'ün ev-i** 'Atatürk's house'; **Atatürk Bulvar-ı** 'the Atatürk Boulevard'. **çoban-ın kız-ı** 'the shepherd's daughter'; **çoban kız-ı** 'the shepherd-girl'.

Suspended affixation operates in izafet too: **halk-ın acı ve sevinçleri** 'the sorrows and joys of the people', i.e. **acı-lar-ı ve sevinç-ler-i**.

A special use of the indefinite izafet with proper names is seen in **Bekir çapkın-ı** 'that rascal of a Bekir'; **Nuri serseri-si** 'that vagabond of a Nuri'; **Ethem hırsız-ı** 'that thief of an Ethem'.

The qualifier may be indefinite in expressions denoting family relationships such as **Bedri eş-i Fatma** 'Bedri's wife Fatima'; **Hasan kız-ı Sevim** 'Hasan's daughter Sevim'; **İsmail oğlu Mehmet** 'Ismail's son Mehmet'. Hence a common type of surname ending in **-oğlu**; cf. our *Johnson* rather than *John's son*. Conversely, villagers use personal names without the third-person suffix after the father's name or family name in the genitive: **Ahmed'in Mustafa** 'Ahmet's son Mustafa'; **Arifler-in Abbas** 'the Arifs' son Abbas'; **Kara Ahmetler-in Leylâ** 'the Black Ahmets' daughter Leyla'. In the last two examples, the 'family name' is the father's name with the plural suffix.

When an adjective, a demonstrative, or an adverb comes between the two elements of an izafet group, the first element must be in the genitive; cf. the English 'committee meeting' but 'the committee's next meeting'. **İstanbul camiler-i** 'the Istanbul mosques' but **İstanbul'un tarihî camiler-i** 'the historic mosques of Istanbul'; **mahkeme karar-ı** 'court decision' but **mahkeme-nin bu karar-ı** 'this decision of the court'; **su donma-sı** 'freezing of water' but **suy-un birdenbire donma-sı** 'the water's suddenly freezing'.[1] This rule does not apply when the intervening adjective is part of a compound noun such as **büyükelçi** 'ambassador' (lit. 'great envoy'): **Türkiye Büyükelçisi** 'the Ambassador of Turkey'. Compare, however, **Türkiye'-nin büyük şehirler-i** 'the great cities of Turkey'.

[1] An adverb may come within an izafet group only, as in this example, when the qualified element is a verbal noun.

ordu subayları	army-officers
bu ordu subayları	these army-officers
bu ordu-nun subayları	the officers of this army
bu ordu-nun bu subaylari	these officers of this army

18. Words indicating nationality. Those formed by suffixing -li (IV, 5) to the name of a country, e.g. **Kıbrıs-lı** 'Cypriot', **Danı-marka-lı** 'Danish', may be nouns or adjectives. All other words indicating nationality, e.g. **Türk, İngiliz, Fransız, Alman**, are nouns and are therefore joined to a following noun by an indefinite izafet: **İngiliz edebiyat-ı** 'English literature'; **Fransız askerler-i** 'French soldiers'. As the singular denotes a class, as well as one member of that class,[1] such expressions must be regarded as meaning not 'the-Englishman his-literature', 'the-Frenchman his-soldiers', but 'the-English their-literature', 'the-French their-soldiers'. A definite izafet may be used instead if it is desired to show a more intimate relationship: **Türk ruh-u** 'the Turkish soul' but **Türk-ün ruh-u** 'the soul of the Turk'.

The colloquialism **Türk iş**, used in self-disparagement when something goes wrong, as we might say 'a typical piece of British muddle', is rather puzzling, since one would expect **Türk iş-i** 'Turkish work'. One explanation is that this expression is not Turkish at all, but German; a relic of the days when German officers were training the Ottoman Army. That is to say, it is an expostulatory *Türkisch!* originally accompanied by a heavenward rolling of the eyes. Alternatively, it might be an imitation, deriving from the same period, of an attempt to say 'Turkish work' on the part of a foreigner unacquainted with the finer points of the language. The former explanation seems more likely.

For 'American' two words exist, **Amerikan** and **Amerikalı**. The former is a noun, used only in izafet, and means 'the body politic of all the Americans'; it bears the same relationship to **Amerikalılar** as 'the English' does to 'the Englishmen' and is used to qualify things, whereas **Amerikalı** is an adjective or noun denoting persons of American nationality: **Amerikan hükümet-i** 'the American government', **bir Amerikan uçağ-ı** 'an American aircraft'; but **bir Amerikalı** 'an American', **Amerikalı subaylar** 'American officers'. 'American Ambassador', however, is **Amerikan Büyükelçi-si**, for **Amerikalı Büyükelçi** would

---

[1] Thus 'Girls' Lycée' is **Kız Lise-si** not **Kızlar Lise-si**.

mean 'Ambassador of American nationality'. The use of **bir Amerikan** for 'an American' is a vulgarism.

A similar pair of words exists for 'Italian': **İtalyan** and **İtalyalı**. The latter, however, is virtually obsolete.

Nouns of nationality may be used in apposition with other nouns, instead of in izafet, when denoting membership of a people rather than of a nation: **Türk liderler-i** and **Türk liderler** both mean 'Turkish leaders', but the first denotes leaders of the Turkish nation whereas the second denotes leaders of the Turkish community in Cyprus. Similarly, **bir Yahudi asker-i** is a soldier of the Jewish nation, an Israeli soldier, while **bir Yahudi asker** is a soldier of any nation who happens to be a Jew. An apposition is also possible if the nationality of the person is not stressed, or if the second element is personified; e.g. **bir Rus jeolog** is a geologist who happens to be a Russian, while **bir Rus sözcü-sü** is 'a Russian spokesman'; **Türk İstanbul** 'Turkish Istanbul'.

There is some fluctuation of usage with the words **Sovyet** and **komünist**; some refer to the Soviet government and the Communist bloc as **Sovyet hükümet-i** and **Komünist blok-u**; others prefer **Sovyet hükümet** and **Komünist blok**.

Names of continents are used as qualifiers of things: **Avrupa başkentler-i** 'European capitals' ('Europe its-capitals'), **Afrika nehirler-i** 'African rivers', **Asya memleketler-i** 'Asian countries'. The forms in **-li** are used only of persons: **Afrikalılar** 'Africans', **Asyalı gazeteciler** 'Asian journalists', **Avrupalı turistler** 'European tourists'.

19. The izafet chain. An izafet group may itself be qualified by a preceding noun: **il seçim kurul-u** 'province election-committee'; **Ankara Kız Lise-si** 'Ankara Girls' Lycée'. Only the last noun in the chain has the third-person suffix, which does double duty: not only does it link **kurul** and **lise** to their immediate qualifiers **seçim** and **kız**; it also links the groups **seçim kurulu** and **kız lisesi** to their qualifiers **il** and **Ankara**.[1]

A definite izafet is also possible in such situations: **hakem-in favl karar-ı** 'the referee's decision of "foul"'; **gün-ün dedikodu konu-su** 'the gossip-topic of the day'. In **Bulgaristan'ın İstanbul Başkonsolosluğ-u** 'the Istanbul Consulate-General

---

[1] It is most important to note that the third-person suffix is not repeated though theoretically one might have expected **Ankara (Kız Lise-si)-si**.

of Bulgaria' the first qualifier is in the genitive because its relationship with the qualified word is closer than is that of the second qualifier.

An izafet group may qualify a following noun: **Diyanet İşler-i Bakanlığ-ı** 'Religion-Affairs Ministry'. Here it will be seen that both qualified nouns have the third-person suffix; that of **İşler** links it to its qualifier **Diyanet**, while that of **Bakanlık** links it to *its* qualifier, the group **Diyanet İşler-i**. Another example: **sene son-u imtihanlar-ı** 'year-end examinations'. The distinction between this pattern and that of **Ankara Kız Lisesi** may be seen by comparing **Ford aile araba-sı** 'the Ford family-car' with **Ford aile-si araba-sı** 'the Ford-family car'. Other possibilities are: **Ford aile-si-nin araba-sı** 'the car of the Ford-family', **Ford'un aile-si-nin araba-sı** 'the car of Ford's family', and **Ford'un aile araba-sı** 'Ford's family-car'.

As a rule (but see §§ 20, 21, 24), any noun in an izafet chain which does not have the third-person suffix is not qualified by a preceding noun. In **Cumhuriyet Halk Parti-si** 'Republican People's Party',[1] as **Halk** has no suffix we know it is not in izafet with **Cumhuriyet**, so the literal meaning is not 'Republic-People Party' but 'Republic People-Party'. In **Türk Dil Kurum-u**, the fact that **Dil** has no suffix shows that it is not qualified by **Türk**, so the phrase means not 'Turkish-Language Society' but 'Turkish Language-Society'. **Türk Dil-i Dergi-si**, however, means 'Turkish-Language Journal'. So too in **makine şerid-i mürekkeb-i**; the second word is qualified by the first and both together qualify the third: 'typewriter-ribbon ink'.

The izafet chain can be extended as required: **İstanbul Üniversite-si Edebiyat Fakülte-si Türk Edebiyat-ı Profesör-ü** 'Istanbul-University Literature-Faculty Turk-Literature Professor', i.e. 'Professor of Turkish Literature of the Faculty of Letters of the University of Istanbul'. **İzmir Örme Sanayi-i İşçiler-i Sendika-sı** 'Izmir Knitting-Industry Workers' Union'.

The rules and examples given should enable the student to unravel any izafet chain, but he may sometimes encounter a definite izafet where he might have expected an indefinite or vice versa; individual authors' ideas of style may vary. It is entirely a matter of taste whether one writes **CHP aday-ı** 'the RPP

[1] Abbreviated to **CHP**. The English translation is abbreviated to RPP.

candidate' or **CHP'nin aday-ı** 'the RPP's candidate'. Most writers keep the number of genitives in an izafet chain down to the minimum that is consistent with intelligibility. For example, in **Bohemya Kırallar-ı saray-ı-nın yeni sâkin-i** 'the new inhabitant of the palace of the Kings of Bohemia' **Kırallar-ı-nın** might have been expected as denoting the owners of the palace. But **saray-ı** had to be in the genitive because the adjective **yeni** separates it from **sâkin-i**, and the juxtaposition of two genitives is avoided as far as possible. The partitive use of the ablative makes it possible to dispense with one genitive, e.g. in **komite üyeler-i-n-den bir-i-nin oy-u** 'the vote of one of the members of the committee'.

20. Place-names consisting in an izafet group. These tend to drop the third-person suffix. **Kadıköy** on the Asiatic shore of the Bosphorus was **Kadı-köy-ü** ('judge-village') barely a generation ago, the **köyü** declining as shown in § 16. Now the **köy** declines as shown in § 6 (*a*). This tendency is doubtless helped by the existence of some other place-names similarly compounded of a noun and **köy** which, if they ever had the third-person suffix, lost it long ago, e.g. **Arnavutköy** and **Bakırköy**, and of some compounded of an adjective and **köy**, e.g. **Yeşilköy**. Another contributory factor may be that the accent in izafet groups is always on the first element, and in place-names is towards the beginning of the word, so that the third-person suffix in the absolute case would tend to be swallowed up. Indeed, the growing practice, frowned on by purists, is for **sokak** 'street' to stand in izafet without the third-person suffix. 'Grocer Street' is properly **Bakkal Sokağ-ı**, but one often hears—and reads—**Bakkal Sokak**. The suffix is secure for the moment in names of roads, squares, hills, and impasses: **Babıâli Cadde-si, Hürriyet Meydan-ı, Fincancılar Yokuş-u, Korsan Çıkmaz-ı**.

European influence has for some years been helping this tendency (which is, however, native in origin), e.g. in names of new office-buildings (**han**), banks, restaurants, and hotels: **Boyacılar Han, Pamuk Bank** (with the western **Bank** instead of the Turkish **Banka**), **Yıldız Lokanta, Paris Otel**, in place of **Hanı, Bankası, Lokantası, Oteli**. Even such gallicisms as **Ristoran Yıldız** and **Otel Paris** have begun to appear in Istanbul.

21. Culinary terms without izafet. The third-person suffix is lacking in some time-honoured names of dishes, such as **şiş kebap** 'skewer roast', **ızgara köfte** 'grill mincemeat', **kuzu pirzola** 'lamb chop', all originally cooks' and waiters' jargon and therefore as untypical of ordinary speech as 'eggs and chips twice'. Analogy with these may help to explain why a recently marketed tomato ketchup is labelled **Domates Ketçap** and not **Ketçapı** (though 'tomato-juice' is **domates suyu** not **su**). The main reason is probably that the manufacturer wishes to familiarize the public with the name **ketçap** and therefore presents it in the absolute form without bothering about grammar.

22. Third-person suffix with substantivizing and defining force. The third-person suffix is used as a syntactic device for creating and defining nouns. The stages in the development of this use are exemplified thus:

(*a*) **iş-in fena-sı şu** 'the bad ⟨part⟩ of the business is this'; **edebiyat-ın iyi-si** 'the good ⟨part⟩ of literature, good literature'; **geceler-in güzel-i, yıldızlı-sı** 'the beautiful night is the starry one' ('of-the-nights, their-beautiful is their-starry').

(*b*) **doğru-su** 'honestly, to tell you the truth' ('the true-of-it'). Here the antecedent is vaguely the matter under discussion.

(*c*) **zengin-i aynı şeyi söylüyor, fakir-i aynı şeyi söylüyor** 'the rich man says the same thing, the poor man says the same thing', lit. 'the-rich-of-it, the poor-of-it'; i.e. of people at large. In **sizden akıllısı yok** 'there is none cleverer than you' **sizden akıllı** 'cleverer than you' is an adjectival phrase, substantivized by the third-person suffix. Cf. **bir-i** 'one of them, someone' (V, 7).

(*d*) **bundan sonra** 'after this'; **bundan sonrası** 'that which is after this, what happens next', lit. 'the after this of it', where the 'it' is the scheme of things entire. Probably under this head is to be sought the explanation for **burası, şurası**, etc. (XII, 12).

23. The Janus construction. By this term is meant the curious facing-both-ways construction wherein, when two people who are related or otherwise closely connected are mentioned in one sentence, each is defined by a third-person suffix linking him to the other:

oğl-u baba-sı-na bir mektup yazdı 'the son wrote a letter to the father', lit. 'his—the father's—son wrote a letter to his—the son's—father'.

kız-ı-nı vermediği için anne ve baba-sı-nı öldürdü (newspaper headline) 'he killed the mother and father because they did not give their daughter ⟨to him in marriage⟩'.

babalar-ı-nın cezasını oğullar-ı çekecek 'the sons will suffer the fathers' punishment' ('their sons . . . their fathers'').

hasta-sı doktor-u-nu arıyor 'the patient is seeking his doctor' ('his patient . . . his doctor').

hoca-sı talebe-si-ne bakar 'the teacher looks after the pupil'.[1]

24. Suffixes with izafet groups. We saw in § 19 that in izafet chains the third-person suffix does double duty and is not repeated. There are two small classes of words which can have two personal suffixes: pronouns such as **bir-i-si** (V, 7; and cf. **şeysi** < **şey-i-si** V, 20) and frozen izafet groups such as **yüzbaşı** 'captain'. Originally this was **yüz baş-ı** 'hundred its-head' but through frequent use has come to be treated as a simple noun, declining like **tarla** (§ 6 (*d*)) and not **ad-ı** (§ 16), e.g. the plural is **yüzbaşılar** not **yüzbaşları**. It can therefore take personal suffixes: **yüzbaşım** 'my captain', **yüzbaşısı** 'his captain', etc. See also XIV, 33. Otherwise no word can have more than one personal suffix.[2] When a third-person possessor of an izafet group is to be indicated, the third-person suffix is not repeated: **yaz** 'summer', **tatil** 'holiday'; **yaz tatil-i** 'summer holiday' or 'his summer holiday'. When a first or second person is the possessor, the third-person suffix of the izafet gives way to the suffix of the first or second person:[3]

---

[1] The last two examples are from Elöve, p. 1053. It must be noted that Elöve, at this point, has quite misunderstood Deny, p. 168.

[2] i.e. in the standard language. Vulgarisms such as **kadın kısm-ı-sı** (for **kısm-ı**) 'the female sex', lit. 'the woman-section', are not unknown.

[3] Fuat Köprülü, on p. 9 of his *Millî Edebiyat Cereyanının İlk Mübeşşirleri* (Istanbul, 1928, in the old alphabet), uses the words **muhtelif eserlerimizde ve bilhassa (Türk edebiyatı tarihi)mizde** 'in our various works and especially in our *History of Turkish Literature*', boldly but solecistically adding the post-vocalic first-person plural suffix **-miz** to the third-person singular suffix **-i** to produce what sounds like the post-consonantal first-plural suffix **-imiz**. He could not have fallen into this error if he had been quoting a title ending in the post-vocalic third-person suffix **-si**, e.g. his *Halk Ebediyatı Ansiklopedisi*.

	summer holiday	summer holidays
my	yaz tatilim	yaz tatillerim
your	yaz tatilin	yaz tatillerin
his	yaz tatili	yaz tatilleri
our	yaz tatilimiz	yaz tatillerimiz
your	yaz tatiliniz	yaz tatilleriniz
their	yaz tatilleri	yaz tatilleri

All the possible ambiguities can be resolved by the use of a noun or personal pronoun in the genitive:

Ahmed'in yaz tatili	Ahmet's summer holiday
onun yaz tatili	his summer holiday
onun yaz tatilleri	his summer holidays
onların yaz tatili	their summer holiday
onların yaz tatilleri	their summer holidays

Similarly, **İzmir büro-su** may mean 'the Izmir office' or 'his Izmir office':

| onun İzmir bürosu | his Izmir office |
| şirket-in İzmir bürosu | the company's Izmir office |

When the suffixes **-li** and **-ci** (IV, 4, 5) are added to an izafet group, the third-person suffix is dropped: **Gece ad-ı** 'the name "Night"'; **Gece ad-lı şiir** 'the poem named "Night"'. **Avrupa fermuar-ı** 'European fastener' (French *fermoir*); **Avrupa fermuar-lı çantalar** 'bags fitted with European fasteners'; **su yol-u** 'water-conduit'; **su yol-cu** 'man responsible for the upkeep of water-conduits'.

25. The vocative use of the third-person suffix. In English a woman may, in the presence of her child, address her brother as 'Uncle', just as the child would do. In Turkish she would address him as **dayı-sı** 'his uncle'. Similarly, if an English-speaking child is being teased by another and runs off calling 'Mother!' the other child may mockingly echo his cry. In Turkish, however, the mocker calls not **Anne!** but **Annesi!**

26. Persian izafet. It was because of the extensive use of this alien grammatical feature, coupled with the borrowing of an immense Arabic and Persian vocabulary, that the literary and administrative language of the Ottoman Empire was largely unintelligible to most of its Turkish subjects. In Persian the qualifier follows the qualified, the opposite of Turkish usage, and the qualified is joined to its qualifier, noun or adjective, by an *i*, as in *koh-i-nur* 'mountain of light' and *koh-i-bozorg* 'great mountain'. This device was used in Ottoman as in Persian, to link Arabic as well as Persian words: **nokta-i nazar** 'point of view'; **Abdülhamid-i sani** 'Abdulhamid the Second'. The linking **i** was usually subjected to the Turkish vowel harmony and was separated from a preceding long vowel by a **y**: **Şura-yı Devlet** 'Council of State'; these words, in Turkish izafet, would be **Devlet Şura-sı**. Analogous violations of normal word-order are found in English: *court martial, blood royal, law merchant*.

As if this was not enough, Ottoman followed Persian in borrowing from Arabic the curse of grammatical gender, from which Turkish and Persian were born free. In Arabic, *dawlat-* 'state, dynasty, empire', whence Turkish **devlet**, is feminine. In Ottoman, therefore, the Arabic adjectives meaning 'high' and 'Ottoman' (*'alī* > **ali**, *'Uthmānī* > **Osmani**) took their Arabic feminine forms in the official name of the Empire: **Devlet-i aliye-i Osmaniye** 'The High Ottoman State'.

The Turkish words for 'some' and 'same', **bazı** and **aynı**, are respectively the Arabic *baʿḍ* 'part' and *ʿayn* 'counterpart' with the Persian izafet, and mean literally 'part of' and 'the counterpart of'.

Purists condemned the use of native Turkish words in Persian izafet, but many Turkish words were in fact so used in Ottoman phraseology, e.g. **ordu** 'army' and **sancak** 'banner' in **ordu-yu hümayun** and **sancağı şerif**,[1] 'Imperial Army' and 'Noble Banner', i.e. the standard of the Prophet. Such phrases were classed as **galat-ı meşhur** 'widely disseminated mistake', i.e. solecism legitimized by usage. The plural of this term, incidentally, was **galatat-ı meşhure**, the adjective being made feminine to agree with the feminine plural noun. An oft-quoted saying runs: **galat-ı meşhur lûgat-i fasihten yeğdir** 'the generally used solecism is better than the chaste locution'.

Nowadays, Persian izafet compounds which have become part of the standard vocabulary are usually spelled as one word: **aksıseda** 'echo' ('reflection of voice'); **hikmetivucüt** *'raison d'être'*; **aklıselim** 'common sense'. These present no difficulty because

---

[1] Originally **sancak-ı şerif**; see I, 10, at end.

they will be found in the dictionary. In less well-acclimatized compounds the rule is to separate the elements, with a hyphen between the first and the vowel of the Persian izafet: **ceride-i çamur** 'organ of the gutter-press' ('newspaper of mud'); **vuzuh-u beyan** 'clarity of exposition'; **muhtac-ı himmet** 'needful of help'; **üful-ü nabehengâm** 'untimely demise'; **mefhum-u muhalif** 'contrary concept, converse'.

Some writers, however, make compounds of either type into two separate words, the first incorporating the vowel of the izafet; **hikmeti vücut, vuzuhu beyan**, etc. Further, some may limit the vowel harmony, rejecting **u** and **ü** and writing, e.g. **vuzuh-ı beyan, üful-i nabehengâm**, as being closer to the original Persian pronunciation.

III

THE ADJECTIVE

1. General observations. The dividing line between noun and adjective is a thin one, but is still worth drawing. If we take as the criterion of a noun the permissibility of using the plural, case, and personal suffixes after it, or the indefinite article **bir** before it, very few of the words classed as adjectives in the dictionary will be excluded. **büyük** 'big, old', **büyüklerim** 'my elders'; **hasta** 'ill', **bir hasta** 'a sick man'; **genç** 'young', **gençlerin** 'of the young'; **Avrupalı** 'European', **Avrupalıya** 'to the European'. The only large class of exceptions, i.e. of adjectives which are not used as nouns, are those formed with the Turkish suffixes **-si**, **-(i)msi**, **-(i)mtrak**, and **-(s)el**, the Arabic **-î**, and the Persian **-ane** and **-varî**, to which may be added recent borrowings like **demokratik** and **kültürel**.

On the other hand, if we take as the criterion of an adjective the permissibility of putting it in the comparative and superlative degrees, vast numbers of nouns will be excluded. In other words, although most adjectives can be nouns, the converse does not hold good.

2. Attributive adjectives. These precede their nouns: **cesur adamlar** 'brave men'; **uzun yol** 'the long road'. Two exceptions:

(*a*) **kare** and **küp**, 'square' and 'cubic', follow names of units of length, as in French from which they are borrowed: **Kıbrıs'ın yüz ölçümü 3.572 mil kare (9.251 kilometre kare) dir** 'the area of Cyprus is 3,572 square miles (9,251 square kilometres)'. **bir metre küp** or **bir metreküp** 'one cubic metre'.

(*b*) **merhum** 'the late' is sometimes used after the name of the deceased instead of before, in imitation of Arabic usage.

3. The indefinite article. bir 'one' may be so termed although the name is not entirely appropriate. One reason is that the noun introduced by **bir** may be in the defined accusative; see XVI, 4.

Another difference from what we understand in English by the indefinite article is that **bir** may introduce a noun in the plural, the effect being vaguer than with a singular noun: **bir şey mırıldandı** 'he mumbled something', but **bir şeyler mırıldandı** 'he mumbled something or other'; **bir zamanlar ben de çocuktum** 'once I too was a child', where **bir zaman** would be too precise: 'at one time'. In this latter example it should also be noted that **bir** is not used before **çocuk**, although the English has 'a child'; its omission is customary with the complement of such verbs as 'to be' and 'to become'.

When it serves as an indefinite article, **bir** usually comes between adjective and noun: **büyük bir tarla** 'a large field', **güzel bir bahçe** 'a beautiful garden'. When it means 'one', it must precede the adjective, just like any other numeral: **iki küçük tarla sattı, bir büyük tarla aldı** 'he sold two small fields, he bought one large field'. This must not be taken to imply that **bir** when it precedes an adjective and noun is always to be translated by 'one'; English idiom may sometimes call for 'a' or 'any' (cf. XVI, 4). The key to understanding this point lies in the basic principle of Turkish syntax: whatever precedes, qualifies. The essential difference between **güzel bir bahçe** and **bir güzel bahçe**, both of which may translate 'a beautiful garden', is that the first means a beautiful member of the class 'garden', the second a member of the class 'beautiful garden'. **güzel bir bahçe** is a beautiful garden as distinct from a less beautiful or even a frankly ugly garden; **bir güzel bahçe** is a beautiful garden as distinct from a beautiful meadow or an ugly forest.

4. Comparison of adjectives. The comparative degree is expressed by putting the second member of the comparison (introduced in English by 'than') in the ablative case: **ağır** 'heavy', **kurşun-dan ağır** 'heavier than lead'. 'Less . . . than' is translated by putting **az** 'little' between the second member in the ablative case and the adjective: **kurşundan az ağır** 'less heavy than lead'. **daha** 'more' may be inserted for emphasis: **kurşundan daha ağır, kurşundan daha az ağır**. It is not essential, however, except in the absence of a second member, e.g. in 'this hammer is cheaper, that one is stronger' **bu çekiç daha ucuz, öteki daha sağlam**, or in such 'floating comparatives' as 'For Whiter Washing' **daha beyaz çamaşır için**.

The ancient comparative suffix -rek, which appears in a few diminutives (IV, 1 (a)), retains its original force in **yeğrek** 'better, best', from **yeğ** 'good'. **yeğ** and **yeğrek**, also spelled **yey, yeyrek**, survive only in proverbs.

The superlative degree is expressed by **en** 'most': **İstanbul en büyük şehrimizdir** 'Istanbul is our greatest city'; **bu toprak en az verimlidir** 'this soil is least fertile'.

5. Arabic and Persian comparatives. The Persian **beter** 'worse' occurs mostly in proverbs. The Arabic elative, which serves as both comparative and superlative, is familiar to us from the slogan *Allāh akbar* 'God is most great'. In Turkish it is currently represented by **elzem** 'essential', the Arabic *alzam*, elative of *lāzim* 'necessary'; **enfes** 'most delightful' (*anfas < nefīs*); **akdem** 'prior' (*aqdam < qadīm* 'ancient'); **ender** 'most rare' (*andar < nādir*); **ehven** 'easiest, very cheaply' (*ahwan < hayn*). Often these words are reinforced in Turkish, as **daha beter** 'worse', **en enfes** 'most delightful'.

6. Intensive adjectives. The only regular[1] use of prefixation is to intensify the meaning of adjectives and, less commonly, of adverbs. The prefix, which is accented, is modelled on the first syllable of the simple adjective or adverb but with the substitution of **m, p, r,** or **s** for the last consonant of that syllable. It is hard to discern any principle governing the choice of consonant, except that **p** is commoner with back vowels than with front vowels. The following list includes the commonest of such formations; the meaning of the intensive is not given when it is obvious from the meaning of the simple word, as **apaçık** 'wide open, manifest' from **açık** 'open', or **yepyeni** 'brand new' from **yeni** 'new'.

açık	open	apaçık	
başka	other	bambaşka	totally different
belli	evident	besbelli	
beyaz	white	bembeyaz	
bok	ordure	bombok	utterly useless
boş	empty	bomboş	
bütün	whole	büsbütün	altogether, entirely

[1] i.e. excluding such curiosities as the recent coinage **ön-görmek** 'to foresee'.

çabuk	quick	çarçabuk	
cavlak	naked, bald	cascavlak	
dızlak	,, ,,	dımdızlak	
doğru	straight	dosdoğru	
dolu	full	dopdolu	
gök	blue	gömgök	
kara	black	kapkara	
katı	hard	kaskatı	
kırmızı	red	kıpkırmızı	
kızıl	,,	kıpkızıl	
kuru	dry	kupkuru	
kütük	drunk	küskütük	
mavi	blue	masmavi	
mor	violet	mosmor	
sarı	yellow	sapsarı	
sıkı	tight	sımsıkı	
siyah	black	simsiyah	
takır	(*imitates tapping*)	tamtakır	quite empty
tamam	complete	tastamam	
taze	fresh	taptaze	
temiz	clean	tertemiz	
toparlak	round	tostoparlak	
uzun	long	upuzun	
yassı	flat	yamyassı	
yeni	new	yepyeni	
yeşil	green	yemyeşil	

Irregular are: **çıplak** 'naked', **çırılçıplak** as well as **çırçıplak**; **sağlam** 'healthy', **sapasağlam**; **yalnız** 'alone', **yapayalnız** as well as **yapyalnız**; **çevre** 'circumference', **çepeçevre** as well as **çepçevre** 'all around'; **gündüz** '(in) daylight', **güpegündüz** 'in broad daylight'; **düz** 'flat', **dümdüz** 'absolutely flat' and **düpedüz** 'downright, openly'; **parça** 'piece', **paramparça** 'broken to bits'. The intensive **sırsıklam** or **sırılsıklam** 'sopping wet' is current, although the simple **sıklam** 'wet' is no longer in use. From **eyü**, an earlier form of **iyi** 'good', comes **epey** 'rather a lot (of)'.

Other such formations are sometimes created in speech without attaining general currency, e.g. **gepegenç** from **genç** 'young'.

IV

NOUN AND ADJECTIVE SUFFIXES

THIS chapter deals with the suffixes whereby nouns and adjectives are derived from other nouns and adjectives.

1. Diminutives. The diminutive suffixes are **-rek**, **-cek**, **-ceğiz**, **-cik**, and **-ce**, of which the first two are no longer productive. Before these suffixes, adjectives invariably and nouns usually drop final **k**.

(*a*) **-rek**, the ancient comparative suffix (III, 4), survives with diminutive force in **acırak** 'rather bitter' (**acı** 'bitter'), **bozrak** 'light grey' (**boz** 'grey'), **küçürek** 'rather small' (**küçük** 'small'), **ufarak** 'rather tiny' (**ufak** 'tiny'), **alçarak** 'lowish' (**alçak** 'low').

(*b*) **-cek** survives in **oyuncak** 'toy' (**oyun** 'game'); in **yavrucak** (also **yavrucuk**), the diminutive of **yavru** 'the young of an animal'; in **büyücek** 'biggish' (**büyük** 'big') and **küçücek** 'very small'. **orayacak** 'all that way' (**oraya** 'thither') is provincial.

(*c*) **-ceğiz**, an extended form of **-cek**, is particularly common with nouns denoting living beings and conveys a sense of affection, sometimes mixed with pity: **adamcağız** 'the poor wee man', **kızcağız** 'the dear little girl', **hayvancağız** 'the poor little creature' (**hayvan** 'animal'), **köyceğiz** 'the dear little village'.

(*d*) **-cik**, the most widely used diminutive suffix, throws the accent on to the first syllable: **Ayşecik** 'little Ayesha'; **Mehmetçik** 'little Mehmet', the affectionate term for the private soldier; **evcik** 'little house'; **alçacık** 'very low, humble' (**alçak**). From **bebek** 'baby' and **köpek** 'dog' come **bebecik** and **köpecik**, less commonly **bebekçik**, **köpekçik**.

A few monosyllables vary slightly from the regular pattern: **az** 'little, few' makes **azacık** and **azıcık** as well as **azcık**; **dar** 'narrow' makes **daracık**; **bir** 'one' makes **biricik** 'unique'.

(*e*) **-ce** has a modifying effect on adjectives: **güzelce** 'quite good' (but not so good as the simple **güzel**), **seyrekçe** 'rather infrequent', **uzunca** 'rather long', **genççe** 'quite young'. It makes a few nouns from verbal nouns in **-me** (X, 7): from **bilme** 'guessing',

bilmece 'riddle'; from **bulma** 'finding', **bulmaca** 'puzzle', especially 'crossword-puzzle'; from **kapma** 'catching', **kapmaca** 'the game of puss-in-the-corner'; from **çekme** 'drawing', **çekmece** 'drawer'. This suffix, which is accented, must not be confused with the enclitic -ce which makes adverbs; see XII, 2. It may be followed by -cik as in **genişçecik** 'pretty wide', **yakıncacık** 'quite near'.

2. Diminutives of personal names. Apart from those formed with -cik, these do not seem reducible to a rule; there is no obvious reason why people named **Mustafa** should be addressed as **Mıstık**. Commonly the first syllable only of the name is retained and to it is added **i, o**, or a syllable ending in **ş**: **Ercüment** > **Erci**; **Neriman** > **Neri**; **Mehmet** > **Memiş** or **Memo**; **Metin** > **Metiş**; **Fatma** > **Fatoş** (also, affectedly, **Fatış** or **Fatı**); **Hasan** > **Hasso**; **Ali** and **Aliye** > **Ališ**; **Cemâl** > **Cemo**; **İbrahim** > **İbo**. The forms in -o are accented on the first syllable. They are more familiar and socially less acceptable than those in -ş; cf. the difference in English between *Bert* and *Bertie* as diminutives of *Albert*.

3. -(i)msi, -(i)mtrak, -si. These three suffixes in some contexts have diminutive effect but basically they mean 'resembling', like English *-ish* in *womanish*. The initial **i** of the first two is lost after vowels.

(*a*) **-(i)mtrak**, the **a** of which is invariable in the best authors, is used with adjectives of colour and taste; **beyazımtrak** 'whitish'; **yeşilimtrak** 'greenish'; **ekşimtrak** 'sourish' (**ekşi** 'sour'); **acımtrak** 'rather bitter'. This suffix is sometimes spelt with what seems to be an epenthetic vowel—e.g. **beyazımtırak**—but may be a survival of an older form.

(*b*) **-(i)msi** is added to nouns and adjectives: **mağara** 'cave', **mağaramsı** 'cavernous'; **duvar** 'wall', **duvarımsı** 'wall-like'; **rapor** 'report', **raporumsu bir yazı** 'a report-like writing', 'a feeble attempt at a report'.

(*c*) **-si** is attached only to nouns and adjectives ending in a consonant, so cannot be confused with the post-vocalic form of the third-person possessive suffix: **erkek** 'male', **erkeksi** 'mannish'; **çocuk** 'child', **çocuksu** 'childish'. But 'foolish', from **budala** 'fool', is **budalamsı**, while **budalası** means 'his fool'.

4. -ci.

This suffix is added to the singular of nouns and occasionally to adjectives and adverbs to denote persons who are professionally or habitually concerned with, or devoted to, the object, person, or quality denoted by the basic word:

iş	work	işçi	workman
süt	milk	sütçü	milkman
diş	tooth	dişçi	dentist
orman	forest	ormancı	forester
Atatürk		Atatürkçü	Ataturkist
halk	people	halkçı	populist or adherent of the People's Party
gürültü	noise	gürültücü	noisy (*of people*)
milliyet	nationality	milliyetçi	nationalist
yol	road	yolcu	traveller
inat	obstinacy	inatçı	obstinate
yalan	falsehood	yalancı	liar, deceiver
kaçak	contraband	kaçakçı	smuggler
şikâyet	complaint	şikâyetçi	complainant
Roentgen	(*discoverer of X-rays*)	röntgenci	radiographer or Peeping Tom, *voyeur*
statüko	*status quo*	statükocu	conservative
sıfır	zero	sıfırcı	schoolteacher who is lavish with zeros
şaka	joke	şakacı	joker
merhum	'the late'	merhumcu	devotee of the late Prime Minister Menderes
eski	old	eskici	old-clothes man
toptan	wholesale	toptancı	wholesaler
beleş	(*slang*) free, gratis	beleşçi	scrounger, parasite
ne	what?	neci	of what profession?

It may be attached to a phrase: **hazır** 'ready', **elbise** 'clothing', **hazır elbiseci** 'dealer in ready-made clothing'; **evet efendim** 'yes, sir', **evet efendimci** 'yes-man'.

In popular speech it is used redundantly with nouns denoting occupation such as **şoför** 'driver', **kasap** (A) 'butcher', **garson** 'waiter': **şoförcü, kasapçı, garsoncu**.
Cf. -ici, XIV, 2.

5. -li. This is added to the singular of nouns to make nouns or adjectives which denote:

(*a*) Possessing the object or quality indicated by the basic word:

şeker	sugar	şekerli	sweet
dikkat	attention, care	dikkatli	attentive, careful
at	horse	atlı	horseman
resim	picture	resimli	illustrated
ümit	hope	ümitli	hopeful
akıl	intelligence	akıllı	intelligent
bulut	cloud	bulutlu	cloudy
gürültü	noise	gürültülü	noisy (*of things*)
rahmet	divine mercy	rahmetli	deceased

(*b*) Possessing the object or quality in a high degree:

çene	jaw	çeneli	talkative
paha	price	pahalı	expensive
hız	speed	hızlı	rapid
sevgi	affection	sevgili	beloved
yaş	age	yaşlı	aged

(*c*) Belonging to a place or institution:

köy	village	köylü	villager, peasant
şehir	city	şehirli	city-dweller
İstanbul		İstanbullu	citizen of Istanbul
Çin	China	Çinli	Chinese
Nicerya	Nigeria	Niceryalı	Nigerian
lise	lycée	liseli	lycée student
Osman	(*founder of the Ottoman dynasty*)	Osmanlı	Ottoman (*member or subject of the dynasty*)

Added to the name of a colour, it makes an adjective or noun meaning dressed in that colour:

| siyah | black | siyahlı | dressed in black |
| kırmızı | red | kırmızılı | dressed in red |

It may be added to a phrase:

uzun boy	long stature	uzun boylu	tall
geniş omuz	broad shoulder	geniş omuzlu	broad-shouldered
orta yaş	middle age	orta yaşlı	middle-aged
kırmızı yanak	red cheek	kırmızı yanaklı	red-cheeked
bir mart tarih-i	the date 1 March	bir mart tarihli	dated 1 March
çamur	mud	çamurlu	muddy
çamurlu yüz	muddy face	çamurlu yüzlü	muddy-faced
kullanış	use (X, 11)	kullanışlı	serviceable
yaygın kullanış	wide use	yaygın kullanışlı	widely used

The suffix appears to be used redundantly in **bombeli** 'convex' < French *bombé* 'convex'. In **şanjanlı** 'shot' (of silk; other forms being **janjanlı** and **cancanlı**) it is not redundant, as the French *changeant* is used as a noun in Turkish in the sense of the quality possessed by shot silk: **şanjan kumaşlar-ı** 'shot fabrics'.

6. ... -li ... -li. Pairs of words of opposite meanings, each with a suffixed **-li**, are used adverbially and adjectivally: **gece-li gündüz-lü çalışmak** 'to work night and day'; **kız-lı erkek-li öğrenci grupları** 'groups of pupils including both girls and boys'. The basic words may be adjectives: **uzak-lı yakın-lı kahkahalar** 'bursts of laughter both far and near'. The **-li** in this use is historically distinct from that discussed in the preceding section.[1]

---

[1] The proof is that these paired forms in **-li** were in use in the ancient language side by side with **-liğ**, from which the single adjectival **-li** subsequently evolved. See Gabain, p. 159.

NOUN AND ADJECTIVE SUFFIXES   IV, 7

7. -siz. This suffix means 'without': **ümitsiz** 'hopeless', **sonsuz** 'endless', **şapkasız** 'hatless', **ğürültüsüz** 'noiseless', **dikkatsiz** 'careless', **tarihsiz** 'undated'. It may be added to pronouns as well as nouns: **onsuz** 'without him', **sensiz** 'without you'. See also XI, 12.

8. -lik.

(a) Added to nouns or adjectives, it makes abstract nouns:

ğüzel	beautiful	ğüzellik	beauty
kolay	easy	kolaylık	ease, facility
iyi	good	iyilik	goodness, good action
asker	soldier	askerlik	military service
çocuk	child	çocukluk	childhood, childish action, childishness
iki	two	ikilik	duality
iş-çi	workman	işçilik	workmanship
kaçak-çı	smuggler	kaçakçılık	smuggling
Atatürk-çü	Ataturkist	Atatürkçülük	Ataturkism
dikkat-li	careful	dikkatlilik	carefulness
dikkat-siz	careless	dikkatsizlik	carelessness

Vulgarly it is added to Arabic abstract nouns: **insaniyet** 'humanity' > **insaniyetlik**; cf. **şoförcü**, etc., § 4, end.

When **-lik** is added to nouns of rank the resulting word is not invariably abstract but exhibits the same ambiguity as 'the President's office' (the office he holds or the office in which he works): **kaymakamlık** may be the rank of lieutenant-governor or his official residence or the district he administers; **kırallık** may be kingship or kingdom or reign.

(b) Added to nouns it makes nouns and adjectives meaning 'intended for or suitable for . . .':

ön	front	önlük	pinafore
tuz	salt	tuzluk	salt-cellar
kira	hire	kiralık	for hire, to let
ğöz	eye	ğözlük	eye-glasses, spectacles

çamaşır	linen	çamaşırlık	laundry
baba	father	babalık	adoptive father, paternity
şehit	martyr	şehitlik	military cemetery, martyrdom
hastane	hospital	hastanelik	hospital-case
mahkeme	law-court	mahkemelik	(person) brought before a court
gelin	bride	gelinlik	marriageable girl, nubile, bridal, wedding-dress, the state of being a bride

(c) Added to numerical expressions it makes nouns and adjectives:

seksen	eighty	seksenlik	octogenarian
on	ten	onluk	tenner; coin or note of ten piastres or pounds
yüz	hundred	yüz liralık	hundred-lira note
yıl	year	yıllık	yearling
		yüz yıllık	hundred-year-old
bir saat	one hour	bir saatlik bir yer	a place one hour's journey away
on iki araba	twelve cars	on iki arabalık bir konvoy	a twelve-car convoy

The numerical expression may be a noun in the locative case:

yüz-de	in a hundred	yüzdelik	percentage
on-da	in ten	ondalık	tithe, ten per cent. commission
Cf. günde	in the day	gündelik	daily wage

(d) Added to adverbs of time:

| şimdi | now | şimdilik | for the present |
| bugün | today | bugünlük | for today |

Although these may be translated as adverbs, as in **bugünlük bu kadar yeter** 'that's enough for today', they really belong in (c) above; **bugünlük** means 'the today-amount'.

9. **-daş**. This suffix is not affected by vowel harmony but it does appear as **-taş** after unvoiced consonants. Added only to nouns, it denotes common attachment to the concept expressed by the basic noun, like English prefixed or suffixed 'fellow'.

vatan	home-land	vatandaş	compatriot, fellow citizen
okul	school	okuldaş	schoolmate
meslek	profession	meslektaş	colleague
din	religion	dindaş	co-religionist
çağ	time, epoch	çağdaş	contemporary (*adj. or noun*)

Two exceptions: **kardeş** not **-daş** is the standard Turkish for 'brother' or 'sister' (< **karın-daş** 'womb-fellow'), and 'namesake' is **adaş** with a single **d** although 'name' is **ad**. The possible explanation for the latter anomaly is that its second element is not **-daş** but **eş** 'mate' and this word **adaş** is the most likely etymon of the suffix **-daş**.

The language reformers have chosen to make this suffix conform to vowel harmony in the neologism **iş-teş-lik** 'co-operation' (**iş** 'work') and in the resurrected **gönül-deş** 'sympathizer' (originally **gönüldaş** from **gönül** 'soul'), apparently through a misunderstanding of the phonetic spelling used by Redhouse.[1]

[1] *A Turkish and English Lexicon* (Constantinople, 1921), p. 1598; cf. *TS*, p. 314.

10. -ģil. This invariable suffix is a provincialism. Added to titles or personal names it denotes 'the house or family of . . .': **Kaymakamģil, Mehmetģil,** also in the plural **Kaymakamģiller, Mehmetģiller.** It may be suffixed, after a personal suffix, to nouns denoting relatives: **teyzemģil** 'my aunt's family', the standard Turkish for this being **teyzemler** (II, 15, end).

It has been used by the reformers to coin names of plant and animal families: **ģül** 'rose', **ģülģiller** 'Rosaceae'; **kedi** 'cat', **kediģiller** 'Felidae'.

11. -(s)el. Arabic words ending in the adjectival suffix -$\bar{\imath}$ have greatly enriched the Turkish vocabulary; witness such words as **tarihî** 'historical', **dinî** 'religious'. Arabic nouns ending in *-at-* drop it before adding the -$\bar{\imath}$, hence **siyasi** 'political' (**siyaset** 'politics'), **iradi** 'voluntary' (**irade** 'will'), **millî** 'national' (**millet** 'nation'). The language reformers, in their desire to purge Turkish of foreign elements, advocated the replacement of this useful suffix by **-sel** or, when added to words ending in **s** or **z**, **-el**. This they employed with foreign borrowings and with Turkish words, both existing and manufactured:

tarih (A)	history	**tarihsel**	historic
siyaset (A)	politics	**siyasal**	political
cebir (A)	algebra	**cebirsel**	algebraic
kimya (A)	chemistry	**kimyasal**	chemical
fizik	physics	**fiziksel**	physical
öz	self	**özel**	private
bilim	science	**bilimsel**	scientific
anayasa	constitution	**anayasal**	constitutional

In this last example the suffix has been reduced to **-l** to avoid the cacophonous **-sasal**.

To justify this innovation the reformers cited such time-honoured words as **uysal** 'compliant' (**uymak** 'to conform'), **kumsal** 'sandy' or 'a sandy tract' (**kum** 'sand') and **yoksul** 'destitute' (**yok** 'non-existent'). The real inspiration of it, however, was in such French words as *culturel* and *social*.[1]

---

[1] *TS* marks as a French borrowing the noun **kültür**, but not the adjective **kültürel**, the implication being that the latter is derived from the former by the addition of the Turkish suffix **-el**.

12. -varî. This Persian suffix, meaning '-like', is still productive in Turkish to a limited extent: **Şekspirvarî** 'Shakespearian'; **Çörçilvarî** 'Churchillian'; **James Bondvarî bir casusluk** 'a James Bond-ish case of espionage'.

13. -cil. This occurs in a few words and has the sense of 'tending towards, accustomed to, addicted to':

ak	white	akçıl	faded
kır	grey	kırçıl[1]	grizzled
adam	man	adamcıl	tame *or* ready to attack man
balık	fish	balıkçıl	heron
tavşan	hare	tavşancıl	eagle
ölüm	death	ölümcül	moribund

Some neologisms have been made with this suffix:

ben	I	bencil	selfish
ana	mother	anacıl	mother-bound
ev	house	evcil	domesticated
kitap	book	kitapçıl	bookish

14. -hane. The Persian *khāne* 'house' is not quite dead as a suffix in Turkish; **pastahane** 'cake-shop' is of more recent origin than **hastahane, eczahane,** and **postahane** and unlike them it keeps its **h** more often than not (I, 11). It is added to a few Turkish words, e.g.: **süthane** 'dairy', **buzhane** 'ice-house', **dikimhane** 'tailoring workshop', **aşhane** 'cook-shop'. Probably ephemeral is **kazıkhane** 'clip-joint' from **kazık** 'swindle'.

15. -ane. This Persian suffix, in which the a is long, serves (*a*) to make adjectives or nouns into adverbs: **mest** 'drunk', **mestane** 'drunkenly'; (*b*) to turn nouns and adjectives indicating persons into adjectives describing things: **şah** 'king', **şahane** 'regal'; **şair** 'poet', **şairane** 'poetic'; **dost** 'friend', **dostane** 'friendly' (as in 'a friendly word'); **müdebbir bir paşa** 'a prudent Pasha', **paşanın müdebbirane hareketi** 'the Pasha's prudent action'. It is mentioned here because it has recently shown itself productive of at least one word: from **bilgiç** 'know-all', **bilgiçane** 'in a know-all fashion'.

[1] Note ç exceptionally following a voiced consonant.

V

PRONOUNS

1. Personal pronouns:

Singular	First	Second	Third
abs.	ben	sen	o
acc.	beni	seni	onu
gen.	benim	senin	onun
dat.	bana	sana	ona
loc.	bende	sende	onda
abl.	benden	senden	ondan
Plural			
abs.	biz	siz	onlar
acc.	bizi	sizi	onları
gen.	bizim	sizin	onların
dat.	bize	size	onlara
loc.	bizde	sizde	onlarda
abl.	bizden	sizden	onlardan

In pre-nineteenth-century texts the usual forms of the third person are: (sing.) ol, anı, anın, ana, anda, andan; (pl.) anlar, anları, etc.

It will be noticed that the table exhibits some anomalies: the **m** in the genitive of the first-person singular and plural; the change from **e** to **a** in the dative of the first- and second-person singular; in the third person the **n** before the case-suffixes of the singular and before the **-lar** of the plural. The most plausible explanations of these anomalies are: the original **beniŋ** and **biziŋ** became **benim** and **bizim** under the influence of the pronominal suffixes **-im** and **-imiz**, helped perhaps by the labial **b**. The original suffix of the dative was not **-e** but **-ge**; **benge, senge** became **beŋe, seŋe** and the influence of this nasal brought about the change to **bana, sana**, a change possibly helped by analogy with the old third-person dative **aŋa** > **ana**; this very form **aŋa** may have been due to the influence of the **ŋ** in an earlier *****oŋa**. The **n** appears in the third person also when **o** takes the suffixes **-siz** (IV, 7) and **-ce** (XII, 2): **onsuz** 'without him', **onca** 'according to him'. The usual explanation of this **n** is that it is the 'pronominal **n**' which appears after the third-person suffix and in the declension of **-ki** and **kendi** (§§ 3, 4).[1]

o is a demonstrative as well as a personal pronoun; see § 5.

siz is the regular polite form for 'you', singular or plural, and **biz** is used colloquially for 'I' (XVI, 3 (c)); they may therefore take the plural suffix in the colloquial—**bizler, sizler**—when referring to more than one person.

In courtly speech, which is steadily becoming rarer, **ben** may be replaced by **bendeniz** 'your slave' (< Persian *banda*; the resemblance to **ben** is coincidental), the full meaning of which has become somewhat abraded, so that a following verb nowadays is usually in the first and not the third person, and, for example, 'my humble opinion' is **bendenizin fikrim** 'your slave's my opinion'. Similarly, **siz** may be replaced by **zat-ı aliniz** or **zat-ı alileri** (Persian izafet), literally 'your high person, their high person', which are followed by a verb in the second-person plural.

2. Uses of the personal pronouns. As they are definite by nature, we may call **ben, sen, o**, etc., the nominative instead of the absolute case; there is no question of their being used as an indefinite accusative.

The persons of verbs are shown by suffixes but a pronoun in the nominative may be used for emphasis: **o ġitti, ben ġitmedim** '*he* went; *I* did not go'.

The pronoun object of a verb is generally omitted if it can be understood from the context: **kitabı dün aldım, daha okumadım** 'I bought the book yesterday; I have not read ⟨it⟩ yet'.

The use of **seni** 'thee' with terms of abuse is conventionally explained by the ellipsis of a verb such as 'I dislike/deplore/warn': **seni ġidi!** 'you scoundrel!'; **seni afacan seni!** 'you cheeky little urchin you!' More precisely, the reason there is no verb is that the speaker does not have in mind any specific verb but only an inarticulate emotion of displeasure of which **seni** is the object.

In the genitive the pronouns can be used predicatively; cf. II, 11: **bu memleket niçin bizim?** 'why is this land ours?'; **bu para benim** 'this money is mine'; **senin olsun** 'keep it' (lit.

---

[1] This is disputed, not quite convincingly, by Elöve, p. 197, note 1.

'let-it-be of-you'); **mesuliyet sizin değil** 'the responsibility is not yours'.

They may reinforce the personal suffixes: **ev-imiz** or **bizim ev-imiz** 'our house'; **sokağ-ınız** or **sizin sokağ-ınız** 'your street'; **ad-ı** or **onun ad-ı** 'his name'.

The genitive of the third-person pronouns can resolve ambiguities which might arise from the various possible senses of, for example, **çocukları** (cf. II, 15):

> **onun çocukları** his children
> **onların çocuğu** their child
> **onların çocukları** their children

Pronouns of the first and second persons in the genitive are also used informally as attributive adjectives, i.e. replacing the personal suffixes: 'our house' can be **bizim ev**; 'your street' **sizin sokak**. The genitive of the third-person pronouns cannot, however, replace the personal suffixes in standard Turkish: **onun ad** instead of **adı** or **onun adı** is a provincialism (cf. **Ahmed'in Mustafa**, p. 43, penultimate paragraph).

3. -ki. The pronominal or 'mixed' suffix **-ki** is exceptional in the matter of vowel harmony, not changing except after **gün** 'day' and **dün** 'yesterday', when it becomes **-kü**. Added to the genitive case of a noun or pronoun, it makes a possessive pronoun: **hizmetçi-nin-ki** 'the one belonging to the servant'; **çocuğ-un-ki** 'the one belonging to the child'; **benimki** 'mine'; **seninki** 'thine'; **onunki** 'his, hers, the one belonging to it'; **bizimki** 'ours'; **sizinki** 'yours'; **onlarınki** 'theirs'. **bu kalem benim değil** and **bu kalem benimki değil** may both be translated 'this pen is not mine'. The former is a simple denial of ownership, the latter implies 'I have a pen but this is not it'.

The noun in the genitive to which **-ki** is suffixed may be in the plural and may have a personal suffix.

arkadaş-ınız-ın-ki   the one belonging to your friend
arkadaş-lar-ınız-ın-ki the one belonging to your friends

Added to an expression of time or place, which may be an adverb or a noun in the locative case, **-ki** makes a pronoun or adjective: **yazın** 'in summer', **köylünün yazınki kazancı** 'the peasant's summer earnings'; **bugün** 'today', **bugünkü gazete** 'today's newspaper', **buğünküler** 'those who are today, people nowadays'; **yarın** 'tomorrow', **yarınki toplantı** 'tomorrow's meeting'; **şimdi** 'now', **şimdiki durum** 'the present situation'; **okul çağ-ı** 'school-age', **okul çağındaki çocuklar** 'children of school-age' ('who are in school-age'). **İzmir'deki büromuz küçük, Adana'daki daha büyüktür** 'our office which-is-in-Izmir is small, the-one-in-Adana is bigger'.

Pronouns in **-ki** may be declined. In the singular the case-endings are preceded by the pronominal **n**, but this does not appear in the plural. Thus the declension of **benimki** is as follows:

	Singular	Plural	
abs.	benimki	benimkiler	mine
acc.	benimkini	benimkileri	mine
gen.	benimkinin	benimkilerin	of mine
dat.	benimkine	benimkilere	to mine
loc.	benimkinde	benimkilerde	in mine
abl.	benimkinden	benimkilerden	from mine

çocuğun boyu babasınınkini geçti 'the child's stature has passed his father's'; **resimlerimiz kardeşlerinizinkilerden kıymetlidir** 'our pictures are more valuable than your brothers''.

4. kendi. As an adjective it means 'own': **kendi oda-m** 'my own room', **kendi kız-ı** 'her own daughter', **kendi memleket-iniz** 'your own country'.

With the personal suffixes it makes the emphatic or reflexive pronouns 'myself', etc.: **kendi-m, kendi-n, kendi or kendi-si, kendi-miz, kendi-niz, kendi-leri**. For 'himself' with reflexive meaning, **kendi** is far more usual than **kendisi**; indeed, purists maintain that **kendisi** should never be used reflexively. Both forms take the pronominal **n** before all case-endings:

abs.	kendi	kendisi
acc.	kendini	kendisini
gen.	kendinin	kendisinin
dat.	kendine	kendisine
loc.	kendinde	kendisinde
abl.	kendinden	kendisinden

As a reflexive pronoun **kendi** is usually repeated, the first time in the absolute with no suffix, the second time with the appropriate personal and case suffix: **kendi kendi-m-i müdafaa ettim** 'I defended myself'; **işi kendi kendi-m-e yapamadım** 'I could not do the job for (or "by") myself'; **kendi kendi-n-i tenkit ediyor** 'he is criticizing himself'; **kendi kendi-miz-den korkmıyalım** 'let us not be afraid of ourselves'.

kendisi and its plural **kendileri** are commonly employed as simple third-person pronouns with no reflexive or emphatic sense: **kendisi evde** 'he is at home'; **kendilerini gördünüz mü** 'have you seen them?'

As **kendi-si** literally means 'his self', it may stand in izafet with a preceding noun in the genitive: **Atatürk'ün kendisi** 'Atatürk himself'; **Meclis'in kendisi** 'the Assembly itself'.

5. Demonstratives:

> **bu** this (close to the speaker)
> **şu** this *or* that (a little further away)
> **o** that (also 'he, she, it')

When used as adjectives these words are invariable. For their declension when used as pronouns see **o** in § 1: sing. **bu, bunu,** etc., **şu, şunu,** etc.; pl. **bunlar,** etc., **şunlar,** etc.

şu means 'the following': **şu teklif** 'the following proposal, this proposal which I am about to mention'; **bu teklif** 'the proposal which has just been mentioned'.

Where we say 'this or that', Turkish prefers 'that or this': **şunu yap, bunu yap** 'do that, do this'.

In archaizing legal language, **işbu** may be found for the adjectival **bu**. Until the last century **şol** was sometimes used for **şu**.

The personal suffixes are not used with the demonstratives except in such stereotyped expressions as **şu-nun bu-nun şu-su bu-su ile alâkadar olmıyan** 'not interested in other people's business' ('the that and this of that one and this one'); **o-nun şu-su bu-su** 'his private concerns' ('his that and this').

Demonstratives precede attributive adjectives: **bu uzun yol** 'this long road'; **şu geniş omuzlu güreşçi** 'that broad-shouldered wrestler'; **o meşhur aktör** 'that famous actor'. They may come within a definite izafet; cf. II, 17, end.

An idiomatic use of **bu** is to place it after a noun: **sanatkâr bu,** literally 'artist this', meaning 'the fellow's an artist; what do you expect?' So **hayat bu** 'that's life for you'; **çocuk bu** 'he's only a child; don't ask too much of him.

From **bu, şu,** and **o** are formed **böyle, şöyle,** and **öyle,** used both as adverbs, 'thus', and adjectives, 'such, this/that kind of':

böyle adamlar such men (as this)
şöyle evler such houses (as those over there)
öyle fikirler such ideas (as those)

The addition of the third-person suffix to these words makes them into pronouns (cf. II, 22): **böylesi** 'this sort of person', **şöylesi** or **öylesi** 'that sort of person'; plural: **böyleleri, şöyleleri, öyleleri.** The singular forms are also used adjectivally—**böylesi adamlar** 'such men'—but this use has not achieved general currency.

beriki, öteki, mean respectively the nearer and the further of two. They may be adjectives or pronouns; being compounded with **-ki,** when used as pronouns they take the pronominal **n** before all cases of the singular: **bu gazeteyi istemiyorum, öteki-n-i ver** 'I don't want this newspaper, give me the other one over there'. **öteki beriki** means 'this one and that one, anybody and everybody'.

6. Interrogatives:

kim who? **hangi** which?
ne what? **kaç** how many?

kim declines like a noun, in singular and plural: **bu çanta kim-in?** 'whose is this bag?'; **kim-i gördün?** 'whom did you see?'; **kimler-e?** 'to what people?'

ne exhibits certain irregularities:

	Singular	Plural
abs.	ne	neler
acc.	ne or neyi	neler or neleri
gen.	neyin or nenin	nelerin
dat.	neye or niye	nelere
loc.	nede	nelerde
abl.	neden	nelerden

There are also alternative forms with the personal suffixes, meaning 'what of mine?', 'what of yours?', etc.:

	Singular		Plural	
1	nem	or neyim	nemiz	or neyimiz
2	nen	neyin	neniz	neyiniz
3	nesi	neyi		neleri

The usual accusative singular is **ne**: **ne yaptın?** 'what have you done?'

neyi is used:

(*a*) For 'what specific thing?'; e.g. if you hear that someone is going to the opera and ask **ne göreceksin?** 'what are you going to see?', you may elicit the facetious reply 'an opera, of course!' There is no danger of this if you ask **neyi göreceksin?** 'what ⟨specific item of the repertoire⟩ are you going to see?' The defined accusative plural **neleri** is similarly used for 'what specific things?'

(*b*) When another interrogative follows, especially one compounded with **ne**: **neyi ve ne zaman yaptın?** 'what have you done, and when?'; **kim neyi kime satıyor?** 'who is selling what, to whom?'

The regular spelling of the dative singular is **neye**: **neye yarar?** 'for-what is it suitable?', while **niye** is generally used in the sense of 'what for, why?', as is the ablative **neden**.

The plural is used in exclamations: **neler gördüm!** 'what things I saw!'

One of the commonest uses of **ne** with personal suffixes is exemplified in **nen var?** 'what's the matter with you?', lit. 'what-of-yours exists, what do you have?', cf. *qu'as-tu?* Note also **neme lâzım?** 'what has it to do with me?', lit. 'to-what-of-mine is it necessary?' **bu okul-un ne-si-sin?** 'you are this school's what?', i.e. 'what is your position in this school?'—**kapıcı-sı-yım** 'I am its janitor'.

ne may be the qualifying element of an indefinite izafet, e.g. **ne ders-i var?** 'what lesson is there?' (on the time-table)—logically, because the answer will also consist in an indefinite izafet: **geometri ders-i**. There is a story of a man who tries to tell a friend that he has seen a ghost, **hortlak**, but in his terror he can only stammer **ho-ho-ho**. His friend asks **ne ho'su?** 'what ho?, the ho belonging to what?'

ne may be an adjective as well as a pronoun: **ne hacet?** 'what need?', **ne inat!** 'what obstinacy!', **ne güzel çiçekler!** 'what lovely flowers!' It may also render 'how' as in **ne güzel!** 'how beautiful!' or **ne malûm?** 'how do you know?' (lit. 'how is it known?'); or 'why' as in **ne karışıyorsun?** 'what are you interfering ⟨for⟩?'

It forms part of many compound interrogatives: **ne zaman, ne vakit** 'when?'; **ne kadar** 'how much?'; **niçin** (< **ne için** 'what for?') 'why?'; **nasıl** (< **ne asıl** 'what basis?') 'how?', 'what sort of . . .?'; **neci** 'of what profession?' With the adverbial suffix -ce (XII, 2), it makes (a) **nece** 'in what language?' and (b) **nice**, meaning originally 'how many?', then 'how many!' and nowadays mostly 'many'.

The elision of the **e** of **ne**, as seen in **niçin** and **nasıl**, is frequent in rapid conversation: **ne olacak?** 'what will happen?' > **nolacak?** or **n'olacak?**; **ne yapalım?** 'what are we to do?' > **napalım?**

hangi and **kaç** are adjectives, the latter always construed with a singular noun: **hangi vilâyet?** 'which province?'; **hangi vilâyetler?** 'which provinces?'; **kaç vilâyet?** 'how many provinces?' With the addition of personal suffixes they become pronouns: **hangi-miz?** 'which of us?'; **kaç-ınız?** 'how many of you?'; **hangi-si?** 'which one of them?'; **hangi-ler-i?** 'which (*pl.*) of them?'

7. Indefinite, determinative, and negative. Most of the pronouns in this category are formed from adjectives by the addition of the third-person suffix, on the pattern of **böyle-si** (§ 5). Thus **bir** 'a, one' is an adjective: **bir adam gitti** 'one man went'; **bir-i** is a pronoun: **adamlardan biri gitti, biri kaldı** 'one of the men went, one remained'. The suffix may be doubled: **bir-i-si**. **biri** and **birisi** may also mean 'someone'; in this use the antecedent of the third-person suffix is people at large, a 'they' as vague as in 'they say': **biri** or **birisi bana seslendi** 'someone called out to me'.

In those words below in which **bir** is the first element it carries the word-accent.

8. **bazı, kimi** 'some' (adjective). The final vowel of **bazı** is the mark of the Persian izafet, so is unaccented. **kimi** in this adjectival sense is a neologism. Both qualify singular or plural nouns: **bazı** or **kimi adam** 'some man'; **bazı** or **kimi adamlar** 'some men'.

Pronouns: **bazısı, bazıları, kimi, kimisi** 'some people'; **bazımız, kimimiz** 'some of us'; **bazınız, kiminiz** 'some of you'.

9. birtakım 'a number of' (lit. 'a set') always qualifies plural nouns: **birtakım köylüler** 'a number of villagers'. Like the indefinite article, it may come between adjective and noun: **küçük birtakım devletler** 'a number of small states'. Note the distinction between **birtakım kitaplar** 'a number of books' and **bir takım kitap** 'a set of books' (for the construction of the latter see XVI, 7).

Pronoun: **birtakımı** 'a number of them'.

10. her (P) 'each, every', with noun in singular: **her gün** 'every day'; **her iki-si** 'both of them'; **her üç-ümüz** 'all three of us'; **her biri, herkes** (P) 'everyone'; **her bir-imiz** 'each one of us'; **her kim** 'whoever'; **her ne** 'whatever'; **her ne kadar** 'however much'; **her hangi** 'whichever'; **her hangi bir** 'any'. Compounds of **her** are sometimes written as one word: **herbirimiz, hernekadar**, etc.

11. hep is an adverb meaning 'altogether, entirely, always'. With the personal suffixes it becomes a pronoun: **hep-imiz** 'all of us', **hep-iniz** 'all of you'. The third person is **hepsi** (< **hep-i-si**, with the suffix doubled) 'all of it, all of them, everyone'.

12. çok as an adverb means 'much, very': **çok konuştuk** 'we talked a lot'; **çok faydalı** 'very useful'. With a noun, singular or plural, it means 'much, many': **çok iş** 'much work'; **çok kişi** 'many persons', **çok yerler** 'many places'.

With possessive suffixes it is a pronoun, meaning 'most' rather than 'many': **çoğ-umuz** 'most of us'; **çoğ-unuz** 'most of you'; **çoğ-u** 'most of it, most of them'. The last is used as an adjective as well as a pronoun, like **böylesi** and **kimi**: **çoğu insanlar** 'most people'; **çoğu zaman** 'most times, most often'.

birçok 'a good deal of' is followed by a noun in singular or plural; pedants say singular only.

Pronouns: **birçoğu** 'a good deal of it, a good many of them'; **birçokları** 'a good many people *or* things'.

13. az as an adverb means 'little : **az içer** 'he drinks little'. As an adjective, with a singular noun, 'few, little': **az kişi** 'few persons'; **az şarap içer** 'he drinks little wine'. See also III, 4.

biraz 'a little': **biraz yürüdük** 'we walked a little'; **biraz ekmek yedim** 'I ate a little bread'; **biraz-ı** 'a little of it'.

14. birkaç 'a few, several', with singular noun: **birkaç gün kaldı** 'he stayed several days'.
Pronoun: **birkaçı** 'several of them', etc.

15. bütün as a noun or as an adjective qualifying a singular noun means 'whole': **millet-in bütün-ü** 'the whole of the nation'; **bütün millet** 'the whole nation'. As an adjective with a plural noun it means 'all': **bütün milletler** 'all the nations'.

16. başka, diğer (P) 'other'.
Pronouns: **başkası, bir başkası, başka biri, diğer biri** 'another of them, someone *or* something else': **kaldır bunu, başkasını getir** 'take this away, bring another'. **benden başka** 'other than me'; **benden başkası** 'someone other than me'.

17. öbür (< **o bir**; I, 31, end) 'the other, the next': **öbür gün** 'the day after tomorrow'; **öbür dünya** 'the next world'.
Pronoun: **öbürü** 'the other one'.

18. birbir or **biribir**, with the appropriate possessive suffix, means 'each other':

1 **birbirimiz** or **biribirimiz**
2 **birbiriniz** or **biribiriniz**
3 **birbiri, biribiri,** or **biribirleri**

birbirimiz-e yardım ediyoruz 'we are helping each other'; **biribiriniz-i seviniz** 'love one another'; **biribirinden güzel kızlar** 'girls each more beautiful than the other'. The Persian **yekdiğer** is an increasingly rarer alternative.

19. aynı 'same'. In view of its origin (II, 26, third paragraph) this ought to be immediately followed by a noun, as in **aynı zaman** 'the same time'. It is, however, regularly used nowadays with an intervening adjective and even predicatively: **aynı uzun yol** 'the same long road'; **hedeflerimiz aynı** 'our aims are the same'. This **aynı**, which is frequently misspelt **ayni**, is accented on the first syllable.

There is another word **aynı**, accented on the last syllable, in which the ı is the Turkish third-person suffix: **ad-ı benim-ki-nin ayn-ı-dır** 'his name is the same as mine', lit. 'his-name of-mine is-its-counterpart'. The suffix is sometimes doubled: **bunun ayn-ı-sı-nı alalım** 'let's buy one just like this' ('of-this its-counterpart'). This, however, is a vulgarism, against which schoolchildren are warned. Oddly, what they are told to put in its place is **tıpkısı** 'its replica', which also contains a doubled third-person suffix. The bare form *tıpık (Arabic *ṭibq*) is never used, though **tıpkı** (the final ı being that of the Persian izafet) is commonly put before a noun or pronoun to reinforce the postposition **ğibi** 'like': **tıpkı babası ğibi** 'just like his father'; **tıpkı onlar ğibi** 'just like them'.

20. **şey**, as well as meaning 'thing', is an all-purpose pronoun, used like French *chose* to take the place of a word or name the speaker cannot for the moment recall. For its syntactic function see XV, 3. When it takes the third-person suffix this is usually doubled, **şey-i-si** (sometimes spelled **şeysi**; cf. **hepsi**, § 11), probably because in ordinary speech **şey-i** is barely distinguishable from **şey**: **şey-in şey-i-si ne oldu—mektub-un zarf-ı?** 'what has become of the what-d'ye-call-it of the what-d'ye-call-it—the envelope of the letter?'

21. **falan, falanca, filân, filânca** 'so and so, such and such' are adjectives and pronouns. **falan tarihte, falanca geldi** 'on such and such a date, so and so came'.

falan and **filân** also mean 'and so on' after nouns, 'or thereabouts' after expressions of time or quantity: **camiler-i falan gezdi** 'he toured the mosques and so on'; **temmuzda filân gelecekler** 'they will come in July or thereabouts'; **on lira falan istiyor** 'he wants ten liras or so'. They may be used together: **Bedri, Orhan, Hâmit falan filân geliyorlar** 'Bedri, Orhan, Hamit and so on and so forth are coming'. **falan festekiz** and **falan feşmekân** are similarly used and convey even less enthusiasm at the prospect.

22. **insan** 'human being' is used for the indefinite 'one': **bu sıcaklıkta insan çabuk yorulur** 'in this heat one gets tired quickly'. See also the use of the impersonal passive in VIII, 54.

23. **hiç**, in origin the Persian for 'nothing', has the same sense in Turkish: **ne yaptın?—hiç** 'what have you done?'—'nothing'. It also functions as an adverb reinforcing negatives: **hiç konuşmaz** 'he doesn't talk at all'. With **bir** it is written as one word: **hiçbir haber yok** 'there is no news at all'.

In positive questions it translates 'ever', in negative questions 'never'; **hiç öyle şey olur mu?** 'does such a thing ever happen?'; **hiç Antalya'da bulunmadınız mı?** 'have you never been in Antalya?'

24. **kimse**, originally 'whoever it is' (**kim+ise**, XX, 7), now means 'person, somebody not clearly specified': **bir kimse sizi arıyordu** 'someone was looking for you'. In conjunction with a negative it means 'no one', like French *personne*: **kimse aldırış etmiyor** 'no one is paying attention'; **kimse-siz çocuklar** 'children who are alone in the world' (IV, 7). Its diminutive is used in the negative sense only: **kimsecik yok** 'there's no one at all'; **kimsecikler kalmamış** 'there are no people left at all'.

VI

NUMERALS

1. Cardinals:

bir	1	on iki	12	kırk	40
iki	2	on üç	13	elli	50
üç	3	on dört	14	altmış	60
dört	4	on beş	15	yetmiş	70
beş	5	on altı	16	seksen	80
altı	6	on yedi	17	doksan	90
yedi	7	on sekiz	18	yüz	100
sekiz	8	on dokuz	19	bin	1,000
dokuz	9	yirmi	20	bir milyon	1,000,000
on	10	yirmi bir	21	bir milyar	1,000,000,000
on bir	11	otuz	30	sıfır	zero

Numbers are compounded by simple juxtaposition: **yüz bir** 'a hundred and one'; **üç milyon dört yüz yirmi bin sekiz yüz doksan altı** 'three million four hundred and twenty thousand eight hundred and ninety-six'.

In the numbers from 11 to 19 inclusive (which may be found written as one word), the accent is on the **on**. In higher numbers the last syllable of the unit is accented.

Whereas 'one hundred' and 'one thousand' are **yüz** and **bin** respectively, 'one million' and 'one milliard' (i.e. an American billion) require **bir**.

In writing figures, a full stop (**nokta**) is used to separate the thousands; thus **beş bin altı yüz otuz iki** is written **5.632**. On the other hand, a comma (**virgül**) is used where English uses a decimal point, so 7·5 ('seven point five') appears as **7,5** (**yedi virgül beş**). Less commonly, the thousands are separated by a comma, and a full stop may be used for the decimal point: **5,632; 7.5**.

In vague assessments of number such as 'two or three', 'five or six', the 'or' is not expressed: **iki üç, beş altı**. For 'three or four', idiom mysteriously prefers **üç beş** to **üç dört**. Cf. the expressions **üç aşağı beş yukarı** 'a little more or less', literally 'three down five up', and **üçe beşe bakmamak** 'not to haggle about the price', literally 'not to look at three ⟨or⟩ five'.[1] Care must be taken not to confuse **on beş** 'fifteen' with **beş on** 'five or ten'.

Care is also necessary with **yüz**, which besides 'hundred' may mean 'cause' or 'face': **iki yüz** 'two hundred'; **iki yüzlü** 'two-faced'; **ikiyüzlülük** 'hypocrisy'; **yüz ölçümü** 'surface-area'; **bu yüzden** 'for this reason'.

kırk is used for an indefinitely high number: **kırkayak** 'centipede' ('forty-feet'); **kırk yılda bir** 'once in a blue moon' ('in forty years').

When case-endings or other suffixes are written after figures, the rules of consonant-assimilation and vowel-harmony must be observed: 'from 2 to 9', **ikiden dokuza, 2 den 9 a**; 'from 3 to 7', **üçten yediye, 3 ten 7 ye**; 'from 6 to 11', **altıdan on bire, 6 dan 11 e**. An apostrophe may precede the suffix: **2'den 9'a**, etc.

For the use of the singular form of the noun after numerals, see II, 2.

2. Classifiers. A numeral is rarely used alone, e.g. in answer to a question; either the noun is repeated or, if the things enumerated are separate entities and not units of measurement, the word **tane** ('seed, grain') is added after the numeral. **kaç saat bekliyorsunuz?—iki saat** 'how many hours have you been waiting?'—'two hours'. **kaç kitap aldınız?—dört tane** 'how many books have you bought?'—'four'. **tane** is often inserted between numeral and noun (unless the latter is a unit of measurement): **beş tane anahtar** 'five keys'; **sekiz tane mendil** 'eight handkerchiefs'. It is also added after **kaç**, especially without a following noun: **kaç tane istiyorsunuz?** 'how many do you want?' If people are being enumerated, **kişi** 'person' is similarly used: **kaç kişi ğeliyor kokteylinize?—kırk altı kişi** 'how many are coming to your cocktail-party?'—'forty-six'.

When enumerating cattle, or vegetables such as onions and cabbage, **baş** 'head' is interposed after the numeral: **elli baş sığır** 'fifty oxen'; **yüz baş koyun** 'a hundred sheep'; **iki baş lâhana** 'two cabbages'. Cf. the English 'fifty head of cattle', but note that Turkish uses simple apposition, with no 'of'.

[1] *TS* defines **birkaç** as '**üçü, beşi ğeçmemek üzere az**' 'few, not exceeding three or five'.

el 'hand' is similarly used when enumerating shots of a firearm or deals of cards: **bir el tabanca attı** 'he fired one pistol-shot'; **bir el poker oynıyalım** 'let's play a hand of poker'. Other such classifiers were used in Ottoman: **aded** 'number' as alternative to **tane**; **kıta** 'piece', of books, documents, ships, and fields; **pare** 'piece', of artillery, ships, and villages.[1]

3. Fractions. The denominator, in the locative case, precedes the numerator: **üçte bir** (lit. 'in-three one') 'one-third'; **yedide dört** 'four-sevenths'; **yüzde yirmi beş** 'twenty-five per cent.'; **yüzde yüz** 'one hundred per cent.' The percentage sign consequently precedes the number: % **25**; % **100**. The numerator is put in definite izafet with the whole, of which the fraction is part: **çocukların beşte üçü** 'three-fifths of the children' ('of-the-children, in-five their-three'); **gelir-im-in yüzde yirmi beşi** 'twenty-five per cent. of my income'. This last example would appear in figures as **gelirimin** % **25 i**. **yekûn yüz ölçüm-ü-nün** % **18,7 si (yüzde on sekiz virgül yedisi) ormanlarla kaplıdır** 'of its total surface-area, 18·7 % is covered with forests'.

buçuk means 'and a half' and is used only after whole numbers and, jocularly, after **az** 'little' and **yarı** 'half': **on buçuk kilometre** 'ten and a half kilometres'; **az buçuk kişi** 'a handful of people'; **yarı buçuk ustalık** 'inadequate craftsmanship'.

yarım is an adjective meaning 'a half-': **yarım saat** 'a half-hour'; **yarım kilo domates** 'half a kilo of tomatoes' (note the apposition). For no clear reason, **yarımda** means 'at half past twelve'. **yarı** is used:

(a) As a noun: **talebeler-in yarı-sı kız** 'half of the pupils are girls'; **gece-nin yarı-sı-nı konuşarak geçirdik** 'we spent half of the night in talking'; **gece yarı-sı** 'midnight'; **gece-nin yarı-sı-n-da** 'in the middle of the night'.

(b) As an adjective meaning 'mid-, at the half-way mark': **yarı gece** 'midnight'; **yarı yol-da bırakmak** 'to leave in the lurch' (lit. 'at mid-way').

(c) As an adverb: **yarı anlamak** 'to half-understand'; **yarı Türkçe, yarı Fransızca konuştuk** 'we spoke half Turkish, half French'.

---

[1] 'An inkling of this system comes to us *via* pidgin English: "one fella man," "five piecee shirt."' Mario Pei, *The Story of Language* (London, Allen & Unwin, 1952), p. 123.

çeyrek (P) 'quarter' is now little used except when telling the time; see XII, 14.

4. Ordinals. The suffix is **-inci** after consonants, **-nci** after vowels:

birinci or **ilk**	1st	**yirminci**	20th
ikinci	2nd	**yirmi birinci**	21st
üçüncü	3rd	**otuzuncu**	30th
dördüncü	4th	**kırkıncı**	40th
beşinci	5th	**ellinci**	50th
altıncı	6th	**altmışıncı**	60th
yedinci	7th	**yetmişinci**	70th
sekizinci	8th	**sekseninci**	80th
dokuzuncu	9th	**doksanıncı**	90th
onuncu	10th	**yüzüncü**	100th
on birinci	11th	**bininci**	1,000th
		milyonuncu	millionth
		milyarıncı	thousand-millionth

As will be seen from '11th' and '21st', the ordinal suffix is attached only to the last member in a compound number.

The suffix should be written in full after figures: **1 inci, 2 nci, 3 üncü**, etc. Some abbreviate it to **ci, cü**, etc.

The suffix also appears in: **kaçıncı** 'how manyeth?'; **sonuncu** 'last' (**son** 'end, last'); **filânıncı** 'so-manyeth'.

Roman numerals are used to indicate centuries and with names of sovereigns and formal events such as congresses and exhibitions; as a rule the ordinal suffix is not then written but a full stop may follow the numeral: **XX** or **XX. asır** 'the twentieth (**yirminci**) century'; **XXVIII** or **XXVIII. İzmir Enternasyonal Fuarı** 'Twenty-eighth (**yirmi sekizinci**) International Fair of Izmir'. The Roman numeral may precede or follow a sovereign's name; 'Selim the Third' may be written **Selim III, III Selim**, or **III. Selim**, all three being read as **üçüncü Selim**.

The first six Arabic ordinals are sometimes used with names of sovereigns in Persian izafet. They are:

evvel	1st		**rabi**	4th
sani	2nd		**hamis**	5th
salis	3rd		**sadis**	6th

All the as are long. **Mehemmed-i sani = ikinci Mehmet**, Muhammad II; **Selim-i salis = üçüncü Selim**, Selim III.

5. *Distributives.* These answer the question **kaçar?** 'how many each?' and are formed by adding to the cardinal the suffix **-er** after a consonant, **-şer** after a vowel:

birer	one each	**sekizer**	eight each
ikişer	two each	**dokuzar**	nine each
üçer	three each	**onar**	ten each
dörder	four each	**on birer**	eleven each
beşer	five each	**yirmişer**	twenty each
altışar	six each	**otuzar**	thirty each
yedişer	seven each	**kırkar**	forty each
		etc.	

As with the ordinals, the suffix is attached only to the last element of compounds: **yirmi üçer** '23 each'; **iki yüz elli dokuzar** '259 each'. With whole hundreds and thousands, however, it is more usual nowadays to attach the distributive suffix to the number preceding the **yüz** or **bin**: **ikişer yüz** rather than **iki yüzer** for '200 each'; **beşer bin** rather than **beş biner** for '5,000 each'. The two foreign borrowings **milyon** and **milyar** never take the distributive suffix: **birer milyon lira** 'a million lira each'; **altışar milyar** 'six thousand million each'.

The distributive of **yarım** is irregular, taking the post-vocalic **-şar** despite its final consonant: **yarımşar** 'half each'. The suffix is not attached to **buçuk** but to the preceding whole number: **yirmi yedişer buçuk** 'twenty-seven and a half each'. **iki kişi birer yıl altışar ay hapis cezasına mahkûm edilmişlerdir** 'two people have each been sentenced to one year and six months' imprisonment' ('one-each year, six-each months'); in figures, **1 er yıl 6 şar ay**.

A frequent idiomatic use of **birer** is seen in: **askerlerimiz, birer aslan gibi düşmana saldırdı** 'our soldiers attacked the enemy like so many lions' (lit. 'like one-each lion, each one like a lion'). Like **bir**, **birer** may come between adjective and noun: **muharririn müşahedeleri bu hususta canlı birer misal teşkil etmektedir** 'the author's observations constitute so many vivid examples in this connexion' (lit. 'vivid one-each example, each one a vivid example').

kaçar 'how many each?' when repeated means 'in lots of how many?' Thus **kaçar kiraz yiyorsunuz?** 'how many cherries each are you eating?' but **kirazları kaçar kaçar yiyorsunuz?** 'how many at a time, at a mouthful, are you eating the cherries?' Cf. XII, 1.

6. Collectives. The suffix **-iz** after consonants, **-z** after vowels, makes numerals denoting twins, triplets, etc.: **iki-z, üç-üz, dörd-üz, beş-iz.** While **altız, yediz,** etc. are theoretically possible, they seem never to be used, for reasons biological rather than grammatical. The collectives are mostly used as nouns but they can qualify a noun, usually in the plural: **ikizler** or **ikiz çocuklar** (rarely **ikiz çocuk**) 'twins'; **üçüzler** 'triplets'; **üçüz kızlar** (rarely **üçüz kız**) 'girl triplets'; **dördüzler** 'quadruplets'; **dördüz kardeşler** 'quadruplet brothers'.

The suffix **-li** may be added, e.g. **beşizli şamdan** 'five-branched candlestick'.

7. Dice numbers. Two dice are employed in the game of **tavla** 'backgammon' and the various possible throws are named in a curious mixture of Turkish and Persian:

1–1 **hepyek**
1–2 **ikibir** or **yekdü**
1–3 **seyek**
1–4 **çarüyek** or **cihariyek**
1–5 **pencüyek**
1–6 **şeşyek**
2–2 **dubara**
2–3 **sebaydü**
2–4 **çarüdü** or **ciharıdü**
2–5 **pencüdü**
2–6 **şeşidü**

3–3 **düse**
3–4 **çarise** or **ciharise**
3–5 **pencüse**
3–6 **şeşüse**
4–4 **dörtçar** or **dörtcihar**
4–5 **beşdört**
4–6 **şeşiçar** or **şeşcihar**
5–5 **dübeş**
5–6 **şeşbeş**
6–6 **düşeş**

Not all these terms are recorded in the dictionaries and some other variant spellings may be found.

Playing-card numbers are formed with **-li**; the ace is **birli**, the deuce **ikili** and so on up to the ten, **onlu**.

VII

POSTPOSITIONS

1. General observations. The functions of some English prepositions are performed in Turkish by the case-suffixes. Those of the rest are performed by postpositions, which follow the word they govern. A few of them can appear as suffixes, but the majority are independent words. Those listed as primary are variously construed with the absolute, genitive, dative, and ablative cases. The only more-or-less current postposition governing the accusative is the obsolescent **mütaakıp** (A) 'following, after': **ziyafet-i mütaakıp** 'after the banquet'. Those listed as secondary postpositions ('postpositional expressions' is another possible term) are nouns in the dative, locative, or ablative case, linked by izafet to the word they govern. An English analogy would be to call 'in' and 'before' primary and 'on the inside of' and 'in front of' secondary prepositions.

2. Primary postpositions with absolute case:

  üzere, üzre on içre in

 üzere is mostly used with the infinitive in **-mek** (X, 2 (*d*)) but may occasionally be found with other substantives: **yol üzere** 'on the road'; **âdeti üzere** 'in accordance with his custom'. **içre** is obsolete except in archaizing poetry: **cihan içre** 'in the world'.

3. Primary postpositions with absolute or genitive case:

ġibi	like	**kadar** (A)	as much as
ile	with	**için**	for

 These take the genitive of the personal pronouns **ben, sen, o, biz,** and **siz,** the demonstrative pronouns **bu, şu,** and **o** and the interrogative **kim**. All other substantives, including pronouns pluralized by **-ler,** appear before these postpositions in the absolute case. Colloquially, however, even the pronouns listed above are used in the absolute case before these postpositions. This is particularly frequent with **kim**; instead of **kiminle, kimin için,** and **kimin gibi** 'with whom?', 'for whom?', 'like whom?', one hears **kimle, kim için,** and **kim gibi,** the last being a more respectable solecism than the first two.

(*a*) **gibi**: **benim gibi bir adam** 'a man like me'; **senin gibi** 'like you'; **bizim gibi** or **bizler gibi** 'like us'; **onlar gibi** 'like them'; **bülbül gibi** 'like a nightingale'. The word may also serve as a noun: **bu gibiler** 'people like these' (lit. 'these likes'); it can also stand in definite izafet with a pronoun—**bu-nun gibi-si** 'the like of this'—or in indefinite izafet with a noun—**bu adam gibisi** 'the like of this man'. A common locution is **öyle gibi-m-e geliyor ki** 'it seems to me as if . . .' ('it so comes to-my-like that . . .'). **gibi-ler-den**, in apposition to a preceding word or clause, means 'on the lines of, of the order of': **köylü, memleketin efendisidir, gibilerden bir nutuk** 'a speech on the lines of "the peasant is the master of the country"'.

(*b*) **kadar** is in origin an Arabic word for 'amount', which helps explain its Turkish use: **bir saat kadar çalıştım** 'I worked for about an hour, as much as an hour' ('an hour amount'); **yirmi, yirmi beş kadar kişi** 'some twenty or twenty-five people'; **taş kadar sert** 'hard as stone' ('stone amount hard'); **o adam kadar zengin** 'as rich as that man'; **fil kadar iri** 'huge as an elephant'. With the genitive of pronouns: **onun kadar zengin** 'as rich as he'; **sizin kadar bir çocuk** 'a child as big (or "old") as you'. When it follows the absolute case of **bu, şu,** or **o**, these function not as pronouns but as demonstrative adjectives, and the resulting **bu, şu,** or **o kadar** may be adverbial as well as adjectival: **o kadar güldük** 'we laughed so much'; **bu kadar para** 'this much money'.

(*c*) **ile** has not only the comitative sense of English 'with' but also denotes the instrument: **kim-in ile gittiniz?** 'with whom did you go?', **vapur ile gittiniz** 'you went by boat'; **bunu zamk ile yapıştırdım** 'I stuck this with glue'. Note also: **kilo ile satmak** 'to sell by the kilogramme'; **para ile satmak** 'to sell for money'. Sometimes it must be translated 'because of'.

It may be suffixed; the **i** is dropped after a consonant and becomes **y** after a vowel, the resulting **-le** or **-yle** being subject to vowel harmony: **kiminle** 'with whom?'; **vapurla** 'by boat';

gümrükçüyle 'with the customs-officer'; karıyla 'with the woman'.

After the third-person suffix it appears as an invariable -yle:

karı-sı	his wife	karısıyle	with his wife
çekiç-i	his hammer	çekiçiyle	with his hammer
göz-ü	his eye	gözüyle	with his eye
omuz-u	his shoulder	omuzuyle	with his shoulder

Although this rule reflects the normal educated pronunciation, many people neglect it, writing **karısıyla, omuzuyla**. Less often, the vowel of the third-person suffix is combined with the postposition to make an invariable **-(s)iyle**: **karısiyle, göziyle, omuziyle**.

Colloquial alternatives to **ile** are **ilen** and **inen**. Instead of **benimle, onunla** 'with me, with him', one hears **benle** or **bennen, onla** or **onnan**, in the informal speech even of educated people.

(d) **için** (for which the older pronunciation **içün** is not uncommon) translates most senses of English 'for': **bunu yurd-un iyiliğ-i için yaptı** 'he did this for the good of the country'; **bunu sizin için aldım** 'I bought this for you'; **yolculuk için hazırlıklar** 'preparations for the journey'; **böyle bir ev için bu kadar para verilir mi?** 'does one pay so much money for such a house?' It also renders 'about' as in 'what do you think about this proposal?' **bu teklif için ne düşünüyorsun?** With the infinitive, rarely with the third-person imperative, it expresses purpose; with the personal participle, cause (XI, 24).

As an archaism it may be found suffixed, as **-çin** or **-çün** after consonants, **-yçin** or **-yçün** after vowels, the forms in **ü** appearing when the vowel of the preceding syllable is rounded: **senin-çin** 'for you'; **onun-çün** 'for him'; **muhabbeti-yçin** 'for love of him'; **komşu-yçün** 'for the neighbour'.

4. Primary postpositions with dative case:

göre, nazaran (A)	according to
doğru	towards
karşı	against
kadar (A), -dek, değin	as far as
dair (A)	concerning

rağmen (A)   in spite of
inat (A)   in despite of
nispeten (A)   in proportion to

The equivalents of these words are italicized in the translations of the examples which follow.

radyo'ya göre, hava güzel olacak '*according* to the radio, the weather is going to be fine'; bu vaziyet-e göre '*in view of* this situation'; yeni ev, tam onlar-a göre 'the new house is *just right for* them'. In the first two examples, göre could be replaced by nazaran. See also the last paragraph on this page, and note bulunduğuna göre on p. 165.

köy-e doğru yürüdük 'we walked *towards* the village'; sabah-a doğru uyandım '*towards* morning I awoke'.

hangi takım-a karşı oynıyacaksınız? '*against* which team are you going to play?'; bu suçlama-ya karşı ne söyleyebildi? 'what could he say *in reply to* this accusation?'; deniz-e karşı oturduk 'we sat *facing* the sea'; sabah-a karşı uyandım '*towards* morning I awoke'.

köy-e kadar yürüdük 'we walked *as far as* the village'; akşam-a kadar konuştuk 'we talked *until* evening'; öğle-ye kadar gelecek 'he will come *by* noon'; bir saat-e kadar gelecek 'he will come *in* an hour'; bir kaç gün-e kadar gelecek 'he will come *in* a few days'. The provincialism -dek or değin is favoured by modernists as a native equivalent of kadar as a postposition with the dative (but not with the absolute or genitive as in (3). -dek is usually suffixed but never changes its vowel: köyedek, akşamadek.[1] The rarer değin is usually written separately: köye değin, akşama değin.

atom bombası-n-a dair bir konferans 'a lecture *about* the atom-bomb'. Modernists prefer üzerine (§ 6) to dair.

gençliğ-i-n-e rağmen büyük bir sanatkârdır '*in spite of* her youth she is a great artist'. The neologism karşın has been proposed as an alternative but has not won general acceptance.

baba-sı-n-a inat okula gitmiyor '*just to spite* his father he doesn't go to school'.

benimki-n-e nispeten sizinki çok pahalı '*in proportion to* mine, *compared with* mine, yours is very expensive'. The modernist alternative is göre. nispeten as an adverb means 'relatively'.

[1] YİK recommends that it be written as a separate word.

A number of adjectives are construed with a dative, e.g. **ait** (A) 'belonging (to)', **mukabil** (A) 'in return (for)', **aykırı** 'contrary (to)'. They are mentioned here because in some contexts they may be parsed as postpositions.

5. Primary postpositions with ablative case:

evvel (A), **önce**	before
sonra	after
beri	since, this side of
bu yana	since
yana	on the side of
içeri	inside
dolayı, ötürü	because of
başka	besides, apart from
itibaren (A)	with effect from

See also § 9, end, and XII, 10.

EXAMPLES: **bugün-den evvel** or **önce** '*before* today'; **toplantı-dan sonra** '*after* the meeting'; **Erzurum'dan sonra yol nasıl?** 'how is the road *beyond* Erzurum?'; **ağustos-tan beri** or **bu yana** '*since* August'; **göl-den beri hava güzel** '*this side of* the lake the weather is fine'; **aydın-ın iyi-si her zaman halk-tan yana-dır** 'the best type of intellectual (II, 22 (a)) is always *on the side of* the people'; **para-dan yana durum kötü** '*as regards money* the position is bad'; **daktilo-dan yana çok talihliyim** 'I am very lucky *as regards* secretarial assistance' ('on-the-side-of typist'); **bir antikacı-dan içeri girdik** 'we went *inside* an antique-dealer⟨'s shop⟩'; **bu-n-dan dolayı gitmedik** '*because of* this we did not go' (the synonymous **ötürü** is far rarer); **bir dayı-dan başka hiçbir akrabası yok** 'he has no relative *apart from* an uncle' (two Arabic synonyms are **maada** and **gayri**, neither very frequent); **perşembe-den itibaren her gün** 'every day, *starting from* Thursday'.

6. Secondary postpositions: I. The words in the following list are all nouns and may be used in any case and with any personal suffix: **ara-mız-da** 'in between us' ('in our interval'); **arka-nız-dan** 'from behind you' ('from your back'); **masa-nın üst-ü-n-ü sildi** 'she wiped the top of the table'. It is only when they are used in izafet with another noun and in the dative, locative, or ablative case that they correspond in function to English prepositions and are called postpositions.

alt	underside	**karşı**	opposite side
ara	interval, space between	**orta**	middle
arka, art	back	**ön**	front
baş	immediate vicinity	**peş** (P)	space behind
dış, hariç (A)	exterior	**üst, üzer-, fevk** (A)	top
etraf (A), **çevre**	surroundings	**yan**	side
iç, dahil (A)	interior		

To these may be added the adverbs of place listed in XII, 10.

EXAMPLES: **topu masa-nın alt-ı-n-a attı** 'he threw the ball under the table'; **ceket-i kol-u-nun alt-ı-n-da, parkta ġeziyordu** 'his jacket under his arm, he was strolling in the park'; **araba-nın alt-ı-n-dan çıktı** 'he emerged from under the car'.

iki ev-in ara-sı-n-a ġirdi 'he entered between the two houses'; **iki evin arasında bekledi** 'he waited between the two houses'; **iki evin arasından çıktı** 'he emerged from between the two houses'.

In such phrases as 'between A and B', 'and' is translated by the postposition **ile**: **Doğu ile** (or **Doğuyla**) **Batı arasında** 'between East and West'; **dağ ile** (**dağla**) **ırmağın arasında** 'between the mountain and the river'. Note that in the first example, where 'East' and 'West' are broad general terms, the izafet is indefinite, while in the second, where a specific river is intended, the izafet is definite, with **ırmak** 'river' in the genitive.

kapı-nın arka-sı-n-a (or **ard-ı-n-a**) **saklandı** 'he hid behind the door' (dative of end of motion); **kapının arkasında (ardında) durdu** 'he stood behind the door' (lit., as in American English, 'in back of the door'); **kapının arkasından (ardından) çıktı** 'he emerged from behind the door'.

masa-sı-nın baş-ı-n-a oturduk 'we sat down at ("to-the-immediate-vicinity-of") his table'; **silâh başına!** 'to arms!'; **vazife başında** 'on duty'; **mikrofon başında şarkı söylemek** 'to sing songs at the microphone'.

baş may be defined by **alt** as in **dam-ın altbaşında** 'immediately under ("in-the-underside-vicinity-of") the roof'. Note also **omuz baş-ım-da duruyor** 'he is standing at my shoulder', lit. 'in-my-shoulder-vicinity', **omuz başım** being an izafet group with the first-person suffix replacing the third; see II, 24.

vilâyet-in sınırlar-ı dış-ı-n-a (or, less commonly, **haric-i-n-e**) **çıkmadı** 'he did not go outside ("to-the-outside-of") the boundaries of the province'; **surlar-ın dışında (haricinde) oturuyorlar** 'they are living outside the city-walls'; **bina-nın dışından (haricinden) bir ses geldi** 'a voice came from outside the building'.

The next two examples well illustrate the difference between the definite and indefinite izafets:

okul-un dışında bir taksi bulunmaktadır 'there is a taxi outside the school'; **bir milyon çocuk okul dışında bulunmaktadır** 'a million children are outside school' (i.e. not attending any school).

etraf is far commoner than its modern replacement **çevre**: **şehr-in etrafında (çevresinde) çok bağ var** 'there are many orchards round the city'.

dahil, on the other hand, is fast going out of use. **deniz-in iç-i-n-e** (**dahil-i-n-e**) 'into the sea'; **acaip bir karışıklık içinde (dahilinde)** 'in a strange confusion'; **küçük kahve-nin içinden (dahilinden)** 'from inside the small cafe'.

We have already met **karşı** as a primary postposition. As a noun it means 'opposite side', so in izafet as a secondary postposition it means 'to/on/from the opposite side of', according to case. It is particularly common in the locative in the sense of '*vis-à-vis*, facing, confronted with': **insan ıztırab-ı karşı-sı-n-da aydın ne diyor?** 'confronted with human affliction, what does the intellectual say?'

kalabalığ-ın orta-sı-n-a 'to the middle of the crowd'; **kalabalığın ortasında** 'in the middle of the crowd'; **kalabalığın ortasından** 'from *or* through the middle of the crowd'.

sahne-nin ön-ü-n-e 'to the front of the stage'; **sahnenin önünde** 'in front of *or* at the front of the stage'; **sahnenin önünden** 'from *or* through the front of the stage'.

kılavuz-un peş-i-n-e düştük 'we began to follow ("we-fell to-the-rear-of") the guide'.

'To run after' is **peşinde** or **peşinden koşmak**.

The hyphen after **üzer** in the list above is to indicate that this word, alone among nouns, is never found without a personal suffix. **düşman-ın üst-ü-n-e** *or* **üzer-i-n-e yürüdüler** 'they marched onto the enemy'; **yayla-nın üstünden** *or* **üzerinden indiler** 'they came down from-on-top-of the plateau'.

üzerine and **üstüne** are commoner than **üzerinde** and **üstünde** for 'on' meaning 'on the subject of': **tarih üzerine** *or* **üstüne araştırmaları** 'his researches on history'. They are also used for 'on top of' in the sense of 'in addition to': **bira üzerine** *or* **üstüne şarap içme** 'don't drink wine on top of beer'. See also **akşamüstü**, etc., XII, 13 (*a*).

fevk is little used nowadays: **kale-nin fevkında** (I, 35, penultimate paragraph) **bir bayrak var** 'there is a flag over the citadel'.

pencere-m-in yan-ı-n-a geldi 'he came beside my window'; **penceremin yanında bir ağaç var** 'there is a tree beside my window'; **penceremin yanından ayrıldı** 'he departed from-beside my window'.

yan may be qualified by **üst**: **kutu-nun üstyanında** 'on the top-side of the box'. It is also compounded with **baş**: **istasyon-un yanıbaşında** 'just beside the station'. The first ı in this word, though originally the third-person suffix, is invariable: **yanı-başımda** 'just beside me'.

art, arka, ön, peş, and **yan** with a personal suffix but no case-ending are compounded with **sıra** 'row' to make postpositions, the **sıra** conveying the sense of close proximity: **kardeş-i-nin ard-ı sıra yürüdü** 'he walked just-behind his brother'; **klâsik dersler-in yan-ı sıra, marangozluk dersleri verilir** 'side by side with the classical lessons, carpentry lessons are given'; **ön-üm sıra yürüdü** 'he walked just-in-front-of-me'. With **kıyı** 'shore' is made the adverb **kıyısıra** 'along the shore'.

The nouns discussed above are also used as adjectives: **alt dudak** 'bottom lip'; **ara kapı** 'communicating door'; **arka bahçe** 'back garden'; **dış ticaret** 'external trade'; **iç ticaret** 'internal trade'. In the official terms for External and Internal Affairs, however, **dış** and **iç** are nouns: **dış işler-i, iç işler-i**.

7. Secondary postpositions: II. The nouns in the first column below (which, as nouns, mean respectively 'truth', 'side', 'cause (or face)', 'regard', 'name') are also used to make postpositions, but differ from the previous group in that in the meanings shown they are used only in the case shown, though they may change for person. That is to say, whereas, for example, 'under' may be

altına, altında, or altından according to context, 'concerning' can only be hakkında in the locative, while 'concerning me' is hakkımda and 'concerning you' hakkınızda, again in the locative.

hak (A)	hakkında	concerning
taraf (A)	tarafından	by, through the agency of
yüz	yüzünden	because of
bakım	bakımından	from the point of view of
nam (P)	namına	in the way of

These are used in indefinite izafet only; i.e. the noun they follow is never in the genitive. Any exceptions to this rule are apparent only, as the examples will show.

inkılâp hakkında bir nutuk söyledi 'he gave a speech about the revolution'; kardeş-i tarafından uzaklaştırıldı 'he was sent away by his brother'; o adam yüzünden kan akacak 'because of that man, blood will flow'; protokol bakımından haklıdır 'from the point of view of protocol, he is right'; para namına bir şeyim yok 'I have nothing in the way of money, nothing you could call money'. If we now make the izafet definite in each example, i.e. if we put the first noun of each into the genitive—inkılâbın, kardeşinin, o adamın, protokolun, paranın—then hakkında, tarafından, yüzünden, bakımından, and namına will revert to their literal meanings: 'in the truth of the revolution'; 'he was sent away from his brother's side'; 'blood will flow from that man's face'; 'he is right from protocol's regard'; 'I have nothing for money's name', the first and the last two being as meaningless in Turkish as in English. bakımından has almost entirely supplanted nokta-ı nazarından (II, 26).

The rule that these postpositions are used only in indefinite izafet is not broken by, for example, o-nun hakkında 'concerning him' any more than it is by benim hakkımda 'concerning me'; the pronoun in the genitive is not in izafet with the following hakk- but merely reinforces its personal suffix.

husus (A) 'particular'—bu hususta itirazım yok 'I have no objection in this particular, in this regard'—is used as a postpositional expression especially with the infinitive of the verb: oraya gitmek hususunda itirazım yok 'I have no objection in-the-matter-of going there'.

8. Secondary postpositions: III. The nouns in the next list are also used only in the case shown (for the ending of **boyunca** see XII, 2) but differ from those in the previous section in that they can be used in definite izafet. In fact, however, they are mostly found in indefinite izafet, even when they follow a defined noun.

boy	length	boyunca	along, throughout
esna (A)	duration	esnasında	in the course of
sıra	row	sırasında	in the course of
zarf (A)	container	zarfında	during
saye (P)	shadow	sayesinde	thanks to
uğur	luck	uğruna, uğrunda	for the sake of
yer	place	yerine	instead of

EXAMPLES: **Kızıl Irmak (Irmağın) boyunca** 'along the Red River'; **Türk tarih-i (tarih-i-nin) boyunca** 'throughout Turkish history'; **muharebe esnasında** 'in the course of the battle'; **muharebe zarfında** 'during the battle'; **Ahmet (Ahmed'in) sayesinde her şey iyi oldu** 'thanks to Ahmet, everything has become all right'; **çocuk, akl-ı (akl-ı-nın) sayesinde kurtuldu** 'the child was saved, thanks to his intelligence'; **vatan-ı (vatan-ı-nın) uğruna/uğrunda can verdi** 'he gave his life for his country's sake'; **asistan, profesör-ü (profesör-ü-nün) yerine derse gitti** 'the assistant went to the class instead of his professor'; **beni eşek yerine alma** 'don't take me for a donkey'.

esna is also used in the locative, defined by a demonstrative: **bu esnada** 'during this time'. So too **bu sayededir ki ...** 'it is thanks to this that ...'

9. leh, aleyh. The Arabic *la-h^u* 'for him' and *'alay-h^i* 'against him' make Turkish secondary postpositions in the locative: **leh-i-n-de** 'for, *pro*' and **aleyh-i-n-de** 'against, *contra*'; **teklif-in lehinde/aleyhinde konuştu** 'he spoke for/against the motion'. Although the final **h** is originally the Arabic third-singular masculine pronoun, these words can be used with the suffixes of any of the three persons: **leh-imiz-de konuştu** 'he spoke for us'; **aleyh-iniz-de konuştu** 'he spoke against you'.

They may also be used in the locative without personal suffixes as primary postpositions following an ablative: **teklif-ten lehte misiniz, aleyhte misiniz?** 'are you for or against the motion?'

10. The preposition ilâ. This, the Arabic *ilā* 'to, towards', is the only preposition used in Turkish as an independent word, as distinct from, for example, the Arabic *bi* in **bilhassa** 'in particular' or the Italian *a* in **alafranga** '*alla franca*, in European style'.

It is employed between numbers: **on beş ilâ yirmi kişi** 'fifteen to twenty people', written in figures **15–20**. Modernists avoid the word and would read these figures as **on beşten yirmiye kadar** 'from 15 to 20' or **on beş ile yirmi arasında** 'between 15 and 20'. Because of the resemblance between **ilâ** and **ile**, the semiliterate trying to show off produces a horrid synthesis: **on beş ilâ yirmi arasında**. Another example of the correct use: **taşlar, yarım ilâ bir metre kalınlığında toprakla örtülür** 'the stones are covered with earth to (lit. 'in') a depth of a half to one metre'.

VIII

THE VERB

1. The stem. The form of the verb which is cited in the dictionaries is the infinitive in **-mek**, e.g. **bilmek** 'to know', **bulmak** 'to find', **görmek** 'to see', **anlamak** 'to understand'. When one is describing the conjugation of the verb it is more convenient to omit this ending and cite only the stem: **bil-, bul-, gör-, anla-**.

2. The verb 'to be'. We shall first deal with those parts of the anomalous and defective verb 'to be' which are used as auxiliaries in the conjugation of all verbs. In the oldest texts the infinitive 'to be' was **ermek**, but the stem **er-**, abraded in the course of time, now appears as **i-**. Some grammarians consequently speak of 'the verb **imek**', but no such form ever existed.

3. The present tense of 'to be'. The forms of the present tense of 'to be' exist only as enclitic suffixes, subject to the fourfold harmony. In origin they are suffixed personal pronouns, with the exception of the third person **-dir**, originally **turur** 'he stands'.[1] **-dir** is placed in parentheses in the following table as a reminder that in Turkish (as in Arabic and Russian), simple 'A is B' equivalences are expressed without a copula. See § 4. When a suffix beginning with a vowel follows a vowel, a **y** is inserted to preserve the identity of both.

Present: 'I am', etc.

Singular

1	-im	-üm	-ım	-um
2	-sin	-sün	-sın	-sun
3	(-dir/tir)	(-dür/tür)	(-dır/tır)	(-dur/tur)

Plural

1	-iz	-üz	-ız	-uz
2	-siniz	-sünüz	-sınız	-sunuz
3	-(dir/tir)ler	-(dür/tür)ler	-(dır/tır)lar	-(dur/tur)lar

[1] This form was already in use as a copula in the eleventh century.

4. Uses of -dir. In writing and in formal speech -dir expresses the copula: **kızın adı, Fatma'dır** 'the girl's name is Fatima'; **enerji kaynaklarımız bol-dur** 'our sources of power are abundant'. It will be noticed that the verb in the latter example is singular; this is customary with inanimate plural subjects and possible with animate plural subjects. See further XVI, 1.

In ordinary speech -dir is not used in such simple 'A = B' sentences; one says **kızın adı Fatma**; **enerji kaynaklarımız bol**. -dir is generally used as a copula in speech as well as in writing:

(*a*) When the predicate is a noun in such a sentence as: **en çok sevdiğim şair Nedim'dir** 'my favourite poet is Nedim', where the omission of -dir might lead to misunderstanding: 'my favourite poet, Nedim, ...'.

(*b*) When the subject is a pronoun understood from the context: **yaman bir adam-dır** 'he is a remarkable man'. There is an alternative, in the colloquial, of using the third-person pronoun instead of -dir: **o, yaman bir adam** or **yaman bir adam, o**.

(*c*) When the subject is a noun which follows the predicate: **yaman bir adamdır, amcanız** 'he is a remarkable man, your uncle'.

(*d*) When the subject is a phrase containing a postposition and the predicate is a noun-clause introduced by **ki** (XIII, 15): **onun sayesinde-dir ki muvaffak olduk** 'it is thanks to him that we have succeeded'; **bundan dolayı-dır ki gitmedim** 'it is because of this that I did not go'. In terms of the equivalent English, one could include these two examples under (*b*) above. This is the only one of the four situations in which the -dir is never omitted.

Otherwise, the use of -dir in informal speech is either for emphasis or, more often, to indicate a supposition. Whereas the written words **vesika kasa-da-dır** mean 'the document is in the safe', the same words in informal speech mean 'the document is surely in the safe, must be in the safe', or, less commonly, a confident 'the document *is* in the safe'; only the tone of voice shows which of the two is intended. If the speaker is stating a simple fact, which he does not think it necessary to emphasize, he will say **vesika kasada**.

The following are possible answers to the spoken question **çocuklar nerede?** 'where are the children?'

bahçede	in the garden
bahçedeler	they are in the garden
bahçededirler	they are in the garden (*emphatic*) *or* they are surely in the garden (*supposition*)
bahçedelerdir	they are surely in the garden (*supposition*)

One manifestation of the rapid closing of the gap between the written and spoken languages is that **-dir** is more and more omitted in writing when it merely expresses the copula.

For **-dir** suffixed to finite verbs see § 42.

5. Examples of the present tense of 'to be'.

'I am, etc., at home'

Singular	*Plural*
1 evde-y-im	evde-y-iz
2 evde-sin	evde-siniz
3 evde(-dir)	evde(-dir)-ler

'I am, etc., Turkish'

Singular	*Plural*
1 **Türk-üm**	**Türk-üz**
2 **Türk-sün**	**Türk-sünüz**
3 **Türk(-tür)**	**Türk(-tür)-ler**[1]

'I am, etc., ready'

Singular	*Plural*
1 hazır-ım	hazır-ız
2 hazır-sın	hazır-sınız
3 hazır(-dır)	hazır(-dır)-lar

'I am, etc., responsible'

Singular	*Plural*
1 sorumlu-y-um	sorumlu-y-uz
2 sorumlu-sun	sorumlu-sunuz
3 sorumlu(-dur)	sorumlu(-dur)-lar

[1] Another possibility is **Türk-ler-dir**, which means not 'they are Turkish' but 'they are the Turks'. See XVI, 3 (*e*).

6. Forms based on i-. The finite forms of 'to be' based on **i-**, namely the past, the conditional, and the inferential, all exist both as independent words and as suffixes. When suffixed, the **i** of the stem is lost after consonants and changes to **y** after vowels, while the remainder of the form is subject both to the fourfold vowel harmony and the alternation **d/t**.

7. The past tense of 'to be'. The base, i.e. the third singular, is **idi** and the other persons are formed by adding to it **-m** for the first and **-n** for the second singular, and **-k** for the first, **-niz** for the second, and **-ler** for the third plural. 'I was', etc.

Suffixed after vowels

Singular
1 idim	-ydim	-ydüm	-ydım	-ydum
2 idin	-ydin	-ydün	-ydın	-ydun
3 idi	-ydi	-ydü	-ydı	-ydu

Plural
1 idik	-ydik	-ydük	-ydık	-yduk
2 idiniz	-ydiniz	-ydünüz	-ydınız	-ydunuz
3 idiler	-ydiler	-ydüler	-ydılar	-ydular

Suffixed after consonants

Singular
1 -dim/tim/düm/tüm/dım/tım/dum/tum
2 -din/tin/dün/tün/dın/tın/dun/tun
3 -di/ti/dü/tü/dı/tı/du/tu

Plural
1 -dik/tik/dük/tük/dık/tık/duk/tuk
2 -diniz/tiniz/dünüz/tünüz/dınız/tınız/dunuz/tunuz
3 -diler/tiler/düler/tüler/dılar/tılar/dular/tular

EXAMPLES: **evde idim** or **evdeydim** 'I was at home'; **Türk idi** or **Türktü** 'he was Turkish'; **hazır idik** or **hazırdık** 'we were ready'; **sorumlu idiniz** or **sorumluydunuz** 'you were responsible'.

8. The present conditional of 'to be': 'if I am', etc. The base is **ise** and the personal endings are the same as those of the past.

Suffixed

Singular	After vowels		After consonants	
1 isem	-ysem	-ysam	-sem	-sam
2 isen	-ysen	-ysan	-sen	-san
3 ise	-yse	-ysa	-se	-sa

Plural

1 isek	-ysek	-ysak	-sek	-sak
2 iseniz	-yseniz	-ysanız	-seniz	-sanız
3 iseler	-yseler	-ysalar	-seler	-salar

EXAMPLES: **evde isem** or **evdeysem** 'if I am at home'; **Türk ise** or **Türkse** 'if he is Turkish'; **hazır isek** or **hazırsak** 'if we are ready'; **sorumlu iseniz** or **sorumluysanız** 'if you are responsible'.

9. The past conditional of 'to be': 'if I was', etc. This expresses open past condition as in: 'if I was right, why did you not agree with me?' For the remote or unfulfilled condition, as in 'if I had been right, would you have agreed with me?', see § 34. The various persons of the present conditional are added to the past base: **idi+isem > idiysem** or, when suffixed, **-ydiysem**. Alternatively, the suffixed third singular of the present conditional is added to the various persons of the past: **idim+se**. The latter alternative is, however, provincial and colloquial, so much so that in writing it occurs only in the suffixed forms **-dimse** or **-ydimse**, etc.; the theoretically possible independent forms **idimse**, etc., seem never to be used.

Comparison with § 7 will show that the following paradigm has been simplified to the extent of three-quarters of its full size by the omission of the suffixed forms (*a*) beginning with **t**, as used after unvoiced consonants, and (*b*) with the rounded vowels **ü** and **u**.

Suffixed

Singular	After vowels		After consonants	
1 idiysem	-ydiysem	-ydıysam	-diysem	-dıysam
2 idiysen	-ydiysen	-ydıysan	-diysen	-dıysan
3 idiyse	-ydiyse	-ydıysa	-diyse	-dıysa

Plural	After vowels			After consonants	
1 idiysek	-ydiysek	-ydıysak		-diysek	-dıysak
2 idiyseniz	-ydiyseniz	-ydısanız		-diyseniz	-dıysanız
3 idiyseler	-ydiyseler	-ydıysalar		-diyseler	-dıysalar

Suffixed

Singular	After vowels		After consonants	
1 -ydimse	-ydımsa		-dimse	-dımsa
2 -ydinse	-ydınsa		-dinse	-dınsa
3 -ydiyse	-ydıysa		-diyse	-ydıysa

Plural				
1 -ydikse	-ydıksa		-dikse	-dıksa
2 -ydinizse	-ydınızsa		-dinizse	-dınızsa
3 -ydiyseler	-ydıysalar		-diyseler	-dıysalar

EXAMPLES: **evde idiysem, evdeydiysem,** or **evdeydimse** 'if I was at home'; **Türk idiyse** or **Türktüyse** 'if he was Turkish'; **hazır idiysek, hazırdıysak** or **hazırdıksa** 'if we were ready'; **sorumlu idiyseniz, sorumluyduysanız** or **sorumluydunuzsa** 'if you were responsible'.

10. The inferential. The inferential present/past **imiş** means 'he is/was said to be' or 'I infer that he is/was although I had not realized it before'. Though some grammarians have termed it the dubitative, in itself it does not imply doubt or uncertainty; e.g. a sentence beginning **Orhan hasta imiş** 'Orhan is said to be ill' may continue 'and we ought to visit the poor man' or 'but I bet he's malingering'. Similarly, a speaker who says **ben gerici imişim** 'I am said to be reactionary' may go on 'and it's true and I'm proud of it' or 'but this is a wicked slander'.

It is formed by adding to the base **imiş**, or the suffixed **-ymiş** or **-miş**, etc., the present suffixes of the verb 'to be', with the exception of **-dir**.

Singular		Suffixed after vowels			
1 imişim	-ymişim	-ymüşüm		-ymışım	-ymuşum
2 imişsin	-ymişsin	-ymüşsün		-ymışsın	-ymuşsun
3 imiş	-ymiş	-ymüş		-ymış	-ymuş

Plural		Suffixed after vowels			
1 imişiz	-ymişiz	-ymüşüz		-ymışız	-ymuşuz
2 imişsiniz	-ymişsiniz	-ymüşsünüz		-ymışsınız	-ymuşsunuz
3 imişler	-ymişler	-ymüşler		-ymışlar	-ymuşlar

Suffixed after consonants

-mişim   -müşüm   -mışım   -muşum
etc.

The **şs** of the second person is sometimes simplified in pronunciation, rarely in writing, to **ş**: **imişin, imişiniz.**

EXAMPLES: **evde imişim** or **evdeymişim** 'I am said to be at home'; **Türk imiş** or **Türkmüş** 'he is said to be Turkish'; **hazır imişiz** or **hazırmışız** 'we are said to be ready'; **sorumlu imişsiniz** or **sorumluymuşsunuz** 'you are said to be responsible'; **kimdir?** 'who is he?'; **kimmiş?** 'who is he supposed to be?'; **günahımız ne imiş?** 'what is our sin said to be?' i.e. 'what are we reported to have done that has offended you?' If told 'the new Minister is a good man', **yeni Bakan iyi bir adam,** one may reply **imiş** or **-mış,** meaning 'so we are told, but I have no first-hand knowledge of him'. **ben mişlere muşlara pek kulak vermem** 'I don't pay much heed to gossip' (lit. 'I do not much give ear to **miş**es and **muş**es').

11. The inferential conditional: 'I gather that if I am/was' or 'if I am/was, as they say', etc. The present conditional endings are suffixed to **imiş** or its suffixed forms.

Singular		Suffixed after vowels			
1 imişsem	-ymişsem	-ymüşsem		-ymışsam	-ymuşsam
2 imişsen	-ymişsen	-ymüşsen		-ymışsan	-ymuşsan
3 imişse	-ymişse	-ymüşse		-ymışsa	-ymuşsa

Plural					
1 imişsek	-ymişsek	-ymüşsek		-ymışsak	-ymuşsak
2 imişseniz	-ymişseniz	-ymüşseniz		-ymışsanız	-ymuşsanız
3 imişseler	-ymişseler	-ymüşseler		-ymışsalar	-ymuşsalar

Suffixed after consonants

-mişsem   -müşsem   -mışsam   -muşsam
etc.

These forms, which are used in reported speech, are not often met with in writing. If someone says to you **hazırsanız yola çıkmalısınız** (§ 30) 'if you are ready you ought to start off', you may report these words thus: **ben hazır imişsem** (or **hazırmışsam**) **yola çıkmalıymışım** 'they are saying that if I am ready I ought to start off'.

12. The negative of 'to be'. This is made by putting after **değil** 'not' the suffixed forms, less commonly the independent forms, given above.

(a) *Present*: 'I am not', etc.

	Singular	*Plural*
1	değilim	değiliz
2	değilsin	değilsiniz
3	değil(dir)	değil(dir)ler

(b) *Past*: 'I was not', etc.

Singular

1	değildim	or	değil idim
2	değildin		değil idin
3	değildi		değil idi

Plural

1	değildik	değil idik
2	değildiniz	değil idiniz
3	değildiler	değil idiler

(c) *Present conditional*: 'if I am not', etc.

Singular

1	değilsem	or	değil isem
2	değilsen		değil isen
3	değilse		değil ise

Plural

1	değilsek	değil isek
2	değilseniz	değil iseniz
3	değilseler (değillerse)[1]	değil iseler

[1] Alternative forms exist for the third-person plural of most tenses and moods. The less common alternative is given in parentheses.

(d) *Past conditional*: 'if I was not', etc.

Singular
1 değil idiysem or değildiysem or değildimse
2 değil idiysen   değildiysen   değildinse
3 değil idiyse   değildiyse   değildiyse

Plural
1 değil idiysek   değildiysek   değildikse
2 değil idiyseniz değildiyseniz değildinizse
3 değil idiyseler değildiyseler değildilerse

(e) *Inferential*: 'I am/was said not to be', 'I infer that I am/was not', etc.

Singular
1 değil imişim or değilmişim
2 değil imişsin   değilmişsin
3 değil imiş   değilmiş

Plural
1 değil imişiz   değilmişiz
2 değil imişsiniz   değilmişsiniz
3 değil imişler   değilmişler

(f) *Inferential conditional*: 'I gather that if I am not', 'if I am not, as they say', etc.

Singular
1 değil imişsem or değilmişsem
2 değil imişsen   değilmişsen
3 değil imişse   değilmişse

Plural
1 değil imişsek   değilmişsek
2 değil imişseniz değilmişseniz
3 değil imişseler değilmişseler

değil alone means 'not' as well as 'is not', usually following the word it negates: **bugün değil, dün geldi** 'he came yesterday, not today' ('this-day not, yesterday he-came'). When it precedes one of two parallel words, it indicates that that one is of less importance than the other: **değil sen, ben de bilmedim** 'never mind about you, even *I* did not know'; **değil parasını, hayatını kurtaramadı** 'never mind about his money, he couldn't save his *life*'. If the positions of **değil** and **parasını** were reversed, the meaning would be 'it wasn't his money, it was his life that he could not save'. **hizmetçiyi değil, beni koğdu** 'it wasn't the servant, it was me he threw out'; **değil hizmetçiyi, beni bile koğdu** 'never mind about the servant, he even threw *me* out'. The sense of 'never mind about' is occasionally expressed by a following **değil**; see XXIV, 32.

13. Interrogative. The interrogative particle is **mi**, which turns the immediately preceding word into a question. It is written separately from the preceding word, but takes its vowel harmony from it: **doğru** 'true', **doğru mu?** 'true?'; **bugün mü?** 'today?'; **yarın mı?** 'tomorrow?' It may even follow and turn into a question a word which is already interrogative; thus the reply to **kim geldi?** 'who came?' may be **kim mi?** 'do you ask "who?"?' (literally '"who?"?'). The forms of the verb 'to be' are appended or suffixed to it, but when **-ler** alone and not **-dirler** is used for the third plural of the present tense the **-ler** precedes **mi**.

(*a*) *Present*: 'am I, etc., at home/Turkish/ready/responsible?'

Singular

1	evde miyim	Türk	müyüm	hazır	mıyım	sorumlu	muyum
2	,, misin	,,	müsün	,,	mısın	,,	musun
3	,, mi(dir)	,,	mü(dür)[1]	,,	mı(dır)	,,	mu(dur)

Plural

1	,, miyiz	,,	müyüz	,,	mıyız	,,	muyuz
2	,, misiniz	,,	müsünüz	,,	mısınız	,,	musunuz
3	evdeler mi / evde midirler	Türkler mi / Türk müdürler		hazırlar mı / hazır mıdırlar		sorumlular mı / sorumlu mudurlar	

(*b*) *Past*: 'was I at home, etc.?'

    evde mi idim   or   evde miydim
    Türk mü idim   Türk müydüm
    hazır mı idim   hazır mıydım
    sorumlu mu idim   sorumlu muydum

---

[1] There is also a noun **müdür** 'administrator, director'. In practice this apparent source of ambiguity gives no trouble.

For the remaining persons see the conjugation of **idim** and its forms when suffixed after vowels, in § 7.

(c) *Inferential*: 'am I said to be at home, etc.?'

> evde mi imişim   or   evde miymişim
> Türk mü imişim   Türk müymüşüm
> hazır mı imişim   hazır mıymışım
> sorumlu mu imişim   sorumlu muymuşum

For the remaining persons see § 10.

Some grammarians complete the paradigm by setting out the interrogative of the conditional; present, past, and inferential: **isem mi, idiysem mi, imişsem mi.** This is unnecessary if it be borne in mind that **mi** functions simply as a question-mark. The 'interrogative of the conditional' of the verb 'to be' occurs only in such contexts as when someone is asked a question like 'what shall we do if he is not at home?' **evde değilse ne yapalım?** and replies 'if he is not at home?' **evde değilse mi?** See also § 34 (*e*).

14. Negative-interrogative. **mi** and the appropriate part of the verb 'to be' are placed after **değil**:

> **evde değil miyim** 'am I not at home?'
> **evde değil mi idim** or **evde değil miydim** 'was I not at home?'
> **evde değil mi imişim** or **evde değil miymişim** 'am I not said to be/am I said not to be at home?'

15. The regular verb. This category includes all verbs other than the verb 'to be'. If we look back over the preceding pages we see that the conjugation of the verb 'to be' may be summarized as follows: there are two distinct sets of personal endings, which we may call Types I and II.

	Type I	Type II
Singular		
1	**-im**	**-m**
2	**-sin**	**-n**
3	**(-dir)**	—

	Type I	Type II
Plural		
1	-iz	-k
2	-siniz	-niz
3	-(dir)ler	-ler

Type I is the present tense, 'I am', etc.; Type II is added to the base of the past tense **idi** and of the conditional **ise**. The Type I endings are suffixed to **imiş** to make the inferential: **imiş-im, imiş-sin**, etc. The conditional, i.e. **ise** plus the Type II endings, is added to the past base **idi** to make the past conditional: **idi-yse-m, idi-yse-n**, etc. Added to the inferential base it makes the inferential conditional: **imiş-se-m, imiş-se-n**, etc. The same principle applies to the conjugation of the regular verb, but two other sets of personal endings are used in addition to Types I and II. Type III is confined to the subjunctive and Type IV to the imperative, which has no first person.

	Type III	Type IV
Singular		
1	-eyim	
2	-esin	—
3	-e	-sin
Plural		
1	-elim	
2	-esiniz	-in, -iniz
3	-eler	-sinler

By the addition of 'characteristics' to the verb-stem the following tense- and mood-bases can be formed:

1. present
2. future
3. aorist
4. **miş**-past
5. necessitative
6. **di**-past
7. conditional
8. subjunctive

The imperative is not included in this list because it has no characteristic. The term 'tense- and mood-bases' is used in preference to 'tenses and moods' because from each base a variety of compound tenses and moods can be formed. Each base is also the third-person singular of its tense or mood. Only the **di**-past, conditional, subjunctive, and imperative have endings of their own; all the other bases are substantival in origin and are verbalized by means of the verb 'to be'. For this reason the device of suspended affixation is regularly used in the verb: just as, in English, there is no need to repeat the part of the verb 'to be' in 'I was sitting in my room [and I was] reading the paper', so in Turkish: **odamda oturuyor[-dum ve] ğazeteyi okuyor-dum.**

The regularity of the verbal system will soon impress itself on the student. Where it has been thought unnecessary to set out a conjugation in full, the first and second persons of the singular and the third person of the plural have been given, as, in this last, alternative forms can occur, with the plural suffix preceding or following the personal suffix. The less common forms of the plural are shown in brackets. When alternative forms exist for a whole conjugation, the most usual is given first. A synopsis of the finite verb will be found on page 136.

The occasional change of **t** to **d** before vowels must be borne in mind; thus the present base of **et-** 'to do' is **ediyor**, of **ğit-** 'to go' **ğidiyor**, of **tat-** 'to taste' **tadıyor**. Most stems in **t**, however, do not undergo this change: **at-** 'to throw', **atıyor**; **tut-** 'to hold', **tutuyor**; **yat-** 'to lie down', **yatıyor**.

16. Present I. The characteristic of the base is an invariable **-yor**, originally an independent verb **yorır**, the aorist of the ancient **yorımak** 'to go, walk'. It is suffixed directly to vowel-stems:

eri- to melt, **eriyor** büyü- to grow, **büyüyor**
tanı- to recognize, **tanıyor** koru- to protect, **koruyor**

Final **e/a** of the stem changes before this suffix in accordance with the rule given in I, 36:

bekle- to wait, **bekliyor** gözle- to observe, **gözlüyor**
anla- to understand, **anlıyor** topla- to collect, **topluyor**

The original final vowel of the stems **bile-** 'to sharpen' and **yıka-** 'to wash' is sometimes preserved in writing—**bileyor, yıkayor**—in order to avoid confusion with the present of **bil-** 'to know' and **yık-** 'to demolish': **biliyor, yıkıyor**.

With consonant-stems, the appropriate close vowel is inserted before -yor:

gel- to come, **geliyor** **gör-** to see, **görüyor**
al- to take, **alıyor** **koş-** to run, **koşuyor**

The accent is on the vowel preceding the **-yor**.

17. Uses of the present. This tense is used for actions either in progress or envisaged: **Antalya'da çalışıyor** 'he is working in Antalya'; **kendisini haftada iki defa görüyorum** 'I see him twice a week'; **yarın gidiyoruz** 'we are going to morrow'. As it can denote actions begun in the past and still going on, it is used in such sentences as **iki sene-dir bu evde oturuyor** lit. 'it is two years he is living in this house' and **burada haziran ayından beri oturuyor** lit. 'he is living here since the month of June', where English has the perfect 'he has been living'. See also § 25.

18. Paradigms of the present. To the base in **-yor** are added the suffixes of the verb 'to be' except **-dir**; both in the written and in the spoken language the addition of **-dir** to the present tense indicates a supposition; see § 42.

(*a*) *Present simple*:

alıyorum	I am taking	**alıyoruz**	we are taking
alıyorsun	you are taking	**alıyorsunuz**	you are taking
alıyor	he is taking	**alıyorlar**	they are taking

(*b*) *Present past*:

alıyordum	I was taking	**alıyorduk**	we were taking
aliyordun	you were taking	**alıyordunuz**	you were taking
alıyordu	he was taking	**alıyorlardı** (**alıyordular**)	they were taking

Theoretically the separate forms **alıyor idim**, etc., might be expected but their use is in fact an Armenianism.

(*c*) *Present conditional*:

alıyorsam	if I am taking	**alıyorsak**	if we are taking
alıyorsan	if you are taking	**alıyorsanız**	if you are taking
alıyorsa	if he is taking	**alıyorlarsa** (**alıyorsalar**)	if they are taking

(d) *Past conditional*: 'if I was taking':

Singular
1 alıyor idiysem   or   alıyorduysam   or   alıyordumsa
2 alıyor idiysen   alıyorduysan   alıyordunsa

Plural
3 alıyor idiyseler   alıyorduysalar   alıyorlardıysa
  (alıyorlar idiyse)

For the full conjugation cf. § 9.

(e) *Inferential*: 'I am/was said to be taking' or 'I gather that I am/was taking':

Singular
1 alıyormuşum
2 alıyormuşsun

Plural
3 alıyorlarmış
  (alıyormuşlar)

See § 10.

(f) *Inferential conditional*: 'if I am/was, as they say, taking' or 'I gather that if I am/was taking':

Singular
1 alıyor imişsem   or   alıyormuşsam
2 alıyor imişsen   alıyormuşsan

Plural
3 alıyor imişseler   alıyormuşsalar
  (alıyorlar imişse)   (alıyorlarmışsa)

See § 11.

(g) *Negative*. The negative suffix is **-me**, added to the verb-stem before the characteristic; its vowel is subject to the rules given in I, 36.

  bekle+me+yor > beklemiyor   he is not waiting
  gör+me+yor   > görmüyor   he is not seeing
  al+ma+yor   > almıyor   he is not taking
  koş+ma+yor   > koşmuyor   he is not running

To the present negative base thus formed, the suffixes of 'to be' are added, just as with the positive base; for example, the negative conjugation of **al-** is exactly as shown in paragraphs (*a*) to (*f*), with the substitution of **almıyor** for **alıyor** throughout.

(*h*) *Interrogative*. The appropriate interrogative form of 'to be' is placed after the present base, positive or negative. As the interrogative particle turns the preceding word into a question, the literal meaning of, for example, **almıyor muydunuz** is 'is it not-taking that you were?'

	'am I taking?'	'am I not taking?'
Singular		
1	alıyor muyum	almıyor muyum
2	alıyor musun	almıyor musun
3	alıyor mu	almıyor mu
Plural		
1	alıyor muyuz	almıyor muyuz
2	alıyor musunuz	almıyor musunuz
3	alıyorlar mı	almıyorlar mı
	'was I taking?'	'was I not taking?'
Singular		
1	alıyor muydum	almıyor muydum
2	alıyor muydun	almıyor muydun
Plural		
3	alıyorlar mıydı (alıyor muydular)	almıyorlar mıydı (almıyor muydular)
	'am/was I said to be taking?'	'am/was I said not to be taking?'
Singular		
1	alıyor muymuşum	almıyor muymuşum
2	alıyor muymuşsun	almıyor muymuşsun
Plural		
3	alıyorlar mıymış (alıyor muymuşlar)	almıyorlar mıymış (almıyor muymuşlar)

19. *Present II*. The base of this tense is the locative case of the infinitive in **-mek**, to which are added the endings of the verb 'to be': **gelmekte-y-im** 'I am (in the act of) coming'; **almakta-sın** 'you are (in the act of) taking'; **gelmekte-ydim** 'I was (in the act of) coming'; **gözlemekte-ymişsiniz** 'you are/were said to be (in the act of) observing'; **koşmakta-larsa** 'if they are (in the act of) running', etc.

The negative is formed with the negative of 'to be' (§ 12): **almakta değilim** 'I am not taking'; **almakta değilsek** 'if we are not taking', etc. For the interrogative and negative-interrogative see §§ 13-14.

This present in **-mekte**, originally a literary formation, is rapidly invading the spoken language. It differs from the present in **-yor** in being used only of actions in progress and never of actions envisaged.

Very rarely the locative of the verbal noun in **-me** is similarly used: **alma-da-y-ım** 'I am in the act of taking'.

Subsequent references to 'the present tense' are to be taken as applying to the present I.

20. Future I. The characteristic is **-ecek**, added directly to consonant-stems: **gel-ecek** 'he will come'; **gör-ecek** 'he will see'; **al-acak** 'he will take'; **bul-acak** 'he will find'.

After vowel-stems a **y** is inserted: **eri-y-ecek** 'it will melt'; **tanı-y-acak** 'he will recognize'. If the final vowel of the stem is **e** or **a**, it is narrowed by the following **y** into **i** or **ı**:

bekle+y+ecek > bekliyecek   he will wait
anla+y+acak > anlıyacak   he will understand

21. Uses of the future. This tense is used, like the English future, to express not only what is going to happen but what the speaker wants to happen: **sigara içmekten vazgeçeceksin** 'you are going to give up smoking cigarettes'; **ister istemez bu işi yapacaksın** 'like it or not, you are going to do this job'. Also as in English, the third person expresses a confident assumption: **şimdi merdivenden çıkan Ahmet olacak** literally 'the one now coming upstairs will be Ahmet', i.e. 'that will be Ahmet coming upstairs now'. The future past, besides expressing past intention—**zaten bunu yapacaktım** 'I was going to do it anyway'—is employed in the apodosis of conditional sentences, both for 'I would do it if . . .' and 'I would have done it if . . .'.

22. Paradigms of the future. To the future base are added the 'to be' endings, as with the present base, except that the written language regularly uses **-dir** in the third person of the future to express a simple future statement and not a supposition.

The change of intervocalic **k** to **ğ** must be borne in mind.

(a) *Future simple*:

	'I shall come'	'I shall take'
Singular		
1	geleceğim	alacağım
2	geleceksin	alacaksın
3	gelecek(tir)	alacak(tır)
Plural		
1	geleceğiz	alacağız
2	geleceksiniz	alacaksınız
3	gelecekler(dir)	alacaklar(dır)

(b) *Future past*:

	'I was about to come, would come'	'I was about to take, would take'
Singular		
1	gelecektim	alacaktım
2	gelecektin	alacaktın
3	gelecekti	alacaktı
Plural		
1	gelecektik	alacaktık
2	gelecektiniz	alacaktınız
3	geleceklerdi	alacaklardı

(c) *Future conditional*:

	'if I am about to come'	'if I am about to take'
Singular		
1	geleceksem	alacaksam
2	geleceksen	alacaksan

Plural

3   geleceklerse   alacaklarsa
    (gelecekseler)   (alacaksalar)

Cf. § 18 (c).

(d) *Future past conditional*: 'if I was about to come':

Singular
1 gelecek idiysem or gelecektiysem or gelecektimse
2 gelecek idiysen   gelecektiysen   gelecektinse

Plural
3 gelecek idiyseler gelecektiyseler geleceklerdiyse
  (gelecekler idiyse)

So, with the changes due to vowel harmony, **alacak idiysem** or **alacaktıysam** or **alacaktımsa**, etc. Cf. § 9.

(e) *Future inferential*: 'I am/was said to be about to come':

Singular
1 gelecek imişim or gelecekmişim
2 gelecek imişsin   gelecekmişsin

Plural
3 gelecekler imiş   geleceklermiş
  (gelecek imişler)   (gelecekmişler)

(f) *Future inferential conditional*: 'if, as they say/said, I am/was about to come':

Singular
1 gelecek imişsem or gelecekmişsem
2 gelecek imişsen   gelecekmişsen

Plural
3 gelecek imişseler   gelecekmişseler
  (gelecekler imişse) (geleceklermişse)

(g) *Future negative*. Compare the present negative in § 18.

gel+me+y+ecek > gelmiyecek   he will not come
al+ma+y+acak > almıyacak   he will not take

The same endings are attached as to the positive base: **gelmiyeceğim, gelmiyecektiniz, gelmiyecek imişseler**, etc.

(*h*) *Future interrogative and negative-interrogative.* These are formed on the same lines as those of the present (§ 18 (*h*)), e.g.

gelecek miyim	am I about to come?
gelmiyecek miyim	am I not about to come?
gelecek miydim	was I about to come?
gelmiyecek miymişim	am I said not to be about to come?

23. *Future II.* The ancient future suffix **-esi** has a restricted use in the modern language. As a finite verb it occurs only in the base-form, i.e. in the third-person singular, and is employed solely for cursing:

ipe gel-esi	may he come to the rope
kör ol-ası	may he become blind
ev-in yıkıl-ası	may your house be demolished
ocak-ları batası	may their hearth sink
ense-n kırılası	may your neck be broken
geber-esi	may he die like a dog
kara toprağa gir-esi	may he enter the black earth

The negative **-me** is narrowed by the buffer **y** before this suffix: **gör-mi-y-esi** 'may he not see'; **ol-mı-y-ası** 'may he not become'.

Provincially, with the inferential forms of 'to be' it does not have this optative force but a different development of the original future meaning: **ceplerinden paralarını çal-ası imişim** 'I am alleged to have stolen their money from their pockets'; **karım benden hoşlan-mı-y-ası imiş** 'my wife is alleged not to like me'. The future sense may not be readily apparent in these two typical examples. The connexion of thought is suggested by the American use of 'I am not about to do it' for 'I am not likely to do it, not the sort of person who would do it'.

Unless otherwise indicated, subsequent references to 'the future tense' apply to the future I.

24. *Aorist.* This term, borrowed from Greek grammar, means 'unbounded' and well describes what the Turks call **geniş zaman**

'the broad tense', which denotes continuing activity. The characteristic is **r**, added directly to vowel-stems:

benze-	to resemble	**benzer**	he resembles
anla-	to understand	**anlar**	he understands
koru-	to protect	**korur**	he protects
de-	to say	**der**	he says
ko-	to put	**kor**	he puts

After consonant-stems, a vowel is added before the **r**. Original monosyllabic stems add **e/a**:

bin-	to mount	**biner**	he mounts
dön-	to turn	**döner**	he turns
et-	to do	**eder**	he does
yap-	to make, do	**yapar**	he makes, does
sun-	to present	**sunar**	he presents

To this rule there are thirteen exceptions; monosyllabic stems which insert **i/ü/ı/u** before the **r**. It will be noted that all but one of these stems end in **l** or **r**:

bil-	to know	**bilir**
ǵel-	to come	**ǵelir**
ver-	to give	**verir**
ǵör-	to see	**ǵörür**
öl-	to die	**ölür**
al-	to take	**alır**
kal-	to remain	**kalır**
san-	to think	**sanır**
var-	to reach	**varır**
bul-	to find	**bulur**
dur-	to stand	**durur**
ol-	to become, be, happen	**olur**
vur-	to strike	**vurur**

Polysyllabic stems add **i/ü/ı/u**:

imren-	to covet	**imrenir**
süpür-	to sweep	**süpürür**
aldat-	to deceive	**aldatır**
konuş-	to speak	**konuşur**

So too do extensions of monosyllabic stems, even if they are themselves monosyllables:

de-	to say	de-n-	to be said	denir	it is said
ye-	to eat	ye-n-	to be eaten	yenir	it is eaten
ko-	to put	ko-n-	to be put	konur	it is put

25. Uses of the aorist. The aorist denotes continuing activity, but to equate, for example, **yapar-ım** with 'I do' and **yapıyor-um** with 'I am doing' is a misleading oversimplification. Fundamentally, **yaparım** means 'I am a doer' and according to context it may represent: 'I habitually do'; 'by and large I am the sort of person who does'; 'I am ready, willing, and able to do'; 'I shall do'. **yapıyorum** means: 'I have undertaken, and am now engaged in, the job of doing'; 'I am doing now'; 'I am doing in the future', i.e. 'I have the job in hand'. **yazarım** and **yazıyorum** may both be translated 'I write'. But more specifically: **yazarım** 'I am a writer; in principle I write (though I may not yet have put pen to paper)'. **yazıyorum** 'I am writing now'; 'as a matter of fact I do write'; 'I write, for example, for four hours every morning' —**her sabah dört saat yazıyorum**—where the broad **yazarım** would be incongruous with the precise expression of time. For 'I love you' the Turk says **seni seviyorum**; if he said **seni severim** that would sound far too vague and without immediacy, corresponding rather to 'I like you'.

The aorist is used in requests: **otur-ur musunuz** 'will you sit down?' The future, **oturacak mısınız**, means 'are you going to sit down?' and the present, **oturuyor musunuz**, 'are you in fact sitting down?'

In promises: **yarın gelir-im** 'I shall come tomorrow'. This carries more conviction than the present **yarın geliyorum** 'I am coming tomorrow' or the future **yarın geleceğim** 'I am going to come tomorrow'. See also § 36.

In stage directions: **Esma gir-er, otur-ur. Osman yerinden kalk-ar** 'Esma enters, sits. Osman rises from his place'.

In proverbs: **it ür-ür kervan geç-er** 'the dogs howl, the caravan moves on'.

As a vivid present: **bir akşam kapı hızla çalın-ır** 'one evening there is a violent ringing at the door'.

The aorist of **ol-** 'to become, happen, be' is used to ask permission: **ol-ur mu** 'is it all right?' (lit. 'does it happen?'); answer

olur 'all right' or **olmaz** 'certainly not' ('it does not happen; it's not on'). Note also **olur mu böyle** 'does it happen thus?' i.e. 'can such things be?'

An instructive example of the difference between the aorist and the present is seen in this cynical remark on traffic hazards in Turkey: **başka memleketlerde kazara ölürler; biz kazara yaşıyoruz** 'in other countries they die by accident; we live by accident'. The force of the aorist **ölürler** is 'I cannot say confidently that anyone abroad is in fact dying at this precise instant, but I am aware that people abroad are liable to die—**kazara**—as the result of accident'. The present **yaşıyoruz** means 'we are in fact living at this moment but—**kazara**—it's more by luck than judgement'.

26. Paradigms of the aorist. To the aorist base the 'to be' endings are suffixed as to **-yor**, i.e. without **-dir** in the third person of the present.

(a) *Aorist present*:

	'I come'	'I see'	'I take'	'I find'
Singular				
1	gelirim	görürüm	alırım	bulurum
2	gelirsin	görürsün	alırsın	bulursun
3	gelir	görür	alır	bulur
Plural				
1	geliriz	görürüz	alırız	buluruz
2	gelirsiniz	görürsünüz	alırsınız	bulursunuz
3	gelirler	görürler	alırlar	bulurlar

The common Anatolian **-ik** instead of **-iz** in the first-person plural occasionally finds its way into literary works: **gelirik, görürük, alırık, buluruk**; sometimes with **h** instead of **k**, to represent the pronunciation kh (see I, 9, end).

(b) *Aorist past*. Besides translating 'I used to do', this tense also occurs in the apodosis of conditional sentences in the sense of 'I would do, would have done'.

Singular
1	gelirdim	görürdüm	alırdım	bulurdum
2	gelirdin	görürdün	alırdın	bulurdun
3	gelirdi	görürdü	alırdı	bulurdu

Plural

1 gelirdik　　görürdük　　alırdık　　bulurduk
2 gelirdiniz　　görürdünüz　　alırdınız　　bulurdunuz
3 gelirlerdi　　görürlerdi　　alırlardı　　bulurlardı
　(gelirdiler)　(görürdüler)　(alırdılar)　(bulurdular)

The separated forms **gelir idim, görür idim**, etc., are an Armenianism.

(c) *Aorist present conditional.* As the conditional suffixes are subject only to the twofold harmony, two examples are sufficient: **görürsem** goes like **gelirsem, bulursam** like **alırsam**.

　　　　　　　'if I come'　　'if I take'
　　Singular
　　1 gelirsem　　alırsam
　　2 gelirsen　　alırsan
　　3 gelirse　　　alırsa

　　Plural
　　1 gelirsek　　alırsak
　　2 gelirseniz　　alırsanız
　　3 gelirlerse　　alırlarsa
　　　(gelirseler)　(alırsalar)

(d) *Aorist past conditional*: 'if I came, used to come', etc. The past conditional forms of 'to be', separate or suffixed, follow the aorist base:

　　Singular
　　1　　gelir/görür/alır/bulur　idiysem
　　2　　　　"　"　"　"　　　idiysen
　　Plural
　　3　　　　"　"　"　"　　　idiyseler
　　　(gelirler/görürler/alırlar/bulurlar idiyse)

Singular
1 gelirdiysem　　görürdüysem　　alırdıysam　　bulurduysam
2 gelirdiysen　　görürdüysen　　alırdıysan　　bulurduysan

Plural

3 gelirdiyseler görürdüyseler alırdıysalar bulurduysalar
 (gelirler-   (görürler-   (alırlar-   (bulurlar-
  diyse)   diyse)   dıysa)   dıysa)

(e) *Aorist inferential*: 'I am/was said to come', etc.

Singular

1 gelirmişim görürmüşüm alırmışım bulurmuşum
2 gelirmişsin görürmüşsün alırmışsın bulurmuşsun

Plural

3 gelirlermiş   görürlermiş   alırlarmış   bulurlarmış
 (gelirmişler) (görürmüşler) (alırmışlar) (bulurmuşlar)

(f) *Aorist inferential conditional*: 'if I am/was said to come', etc.

Singular

1   gelir/görür/alır/bulur imişsem
2   ,,   ,,   ,,   ,,   imişsen

Plural

3   ,,   ,,   ,,   ,,   imişseler
      (gelirler/görürler/alırlar/bulurlar imişse)

Singular

1 gelirmişsem görürmüşsem alırmışsam bulurmuşsam
2 gelirmişsen görürmüşsen alırmışsan bulurmuşsan

Plural

3 gelir-   görür-   alır-   bulur-
  mişseler   müşseler   mışsalar   muşsalar
 (gelirler-   (görürler-   (alırlar-   (bulurlar-
  mişse)   mişse)   mışsa)   mışsa)

(g) *Negative*. The aorist is unique in that its negative is not formed by inserting -me before the characteristic r of the positive; instead, the negative has a characteristic of its own, -mez, and this is abraded to -me in the first persons. Further, whereas in other negative bases it is the syllable before the -me that is accented, in the negative of the aorist the -me or -mez itself is accented except in the third-person plural, where the accent is on the **-ler**. As **-mez** is subject to the twofold harmony, two examples are sufficient.

'I do not come' 'I do not take'

Singular
1 gelmem   almam
2 gelmezsin almazsın
3 gelmez   almaz

Plural
1 gelmeyiz   almayız
2 gelmezsiniz almazsınız
3 gelmezler   almazlar

An older form of the first-person plural was in **-mezük** instead of **-meyiz**. It survives in **istemezük** 'we don't want ⟨it⟩', used to typify opposition to progress.

The other tenses and moods are formed like their positive counterparts but on the base **-mez/maz**, e.g.

Aorist past negative: 'I used not to come/take, would not come/take':

Singular
1 gelmezdim almazdım
2 gelmezdin almazdın

Plural
3 gelmezlerdi almazlardı
  (gelmezdiler) (almazdılar)

(*h*) *Interrogative*. The interrogative and negative-interrogative conjugations follow the usual pattern:

gelir miyim do I come?   **alır mıyım** do I take?
gelmez miyim do I not come? **almaz mıyım** do I not take?
gelir miydim used I to come? **alır mıydım** used I to take?
etc.   etc.

The negative-interrogative is used colloquially as a vivid present: **terbiyesiz herif ayağıma basmaz mı** 'the mannerless

fellow goes and steps on my foot', lit. 'does he not step on . . .?'
i.e. 'is he the sort of man who would not step on . . .?', a rhetorical
question expecting the answer 'no'.

27. miş-past. This base is formed by adding **-miş** to the stem:
gelmiş, görmüş, almış, bulmuş. Two distinct functions are
combined in it.

It is first a past participle, describing present state arising out
of past action. If you say **kar yağmış** 'snow has fallen', it may be
that you yourself saw the snow falling, but that is not what you
are concerned with. What you are reporting is not what happened but what is now the case: that there is fallen snow.

But precisely because the perfect participle does not indicate
that the speaker has seen the action take place, it has come to be
used as a finite verb to convey that the information it gives is not
based on having witnessed the action but on hearsay or on
inference from observed facts. In this respect the **miş**-past
resembles the inferential of the verb 'to be', except that (i)
whereas **imiş** refers to past or present time, the **miş**-past is
exclusively a past tense; (ii) **imiş** mostly conveys that the
information given is based on hearsay, less often that it is based
on inference; (iii) whereas **imiş**, whether pronounced as a separate
word or suffixed, is enclitic, the **-miş** of the **miş**-past is accented.

There is no inferential connotation when it is conjugated with
-dir or with the past or conditional forms of the verb 'to be'. The
addition of **-dir** to the base makes a definite past tense: **gelmiştir**
'he came, has come'; this is the normal past tense, third person,
in the language of the media, for which the normal spoken
language employs the past tense in **-di**.

Thus with the Type I endings the following distinction can
be made:

(a) *Inferential past*: 'I gather that I have come', etc.

Singular

1 gelmişim	görmüşüm	almışım	bulmuşum
2 gelmişsin	görmüşsün	almışsın	bulmuşsun
3 gelmiş	görmüş	almış	bulmuş

Plural

1 gelmişiz	görmüşüz	almışız	bulmuşuz
2 gelmişsiniz	görmüşsünüz	almışsınız	bulmuşsunuz
3 gelmişler	görmüşler	almışlar	bulmuşlar

As with **imiş**, the **ş**s of the second person is frequently reduced to **ş** in speech and sometimes in informal writing.

(b) *Definite past*: 'he came, has come; they came, have come', etc. (third person only):

Singular

3 gelmiştir	görmüştür	almıştır	bulmuştur

Plural

3 gelmişlerdir	görmüşlerdir	almışlardır	bulmuşlardır
(gelmiştirler)	(görmüştürler)	(almıştırlar)	(bulmuşturlar)

28. Pluperfect. The addition of the Type II endings makes the pluperfect; **gelmiş-tim** literally means 'I-was having-come'.

'I had come'   'I had seen'   'I had taken'   'I had found'

Singular

1 gelmiştim	görmüştüm	almıştım	bulmuştum
2 gelmiştin	görmüştün	almıştın	bulmuştun

Plural

3 gelmişlerdi	görmüşlerdi	almışlardı	bulmuşlardı
(gelmiştiler)	(görmüştüler)	(almıştılar)	(bulmuştular)

The pluperfect is used more frequently than its English counterpart to show that one past event preceded another, e.g. **şehre saat 10'da varmıştık, bürosuna saat 3'te gittik** 'we [had] arrived in the city at 10 and went to his office at 3'.

29. Other paradigms of the **miş**-past.

(a) *Conditional*: literally 'if-I-am having-come', etc.

'if I have come'   'if I have seen'   'if I have taken'   'if I have found'

Singular

1 gelmişsem	görmüşsem	almışsam	bulmuşsam
2 gelmişsen	görmüşsen	almışsan	bulmuşsan

Plural

3 gelmişlerse görmüşlerse almışlarsa bulmuşlarsa
(gelmişseler) (görmüşseler) (almışsalar) (bulmuşsalar)

(b) *Pluperfect conditional*: 'if I had come', etc. The independent **idiysem**, etc. (§ 9) is put after the base:

Singular

1   gelmiş/görmüş/almış/bulmuş idiysem
2   ,,   ,,   ,,   ,,   idiysen

Plural

3   ,,   ,,   ,,   ,,   idiyseler
(gelmişler/görmüşler/almışlar/bulmuşlar idiyse)

The one-word forms **gelmiştiysem**, etc., are not literary.

(c) *Inferential*: 'I am said to have come', etc. The base in **-miş** is followed by the inferential of 'to be'; see § 10. In this tense, for reasons of euphony, the independent **imişim**, etc., are used very much more often than the suffixed forms. It must be emphasized that the inferential element here comes from the **imiş** and not from the base, which in this tense, as in the pluperfect and conditional, functions simply as a past participle. Thus **gelmiş imiş** or **gelmişmiş** means literally 'he-is-said-to-be having-come', just as **gelmişti** means 'he-was having-come' and **gelmişse** 'if-he-is having-come'.

Singular

1   gelmiş/görmüş/almış/bulmuş imişim
2   ,,   ,,   ,,   ,,   imişsin

Plural

3   ,,   ,,   ,,   ,,   imişler
(gelmişler/görmüşler/almışlar/bulmuşlar imiş)

Singular

1 gelmiş-   görmüş-   almış-   bulmuş-
   mişim   müşüm   mışım   muşum
2 gelmiş-   görmüş-   almış-   bulmuş-
   mişsin   müşsün   mışsın   muşsun

Plural

3	gelmiş- lermiş (gelmiş- mişler)	görmüş- lermiş (görmüş- müşler)	almış- larmış (almış- mışlar)	bulmuş- larmış (bulmuş- muşlar)

(*d*) *Inferential conditional*: 'if I am said to have come', etc. Here again the separate **imişsem**, etc., are commoner than the suffixed forms.

Singular

1   gelmiş/görmüş/almış/bulmuş imişsem
2   ,,   ,,   ,,   ,,   imişsen

Plural

3   ,,   ,,   ,,   ,,   imişseler
(gelmişler/görmüşler/almışlar/bulmuşlar imişse)

The rare suffixed forms are like the **miş**-past conditional (see (*a*) of this section) but with **-mişmiş**, etc., replacing **-miş**.

(*e*) *Negative*. **-me** is added after the stem: **gelmemiş, görmemiş, almamış, bulmamış**. To this negative base the endings are attached as to the positive base, except that as the negative **-me** is subject only to the twofold harmony the suffixes following it appear only in two forms: **gelmemişim, bulmamışım** 'I gather that I have not come/found'; **gelmemişlerdir, bulmamışlardır** 'they have not come/found'; **gelmemiştik, bulmamıştık** 'we had not come/found'; **gelmemiş imişim, bulmamış imişim** 'I am said not to have come/found'.

(*f*) *Interrogative*. This and the interrogative-negative are as usual, with **mi** preceding the personal endings except **-ler**, which it follows: **gelmiş miydik** 'had we come?'; **gelmemiş miydik** 'had we not come?'; **görmüş imişler mi** 'are they said to have seen?'; **almamış imişler mi** 'are they said not to have taken?'.

30. Necessitative. The characteristic is **-meli**, which may be used impersonally: **gelmeli** 'one ought to come'; **almalı** 'one ought to take'. It may also be conjugated with the present, past, and inferential of 'to be', but not with the conditional; see (*f*) below. In origin it is the verbal noun suffix **-me** with **-li** (IV, 5).

(a) *Present*: 'I ought to come', 'I ought to take':

Singular
1. gelmeliyim almalıyım
2. gelmelisin almalısın
3. gelmeli(dir) almalı(dır)

Plural
1. gelmeliyiz almalıyız
2. gelmelisiniz almalısınız
3. gelmeli(dir)ler almalı(dır)lar

(b) *Past*:

'I had to come, 'I had to take,
should have come' should have taken'

Singular
1. gelmeli idim almalı idim
2. gelmeli idin almalı idin

Plural
3. gelmeli idiler almalı idiler

Singular
1. gelmeliydim almalıydım
2. gelmeliydin almalıydın

Plural
3. gelmeliydiler almalıydılar
   (gelmelilerdi) (almalılardı)

(c) *Inferential*: 'they say/said I ought to . . .'. The separate **gelmeli imişim, almalı imişim** is rare.

Singular
1. gelmeliymişim almalıymışım
2. gelmeliymişsin almalıymışsın

Plural
3. gelmeliymişler almalıymışlar
   (gelmelilermiş) (almalılarmış)

(d) *Negative*. The negative base, **gel-me-meli, al-ma-malı**, like the positive base, is used impersonally, 'one ought not to come/take', as well as with the present, past, and inferential endings: gel-me-meli-siniz 'you ought not to come'; gel-me-meli-ydik 'we ought not to have come', gel-me-meli-ymiş-siniz 'they say/said you ought not to come'.

(*e*) *Interrogative*: gelmeli mi 'should one come?' or 'should he come?'; gelmeli miydiniz 'should you have come?'; gelme-meli miydik 'should we not have come?' etc.

(*f*) *Conditional*. In place of the conditional forms of the necessitative, a periphrasis is used, with the conditional forms of the verbs gerekmek or icabetmek 'to be necessary', or, particularly for the present conditional, the adjectives lâzım or gerek 'necessary' and the conditional forms of 'to be', following the -me verbal noun of the required verb with the appropriate personal suffix:

gelme-m gerekirse
,, icabederse
,, lâzım-sa
,, gerek-se
 if I ought to come (lit. 'if my-coming is necessary')

gelme-miz gerektiyse
,, icabettiyse
 if we had to come (lit. 'if our-coming was necessary')

The future necessitative is expressed by a similar periphrasis: gelme-si gerekecek 'he will have to come' ('his-coming will-be-necessary').

31. di-past. This tense corresponds to both the English simple past and perfect with 'have'. Its characteristic is **-di** (**-ti** after unvoiced consonants), to which are added the Type II endings.

	'I came, have come'	'I saw, have seen'	'I did, have done'	'I found, have found'
Singular				
1	geldim	gördüm	yaptım	buldum
2	geldin	gördün	yaptın	buldun
3	geldi	gördü	yaptı	buldu
Plural				
1	geldik	gördük	yaptık	bulduk
2	geldiniz	gördünüz	yaptınız	buldunuz
3	geldiler	gördüler	yaptılar	buldular

32. Uses of the di-past. This is the tense used in speech when relating past events positively known to the speaker. If one has witnessed the arrival of a tourist-ship, one may report the event in the words **bir turist vapuru geldi**. The newspapers will say **bir turist vapuru gelmiştir**, although in the headline they will use the synonymous but shorter **geldi**. Someone who has learned of the event from an eyewitness or from the newspapers will report it as **bir turist vapuru gelmiş**.

33. Other paradigms of the di-past.

(a) *Pluperfect.* From the **di**-past two pluperfect tenses are made, on the pattern of the two past conditionals of the verb 'to be'; see § 9. The first is commoner than the second, but not so common as the pluperfect in **-miş-ti** (§ 28).

'I had come'   'I had seen'

Singular
1 geldiydim or geldimdi   gördüydüm or gördümdü
2 geldiydin   geldindi   gördüydün   gördündü
3   geldiydi   gördüydü

Plural
1 geldiydik   geldikti   gördüydük   gördüktü
2 geldiydiniz geldinizdi   gördüydünüz gördünüzdü
3 geldiydiler geldilerdi   gördüydüler gördülerdi

'I had done'   'I had found'

Singular
1 yaptıydım or yaptımdı   bulduydum or buldumdu
2 yaptıydın   yaptındı   bulduydun   buldundu
3   yaptıydı   bulduydu

Plural
1 yaptıydık   yaptıktı   bulduyduk   bulduktu
2 yaptıydınız yaptınızdı   bulduydunuz buldunuzdu
3 yaptıydılar yaptılardı   bulduydular buldulardı

(b) *Conditional.* As in the **di**-pluperfect, there are two possible forms, the first being more frequent. Only the conjugation of stems with rounded vowels has been shown; for the conjugation of **geldiysem/geldimse** and **yaptıysam/yaptımsa**, cf. § 9.

VIII, 33 THE VERB

'if I saw, have seen'   'if I found, have found'

Singular
1 gördüysem or gördümse   bulduysam or buldumsa
2 gördüysen   gördünse   bulduysan   buldunsa
3   gördüyse   bulduysa

Plural
1 gördüysek   gördükse   bulduysak   bulduksa
2 gördüyseniz   gördünüzse   bulduysanız   buldunuzsa
3 gördüyseler   gördülerse   bulduysalar   buldularsa

(c) *Pluperfect conditional.* The separate **idiysem**, etc., and not the suffixed forms are used.

'if I had come'

Singular
1 geldi idiysem or geldim idiyse
2 geldi idiysen   geldin idiyse

Plural
3 geldi idiyseler   geldiler idiyse

(d) *Negative.* The same endings are attached to the negative stem: **gelmedim** 'I have not come', **almadınız** 'you did not take', **yapmadıydık** or **yapmadıktı** 'we had not done', **görmediyse** 'if he has not seen', **bulmadı idiyseler** 'if they had not found', etc.

(e) *Interrogative.* As the **di**-past is conjugated with the Type II endings and not the present of 'to be', the interrogative particle follows the whole word: **geldim mi?** 'did I come?'; **yaptınız mı?** 'did you do?'; **gördük mü?** 'did we see?'; **buldular mı?** 'did they find?', etc.

There are alternative forms for the interrogative of the pluperfect:

'had I come?'

Singular
1 geldi miydim or geldim miydi
2 geldi miydin   geldin miydi
3   geldi miydi

Plural

1 geldi miydik   geldik miydi
2 geldi miydiniz   geldiniz miydi
3 geldi miydiler   geldiler miydi

Negative-interrogative: **gelmedim mi?** 'did I not come?'; **yapmadınız mı?** 'did you not do?'; **görmedik mi?** 'did we not see?'; **bulmadılar mı?** 'did they not find?'; **gelmedi miydim?** or **gelmedim miydi?** 'had I not come?', etc.

34. Conditional. As we have seen, the various bases so far examined, except **-meli**, have conditional moods, formed by adding the conditional of 'to be':

(i) To the base:

geliyorsam   if I am coming
geleceksem   if I am going to come
gelirsem   if I come
gelmişsem   if I have come
geldiysem   if I came, have come

(ii) To the past or inferential forms of 'to be' added to the base:

geliyorduysam   if I was coming
geliyormuşsam   if I am/was said to be coming
gelecektiysem   if I was going to come
etc.

In addition, every verb has its own conditional base, the characteristic being **-se**, which expresses (*a*) remote condition: **gelse** 'if he were to come'; (*b*) wish: **gelse!** 'if only he would come!' As with the conditional of 'to be', the Type II endings are used.

(*a*) *Conditional present*: 'if I were to come', 'if I were to take':

Singular
1 gelsem   alsam
2 gelsen   alsan
3 gelse   alsa

Plural
1 gelsek   alsak
2 gelseniz   alsanız
3 gelseler   alsalar

(b) *Conditional past.* With the past endings of 'to be', the conditional base expresses (a) unfulfilled conditions: **gelseydi** 'had he come'; (b) hopeless wishes relating to past time: **gelseydi!** 'if only he had come!' This must be carefully distinguished from the **di**-past conditional:

past conditional: **gel+di+ise > geldiyse** 'if he came'
conditional past: **gel+se+idi > gelseydi** 'if (only) he had come'

Singular

1. gelseydim   alsaydım
2. gelseydin   alsaydın
3. gelseydi   alsaydı

Plural

1. gelseydik   alsaydık
2. gelseydiniz   alsaydınız
3. gelseydiler   alsaydılar
   (gelselerdi)   (alsalardı)

(c) *Conditional inferential.* With the inferential endings of 'to be', remote conditions and wishes can be quoted: **gelseymiş** 'they say that if he were to come' or 'they are saying "if only he would come!"'

Singular

1. gelseymişim   alsaymışım
2. gelseymişsin   alsaymışsın
3. gelseymiş   alsaymış

Plural

1. gelseymişiz   alsaymışız
2. gelseymişsiniz   alsaymışsınız
3. gelseymişler   alsaymışlar
   (gelselermiş)   (alsalarmış)

The separate forms **gelse imişim**, etc., are also found.

(d) *Negative*: **gelmesem** 'if I were not to come'; **gelmeseydim** 'if only I had not come!'; **gelmeseymişim** 'they are saying of me "if he were not to come" *or* "if only he would not/had not come!"'

(e) *Interrogative.* Besides asking for confirmation of what you think you have heard (cf. § 13 end)—**ġelse mi?** '"if he were to come!" do you say?'—the interrogative of the conditional may express indecision: **ġitsem mi?** 'should I go?' See XX, 10.

The interrogative of the conditional inferential has alternative forms:

Singular

1 ġelse miymişim or ġelsem miymiş
2 ġelse miymişsin   ġelsen miymiş

Plural

3 ġelse miymişler   ġelseler miymiş

Negative-interrogative: **ġelmese miymişim** or **ġelmesem miymiş**, etc., 'are they saying of me "if only he wouldn't come!"?'

35. Subjunctive. This mood, which some grammarians prefer to call the subjunctive-optative, uses the Type III endings. It is not strictly correct to speak of the third-person singular as the base of this mood, as the elements of the first-person endings are inseparable.

Present

Singular

1 ġeleyim   alayım
2 ġelesin   alasın
3 ġele   ala

Plural

1 ġelelim   alalım
2 ġelesiniz alasınız
3 ġeleler   alalar

The accent is on the last syllable of the first persons and of the third-person plural. The other endings are accented on the -e/a.

The endings of the second persons are those of Type I. The final syllable of the first singular, though it looks like the 'I am' ending, is not; this is evident from the fact that it is accented, which the 'to be' endings never are. Historically, this ending seems to be a hybrid; in the sixteenth century the ending of the first singular of the subjunctive was **-eyin**, and of the imperative **-eyem**. The 'I am' ending has, however, doubtless influenced its development. The first plural is historically part not of the subjunctive but of the imperative, the old first-plural ending of the subjunctive being -evüz.

The Anatolian forms of the first singular and plural sometimes find their way into print; they are in -em and -ek respectively: **gidem** 'let me go, I'll go'; **gidek** or **gideh** 'let's go'.

With vowel-stems, a **y** is inserted before the Type III endings. This narrows the preceding vowel, but the phonetic change is not invariably reflected in writing. Thus the subjunctive of **söyle-** 'to tell' and **başla-** 'to begin' is:

Singular

1	söyliyeyim	başlıyayım
2	söyliyesin	başlıyasın
3	söyliye	başlıya

Plural

1	söyliyelim	başlıyalım
2	söyliyesiniz	başlıyasınız
3	söyliyeler	başlıyalar

Note that the **y** of the first-person ending **-eyim/ayım** does not narrow the preceding vowel in writing. In the colloquial, however, the first singular of vowel-stems often loses the stem-vowel and the buffer **y**; thus **söyliyeyim, başlıyayım** are heard as **söyleyim, başlayım**, while **yapayım** 'let me do' is heard as **yapiim** or **yapim**.

36. Uses of the subjunctive.[1] The subjunctive expresses concepts envisaged by the subject or the speaker; it makes no statement about facts, except that the first singular is used colloquially with future meaning: **yarın geleyim** 'let me come tomorrow' > 'I may come tomorrow' > 'I'll come tomorrow'. The first plural means 'let us come', etc. The second persons are used in formal speech to relay requests and commands: **babam dedi ki, yarın bize gelesiniz** 'my father said that you-should-come to us tomorrow'. The third singular is used colloquially to ask cautious questions—

---

[1] It is a pity that the subjunctive is dying in England, though it seems more healthy in America. It will be a sad day when we forget the distinction between 'I insist that the claimant is adequately compensated' and 'I insist that the claimant be adequately compensated'.

evde mi ola 'might he be at home?'—and in a few set expressions: from **rasgel-** 'to chance', **rasgele** 'may it fall out well, good luck!'; **kolay gele** 'may it come easy', a form of greeting used when entering a place where someone is working; **geçmiş ola** 'may it be past', said when hearing of someone's illness. In these last two expressions, however, Istanbul idiom favours the imperative: **kolay gelsin, geçmiş olsun**. The third plural is virtually obsolete. In subordinate clauses (see Chapter XIX) the third persons are almost invariably replaced nowadays by the third persons of the imperative.

37. Other paradigms of the subjunctive.

(*a*) *Subjunctive past*. Formed by suffixing to the third-person singular of the subjunctive the past endings of 'to be':

Singular

1 geleydim   başlıyaydım
2 geleydin   başlıyaydın
3 geleydi   başlıyaydı

Plural

1 geleydik   başlıyaydık
2 geleydiniz   başlıyaydınız
3 geleydiler   başlıyaydılar
  (gelelerdi)   (başlıyalardı)

It expresses unfulfillable past wishes: 'would that I had come/begun', usually reinforced by **keşki: keşki bacağı kırılaydı** 'would that his leg had broken!' It is also used (i) as an alternative to the conditional past and (ii) as the past tense of the imperative.

EXAMPLES: (i) **bileydim buraya kadar gelmezdim** 'had I known, I would not have come this far'. See, however, Chapter XXI, last paragraph.

(ii) **söyliyeydiniz** 'you should have said'. There is an idiomatic use of the third-person singular imperative of **var-** 'to come' with that of another verb, usually in the negative, e.g. **varsın demesin,** lit. 'let him come let him not say', meaning 'he might as well not say, it doesn't matter whether he says or not'. One way of expressing the past tense of this is **vara demiyeydi** 'he might as well not have said'. Note the suspended affixation of the

-**ydi,** which belongs both to **vara** and **demiye.** For an alternative way see § 40 and XXIV, 31 (the latter positive not negative).

(b) *Subjunctive inferential*

Singular

1 geleymişim   başlıyaymışım
2 geleymişsin   başlıyaymışsın
3 geleymiş   başlıyaymış

Plural

1 geleymişiz   başlıyaymışız
2 geleymişsiniz   başlıyaymışsınız
3 geleymişler   başlıyaymışlar
  (gelelermiş)   (başlıyalarmış)

These forms, which are not of frequent occurrence, are used to quote the present and past subjunctive—**geleymiş** 'people are/were saying "would that he had/might come!"'—and as an alternative to the conditional inferential.

(c) *Negative.* The buffer **y** narrows the **-me/ma** to **-mi/mı**: **gelmiye, başlamıya, gelmiyeydim, başlamıyaymış,** etc. YİK recommends that this change should not be shown in writing.

(d) *Interrogative.* The particle **mi** follows the endings of the present. In practice, only the first persons are likely to be encountered: **geleyim mi, gelelim mi** 'should I/we come?'; **gelmiyeyim mi, gelmiyelim mi** 'should I/we not come?' The interrogative of the past is not in use.

The interrogative of the inferential is rare, as one might expect from its meaning: 'do/did they say "would that he might come!"?'

Singular

1 gele miymişim   başlıya mıymışım
2 gele miymişsin   başlıya mıymışsın

Plural

3 gele miymişler   başlıya mıymışlar
  (geleler miymiş)   (başlıyalar mıymış)

38. Synopsis of the verb. The accompanying table shows the first-person singular of all tenses and moods of **gelmek** 'to come' except the imperative (for which see the next section), the present

THE VERB

	Base	Simple I	Past idi+II	Conditional ise+II	Past conditional idi+ise+II	Inferential imiş+I	Inferential conditional imiş+ise+II
Present	geliyor 16	geliyorum 18 (a)	geliyordum 18 (b)	geliyorsam 18 (c)	geliyorduysam 18 (d)	geliyormuşum 18 (e)	geliyormuşsam 18 (f)
Future	gelecek 20	geleceğim 22 (a)	gelecektim 22 (b)	geleceksem 22 (c)	gelecektiysem 22 (d)	gelecekmişim 22 (e)	gelecekmişsem 22 (f)
Aorist	gelir 24	gelirim 26 (a)	gelirdim 26 (b)	gelirsem 26 (c)	gelirdiysem 26 (d)	gelirmişim 26 (e)	gelirmişsem 26 (f)
miş-past	gelmiş 27	gelmişim 27 (a)	gelmiştim 28	gelmişsem 29 (a)	gelmiş idiysem 29 (b)	gelmiş imişim 29 (c)	gelmiş imişsem 29 (d)
Necessitative	gelmeli 30	gelmeliyim 30 (a)	gelmeliydim 30 (b)	..	..	gelmeliymişim 30 (c)	..
di-past	geldi 31	geldim 31	geldiydim 33 (a)	geldiysem 33 (b)	geldi idiysem 33 (c)	..	..
Conditional	gelse 34	gelsem 34 (a)	gelseydim 34 (b)	..	..	gelseymişim 34 (c)	..
Subjunctive	gele 35	geleyim 35	geleydim 37 (a)	..	..	geleymişim 37 (b)	..

Personal endings

	Type I	Type II	Type III
Singular			
1	-im	-m	-eyim
2	-sin	-n	-esin
3	(-dir)	..	-e
Plural			
1	-iz	-k	-elim
2	-siniz	-niz	-esiniz
3	-(dir)ler	-ler	-eler

Participles

Present	gelen
Future	gelecek
Aorist	gelir
miş-past	gelmiş
di-past	geldik

Verbal nouns

gelmek
gelmeklik
gelme
geliş
gelmezlik
gelmemezlik

II (§ 19), and the future II (§ 23). The participles and verbal nouns are also shown. See also the periphrastic tenses and moods in IX, 11.

The Roman numerals in the table refer to the types of personal endings set out in § 15, which for convenience are repeated beneath the table. The Arabic numerals refer to the relevant sections of the present chapter.

The suffixed forms of the verb 'to be' have been shown except where literary usage has a strong preference for the separate forms.

39. Imperative.

Singular

| 2 | gel | gör | al | bul |
| 3 | gelsin | görsün | alsın | bulsun |

Plural

2	gelin	görün	alın	bulun
	geliniz	görünüz	alınız	bulunuz
3	gelsinler	görsünler	alsınlar	bulsunlar

It will be seen that the imperative of the second singular is identical with the stem; cf. the English imperative, which is identical with the infinitive without 'to'. Of the second person plural forms, the longer is the more polite. Care should be taken not to confuse the third-person suffix of this mood with the second-person singular of the Type I endings; if **-sin** is added to a stem, it makes the third-singular imperative: **gel-sin** 'let him come'; if added to a base, it makes the second-singular present: **gel-ecek-sin** 'you-are about-to-come'; **gel-miş-sin** 'you-are having-come'. An ancient suffix of the second person imperative was **-gil**: **bilgil** 'know!'

The interjections **haydi** 'come on!', **aman** 'mercy!' and **destur** 'mind out of the way!' (*dastūr* (P) 'permission') are sometimes given one or other of the endings of the second-plural imperative when more than one person is being addressed: **haydiniz, amanın, desturun.**

In the second-person plural of the imperative, the buffer **y** does not narrow a preceding vowel: this rule applies to the second-person plural of the positive imperative of vowel-stems and of the negative imperative of all stems. Thus the positive imperative of **söyle-** 'to tell' and **başla-** 'to begin' is as follows:

Singular
2 söyle   başla
3 söylesin   başlasın

Plural
2 söyleyin   başlayın
  söyleyiniz   başlayınız
3 söylesinler başlasınlar

The negative imperative of **gel-** and **bul-**:

Singular
2 gelme   bulma
3 gelmesin   bulmasın

Plural
2 gelmeyin   bulmayın
  gelmeyiniz   bulmayınız
3 gelmesinler bulmasınlar

The interrogative: **gelsin mi** 'is he to come?'; **söylemesin mi** 'is he not to tell?' An idiomatic use of the third-person negative interrogative is to state a fact, with an implication of surprise: **satıcı iki gazete için benden on yerine sekiz kuruş almasın mı?** 'would you believe it, for two newspapers the shopkeeper took eight piastres from me instead of ten' (lit. 'is the seller not to take . . .?'). Cf. § 26 (*h*).

Such colloquial imperatives as **durundu** 'hey, stop!' and **bakındı** 'hey, look here!' (accented on the first syllable and used as singular or plural) are compounded of the second-singular imperative—**dur, bak**—and the adverb **imdi** 'now'. A less likely theory derives them from the second plural—**durun, bakın**—and the interjection **di** seen in **haydi** 'come on!'

40. -sindi. This ending is the third-person imperative **-sin** with the third-person past of the verb 'to be': **onun gizli fikirlerini halk ne bilsindi?** 'how should the common people know his secret thoughts?' **bu yazıya Bakan kızmasın da, kim kızsındı?** 'who should have been angry at this article, if not the

Minister?' (for this form of conditional sentence see XX, 2). Some Turkish grammarians reject this explanation and regard the -di as identical with the last syllable of **haydi**. This is because they will not admit the possibility of a past imperative, on the grounds that the function of the imperative is to give an order and one cannot give an order in the past. This somewhat mechanistic objection can be disposed of if we regard the -di as putting not the imperatives **bilsin, kızsın**, but the whole sentences into the past. For another example see XXIV, 31.

41. -sin için. An uncommon method of expressing purpose is to use the postposition **için** after the third-person imperative, as in **bu gömleği ona giysin için verdim** 'I gave him this shirt for him to wear', where **giysin diye** (XI, 2) would be more normal. For an example where it has clearly been used to avoid the repetition of **diye**, see XXIV. 25.

42. -dir suffixed to finite verbs. As we have seen in § 4, **-dir** indicates supposition or, less commonly, emphasis when used as a third-person copula. It may also be suffixed to verbs (except the di-past, conditional, subjunctive, and imperative) in any person, including the first and second persons of the verb 'to be'. In such situations it generally does not so much emphasize the verb as weaken it, the implication being that the speaker is stating as a fact something of which he has no positive knowledge but only a strong feeling or impression: **çocuk gibi-y-im-dir ya . . . elbette çocuk gibi-y-im** 'I-really-think-I-am-like a child, you know . . . indeed I am like a child'. Here the speaker begins by stating his feeling and then, his conviction becoming firmer, he states it as a fact.

şair-im   I am a poet
şair-im-dir   I am surely a poet, I think I must be a poet
şiir yazıyorum   I am writing poetry
şiir yazıyorum-dur   surely I am writing poetry! *or* why, I must be writing poetry!
biliyorsunuz   you know
biliyorsunuzdur   you surely know, I presume you know
uyu-muş-um   I gather that I have slept
uyu-muş-um-dur   I must have slept

oku-muş-sunuz you have read, I gather
oku-muş-sunuz-dur you are bound to have read
İngiliz polisinin methini hepimiz duy-muş-uz-dur we have all certainly heard the praises of the English police

It must be remembered, however, that the addition of **-dir** to the third person of the **miş**-past makes a definite past tense (§ 27), while its addition to the third person of the future I makes a definite future (§ 22).

Note the differences between the three possible ways of translating 'I have written him a letter':

> **ona bir mektup yazdım**
> **ona bir mektup yazmışım**
> **ona bir mektup yazmışımdır**

The first is a statement of fact; I remember writing the letter. The second is an inference; I do not remember writing the letter, but I have found the carbon copy in my file. The third, which might be translated 'I must have written . . .', suggests that I do not remember writing, nor have I any evidence that I wrote, but after all it is over three months since I received his letter and I presume I must have done something about it.

The nuances in the three possible ways of saying 'my friend is waiting for me' should also be noted:

> **arkadaşım beni bekliyor**
> **arkadaşım beni bekliyormuş**
> **arkadaşım beni bekliyordur**

The first states a fact; I can see him there at the corner. The second is based on hearsay; someone has seen him waiting and told me so. The third is a supposition—'I'm sure he is waiting'—based on the knowledge that my friend is always punctual, that he said he would wait from five o'clock, and that it is now five past five.

The context shows when **-dir** has the emphasizing function: **çocuğa anlatın, mutlaka ilâcı içmelidir** 'explain to the child, he really must drink the medicine'; **babanın sözünden çıkmıyacaksınızdır, değil mi?** 'it-is-a-fact-that-you-will-not-depart from your father's word, is it not?' i.e. 'you will positively not disobey your father, will you?'

43. -dir with a following verb. bir kıyamet-tir koptu, literally 'it is a resurrection broke out', may be translated 'all hell broke loose'. In such sentences, in which the noun is always preceded by **bir**, the **-dir** is the main verb, with the following verb subordinate. This is a vivid form of expression drawing attention to a sudden startling event or a remarkable state of affairs, the latter when the second verb is **git-**, denoting continuity (XI, 35 (g)).

sol kulağıma bir tokat-tır indi 'such a slap came down on my left ear!'

bütün gün evde bir konuşma bir patırdı bir gürültü-dür gider 'there is a perpetual talking and scurrying and noise in the house all day long'.

Comparable is the colloquial use exemplified in: **Adalarda yaz mevsimi bir hayat-tır hiç sorma** 'on the Islands, the summer season is such a marvellous life!' ('is a life don't ask!').

44. Summary of the forms of 'to be'. The following summary of the forms of the verb **olmak** 'to become, be, happen, mature', and of the verb 'to be', shows how the former supplements the deficiencies of the latter.

-im	I am	**oluyorum**	I am becoming, I tend to be
		olmaktayım	I am becoming
		olacağım	I shall be, shall become
		olurum	I become, shall be
imişim	I am/was said to be	**olmuşum**	I infer that I have become
		olmalıyım	I ought to become, ought to be
idim	I was	**oldum**	I became, have become
isem	if I am	**olsam**	if I were, if I should become
		olayım	let me be, let me become

		ol, olun,	
		olunuz	be! become!
		olsun	let him be
		olası	may he be
		olabilirim	I can be, become (§ 55 (a))
		olamam	I cannot be, become (§ 55 (b))
iken	while being	olurken	while becoming (XI, 34)

The stem **ol-** also supplies the participles and verbal nouns of 'to be'.

Particularly to be noted is the difference between **idim** and **oldum**. Here is a pair of examples in the third person: **bir zelzele idi** 'it was an earthquake'; **bir zelzele oldu** 'an earthquake occurred'.

45. var, yok. These words are adjectives meaning respectively 'existent' and 'non-existent'. They take the place of English 'there is/are' and 'there is/are not' and of the verb 'to have'.

köşede bir kahve var 'there is a café on the corner'. **bu köyde postahane var mı?** 'is there a post-office in this village?' Answer: **var** 'there is' or **yok** 'there is not'. **bıçak var mı sizde?** 'have you a knife on you?' Answer: **var** 'I have' or **yok** 'I have not'. **bıçağınız yok mu?** 'do you not have a knife?' ('your-knife non-existent?'). **çocuk yetim değil, babası var** 'the child is not an orphan, he has a father'.

The subject of English 'have' is put in the genitive in Turkish: **çocuğun babası var** 'the child has a father'. This is not an ordinary izafet group; it literally means not 'the child's father exists' but 'his father exists—the child's'. The distinction may seem slight but has practical consequences; see XVI, 6. **imparatorun elbisesi yok** 'the emperor has no clothes'; **benim şüphem yok** 'I have no doubt'.

The present tense of 'to be' and the forms based on **i-** may be used in conjunction with **var** and **yok**: **Burada yalnız mı-y-ız? dedi. Hayır, ben de var-ım, dedim** '"Are we alone here?" said he. "No, I-am-present too", I said'.

ev-in bahçe-si var			the house has a garden
,,	,,	vardı	the house had a garden
,,	,,	vardır	*formal*: the house has . . .; *informal*: the house surely has . . .
,,	,,	varsa	if the house has . . .
,,	,,	varmış	the house is said to have . . .
,,	,,	varken	while the house has . . .

For the negative of all these, **var** is replaced by **yok**: **yoktu, yoktur, yoksa, yokmuş, yokken**.

For other verbal forms, i.e. for those missing from the left-hand column in § 44, **ol-** and its negative **olma-** take the place of **var** and **yok**:

evin bahçesi olacak			the house will have a garden
,,	,,	olmıyacak	the house will not have a garden
,,	,,	olsaydı	if the house had a garden
,,	,,	olmasaydı	if the house had no garden
,,	,,	olsun	let the house have a garden
,,	,,	olmasın	let the house not have a garden
,,	,,	olmalı	the house ought to have a garden

The attributive use of **var** is confined to such expressions as **var kuvvet-i-yle dayandı** 'he resisted with all his strength' ('with his existent strength') and of **yok** to **yok yere** 'vainly' ('to non-existent place').

Care is necessary to distinguish **vardı** < **var**+**idi** from **vardı** the third-singular **di**-past of **varmak** 'to arrive'. As the suffixes of 'to be' are enclitic, no confusion is possible in speech; 'there was' is **várdı**, 'he arrived' is **vardí**. In writing, the context should obviate ambiguity: **köyde bir misafir vardı** 'there was a guest in the village'; **köye bir misafir vardı** 'a guest arrived in the village'.

46. Extended stems. The suffixes treated in the next seven sections are used to make reciprocal, causative, repetitive, reflexive, and passive verb-stems, to which are added the tense- and mood-endings set out above.

47. The reciprocal or co-operative verb. The addition of **-iş-** to a consonant- or of **-ş-** to a vowel-stem shows that the action is

done by more than one subject, one with another or one to another:

anla-	to understand	**anlaş-**	to understand one another
benze-	to resemble	**benzeş-**	to resemble one another
döv-	to beat	**dövüş-**	to fight one another
koş-	to run	**koşuş-**	to make a concerted rush *or* to run in all directions
sev-	to love	**seviş-**	to love one another
uç-	to fly	**uçuş-**	to fly about together

The precise meaning of such a verb cannot always be deduced logically; thus **tutuş-**, from **tut-** 'to hold', means not only 'to hold each other' and 'to hold mutually'—**el tutuştular** 'they held hands'—but also 'to catch fire'. **yatış-** is not 'to lie down together' but 'to subside'; **yapış-** is not 'to do together' but 'to adhere'; **geliş-** is not 'to come together' but 'to develop'; **kalkış-** is not 'to rise together' but 'to attempt something beyond one's powers'. A grammar cannot be a substitute for a dictionary.

48. The causative verb. This is formed by adding one or other of the suffixes listed below to the stem, original or reciprocal.

(*a*) **-dir-**. This suffix is etymologically and functionally distinct from the suffix meaning 'is', but is identical with it in its phonetic metamorphoses (see § 3). It is the commonest causative suffix, but is not used with polysyllabic stems ending in a vowel or **l** or **r**.

don-	to freeze (intr.)	**dondur-**	to freeze (tr.)
dön-	to turn (intr.)	**döndür-**	to turn (tr.)
inan-	to believe	**inandır-**	to persuade
öl-	to die	**öldür-**	to kill
sev-iş-	to love one another	**seviştir-**	to make to love one another
ye-	to eat	**yedir-**	to feed

The final sentence of the previous section is applicable to this section too; e.g. **al-dır-** means 'to cause to take' but also 'to pay attention'; **koş-tur-** is 'to cause to run' but also 'to run about in a panic'. See further § 51.

(b) -ir-. This is used with some twenty monosyllables, of which the commonest are:

aş-	to pass, surpass	aşır-	to cause to pass over
bat-	to sink (intr.)	batır-	to sink (tr.)
bit-	to finish (intr.)	bitir-	to finish (tr.)
doğ-	to be born	doğur-	to give birth to
doy-	to be satiated	doyur-	to satiate
duy-	to feel, hear	duyur-	to divulge
düş-	to fall	düşür-	to make fall, drop
geç-	to pass (intr.)	geçir-	to pass (tr.)
göç-	to migrate	göçür-	to cause to migrate
iç-	to drink	içir-	to make drink
kaç-	to escape	kaçır-	to let escape, lose
piş-	to cook (intr.)	pişir-	to cook (tr.)
şiş-	to swell (intr.)	şişir-	to inflate
taş-	to overflow	taşır-	to make overflow
yat-	to lie down	yatır-	to lay down, deposit

(c) -t- is used with polysyllabic stems ending in a vowel or l or r:

anla-	to understand	anlat-	to explain
bekle-	to wait	beklet-	to keep waiting
düzel-	to be put in order	düzelt-	to arrange
küçül-	to become small	küçült-	to belittle
otur-	to sit	oturt-	to seat
söyle-	to speak	söylet-	to make speak

(d) -it- is used after a few monosyllabic stems, mostly ending in k, e.g.:

ak-	to flow	akıt-	to shed
kok-	to smell (intr.)	kokut-	to make smell
kork-	to fear	korkut-	to frighten
sap-	to deviate	sapıt-	to send astray
sark-	to lean down	sarkıt-	to suspend
ürk-	to start with fear	ürküt-	to startle

(e) -er- occurs only in these words:

çık-	to go out, go up	çıkar-	to remove, raise
çök-	to collapse	çöker-	to cause to collapse, make kneel

git-	to go	gider-	to remove
kop-	to break off, break out (intr.)	kopar-	to break off (tr.), cause to break out
on-	to prosper	onar-	to repair (also ondur- 'to improve' (tr.))

(*f*) Irregular are:

em-	to suck	emzir-	to suckle (also emdir- 'to cause (e.g. a pump) to suck')
gel-	to come	getir-	to bring
gör-	to see, perform (a duty, task, etc.)	göster-	to show (also gördür- 'to make perform')
kalk-	to rise	kaldır-	to raise, remove

49. Doubly causative verbs. The causative **-t-** may be suffixed to **-dir-, -ir-,** and **-er-**; the causative **-dir-** may be suffixed to **-t-** and **-it-**, sometimes with no change of meaning. Thus from **de-** 'to say' the causative 'to make say' is **dedir-** or **dedirt-**; from **kon-** 'to settle', **kondur-** or **kondurt-**. More often, however, both suffixes have their full value:

öl-	to die
öldür-	to kill
öldürt-	to have someone killed
piş-	to cook (intr.)
pişir-	to cook (tr.)
pişirt-	to get something cooked

Causatives of the third and fourth degree are theoretically possible but are rarely if ever found outside the pages of grammar-books, e.g. **öl-dür-t-tür-t-** 'to get someone to get someone to get someone to make someone die', i.e. to kill through the agency of three intermediaries.

50. Syntax of the causative. When a transitive verb is made causative, the object of the basic verb remains in the accusative, while the object of the causative element of the verb is put in the dative: **mektub-u imzala-dım** 'I signed the letter'; **mektub-u**

müdür-e imzala-t-tım 'I got the director to sign the letter' ('to-the-director I-was-the-cause-of-signing'). salon-un duvarlar-ı-n-ı boya-y-acaktım 'I was going to paint the walls of the drawing-room'; salonun duvarlarını bir amele-y-e boya-t-acaktım 'I was going to get a workman to paint', etc.

When the object of the causative element is not expressed, English idiom usually demands a passive participle: mektubu imzalattım 'I got the letter signed'; salonun duvarlarını boyatacaktım 'I was going to have the walls of the drawing-room painted'.

When an intransitive verb is made causative, the subject of the basic verb becomes the object: rakib-i öl-dü 'his rival died'; rakib-i-n-i öl-dür-dü 'he killed his rival'. This causative verb, having an object, can now be treated like any other transitive verb: rakibini kiralık bir kaatil-e öl-dür-t-tü 'he got a hired murderer to kill his rival' ('to a hired murderer he-was-the-cause-of-making-die his-rival'). çocuk doğ-du 'the child was born'; anne, çocuğ-u doğ-ur-du 'the mother bore the child'; ebe, anne-y-e çocuğ-u doğ-ur-t-tu 'the midwife helped the mother to bear the child'; ebe, anne-y-i doğ-ur-t-tu 'the midwife brought the mother to birth'.

When an originally transitive verb is made doubly causative, the second intermediary may be expressed with the help of vasıta (A) 'means', or tavassut (A) or the neologism aracılık 'mediation': mektub-u müdür-e kâtib-in vasıtasiyle/tavassutiyle/aracılığıyle imzala-t-tır-dım 'through the agency of the secretary, I got the director to sign the letter'. The second intermediary need not be mentioned: mektubu müdüre imzalattırdım 'I got someone to get the director to sign the letter'.

Verbs construed with the dative retain the dative when they are made causative, the object of the causative element being put in the accusative: söz-üm-ün doğru-luğ-u-n-a inan-ır mısınız? 'do you believe in the truth of my statement?'; sözümün doğruluğuna siz-i nasıl inan-dır-ayım? 'how am I to make you believe in the truth of my statement?'. çocuk, okul-a başla-dı 'the child started school'; çocuğ-u okul-a başla-t-tık 'we made the child start school'. herkes kendisi-n-e acın-ıyor 'everyone is sorry for him'; herkes-i kendisine acın-dır-ıyor 'he is making everyone sorry for him'.

The causative means not only 'to make someone do something'

but also, voluntarily or involuntarily, 'to let someone do something': **orman-lar-ımız-ı keçi-ler-e ye-dir-iyor-uz** 'we are letting the goats eat our forests'. **para-m-ı tramvay-da çal-dır-mış-ım** 'I have had my money stolen on the tram'. **tren-i kaç-ır-dık** 'we missed the train' ('we let the train escape').

The negative of the causative is commonly used for 'not to permit': **bizi söyle-t-me-diler** 'they did not let us speak'.

51. The repetitive verb.

(a) Stems extended by the suffixes **-(i)ş-** plus **-dir-** are not necessarily reciprocal and causative. In the following verbs the ending **-(i)ştir-** conveys repeated and intensive action:[1]

ara-	to seek	araştır-	to research, investigate
at-	to throw	atıştır-	to gobble up
çek-	to pull	çekiştir-	to slander
serp-	to sprinkle	serpiştir-	to scatter about
sor-	to ask	soruştur-	to make inquiries
sür-	to smear	sürüştür-	to put on make-up
tak-	to attach	takıştır-	to dress up
ver-	to give	veriştir-	to be abusive

For additional emphasis the simple verb may be used before the repetitive verb: **tak takıştır, sür sürüştür** (imperative second sing.) 'doll yourself up in your best clothes and put on your full war-paint'.

(b) **-(e)kle-** can no longer be regarded as a live suffix (with the reservation that the language reformers may at any time decide to resurrect it) but is found in a small number of verbs, e.g.:

dürt-	to prod	dürtükle-	to keep prodding
it-	to push	itekle-	to manhandle
sür-	to drive	sürükle-	to drag
uyu-	to sleep	uyukla-	to keep dozing off

(c) **-ele-** is even rarer:

şaş-	to be bewildered	şaşala-	to be bewildered
gev-	(*obsolete*) to chew	gevele-	to chew over, beat about the bush

---

[1] This is not an exhaustive list but includes the commonest of such verbs. See Tahsin Banguoğlu, 'Türkçede Tekerrür Fiilleri', in *TDAYB* 1956, 111–23.

52. The reflexive verb.

The suffix is -(i)n-:

bul-	to find	bulun-	to find oneself, be
döv-	to beat	dövün-	to beat one's breast
ġiy-	to put on, wear	ġiyin-	to dress oneself
sal-	to throw	salın-	to oscillate
söyle-	to speak	söylen-	to grumble to oneself
yıka-	to wash	yıkan-	to wash oneself

Like the 'middle voice' of ancient Greek, this suffix denotes action done not only to oneself but also for oneself:

et-	to do	edin-	to acquire
ġeç-	to pass	ġeçin-	to make a living, get along
kaç-	to run away	kaçın-	to abstain
kalk-	to rise	kalkın-	to progress, recover
yap-	to make	yapın-	to make for oneself *or* to have (e.g. a suit) made

Both senses may occur in the same verb: **aran-**, from **ara-** 'to seek', means 'to search one's mind' and 'to seek something for oneself': **aranıyorsun** 'you're asking for it!' (i.e. a good hiding). **taşın-**, from **taşı-** 'to carry', means both 'to turn things over in one's mind' and 'to move oneself, move house'.

As with verbs in **-iştir-**, the meaning of the reflexive is not always guessable: e.g. **ġör-ün-** is not 'to see oneself' but 'to seem, to appear'; **sev-in-** is not 'to love oneself' but 'to rejoice'.

53. The passive verb.

This is formed by adding **-il-** after all consonants except **l**: **sev-il-** 'to be loved'; **ġör-ül-** 'to be seen'; **yap-ıl-** 'to be made'; **tut-ul-** 'to be held'.

Stems in **l** or a vowel form their passive identically with the reflexive:

al-	to take, buy	al-ın-	to be taken, bought
oku-	to read	oku-n-	to be read
kapa-	to shut	kapa-n-	to be shut

Thus, for example, **söylen-** is both the reflexive and the passive of **söyle-** and means either 'to grumble' or 'to be spoken';

yıkan- is either 'to wash oneself' or 'to be washed'. In cases where ambiguity might arise, the passive can be indicated by adding -il- to the -(i)n, or the reflexive can be shown by using the reflexive pronoun (V, 4): **çocuğu yıkadı** 'she washed the child'; **çocuk yıka-n-dı** 'the child washed himself *or* was washed'; **çocuk yıka-n-ıl-dı** 'the child was washed'; **çocuk kendi kendini yıka-dı** 'the child washed himself'.

Just as some verbs take a doubly causative suffix for no obvious reason, so some take a doubly passive suffix; e.g. the passive of **de-** 'to say' is **de-n-il-** as well as **de-n-**.

The passive of **anla-** 'to understand' is irregular: **anla-ş-ıl-** 'to be understood'.

54. Uses of the passive. It is not used as much as its English equivalent for the sake of elegant variation; e.g. instead of 'he was rebuked by his father', a Turk is more likely to say 'his father rebuked him'. Another difference from English idiom is that when **başla-** 'to begin' is construed with a passive verb it is put in the passive itself: **bu iş-i yap-ma-y-a**[1] **başlıyorlar** 'they are beginning to do this job'; **bu iş yap-ıl-ma-y-a başla-n-ıyor** 'this job is beginning to be done' ('is-being-begun to-be-done').

But the most remarkable feature of the Turkish passive is its impersonal use: **niçin yalan söyle-n-ir?** 'why are lies told?', i.e. 'why do people tell lies?' In this example the passive verb appears to have[2] a subject, but impersonal passives are also regularly formed from intransitive verbs and then have no conceivable grammatical subject; indeed, the example **bu . . . başlanıyor** above may be explained under this head. **bahşis at-ın diş-i-n-e bak-ıl-maz** 'one does not look at the teeth of a gift horse' ('looking-is-not-done to the tooth . . .'). **o zamanlar Karaköy'- den Harbiye'ye taksi ile iki lira-y-a ģid-il-ir-di** 'in those days, one used to go from K. to H. by taxi for two liras'. **ģidilirdi** is the aorist past passive, lit. 'going-used-to-be-done'; cf. Virgil's *sic itur ad astra* 'thus does one go to the stars' **yıldızlara böyle ģidilir.** The sentence **bu ilâç-la kimse iyi olmaz** 'with this medicine no one becomes well' may be expressed impersonally thus: **bu ilâçla iyi ol-un-maz**, using the aorist present negative

---

[1] This is the dative of the verbal noun; see X, 7.
[2] 'appears to have', because the verb is in fact impersonal; the literal meaning of **yalan söylenir** is not 'lies are told' but 'lie-telling is-done'.

passive of **ol-**; 'becoming-is-not-done'. **yerli-ler-le çabuk arkadaş ol-un-ur** 'one quickly becomes friends with the natives'.

This passive of **ol-** is used to form the passive of verbs compounded of **et-** 'to do' and a verbal noun (see § 57) and is commoner in this use than the passive of **et-**: **onu tenkit ediyorlar** 'they are criticizing him'; **tenkit olunuyor** or **tenkit ediliyor** 'he is being criticized'.

55. The potential verb. This might, on grounds of structure, have been treated together with compound verbs such as **ol-a-ġel-** and **ġel-i-ver-** (see XI, 35) but is singled out here because of its great frequency and its anomalous negative.

(*a*) The positive is formed by adding the appropriate part of **bil-** 'to know' to the required stem, original or extended, plus -e/a[1] (with the usual buffer **y** after vowel-stems): **ġel-ir-im** 'I come'; **ġel-e-bil-ir-im** 'I can come'. **ġel-di-yse** 'if he came'; **ġel-e-bil-di-yse** 'if he was able to come'. **anlı-y-acak-mış-ım** 'I gather that I shall understand'; **anlı-y-a-bil-ecek-miş-im** 'I gather that I shall be able to understand'. The verb in the next example is the aorist present interrogative of the potential passive of **ol-**, used impersonally: **hakikaten bedbaht ol-un-a-bil-ir mi?** 'is it possible to be truly unhappy?'

(*b*) The negative, i.e. the impotential, is formed by adding to the stem the suffix **-eme/ama**, the first vowel of which is accented and the second may be narrowed to **i/ı** by a following **y**. This was originally the negative of a now obsolete verb **umak** 'to be powerful, able'. **ġel-me-mek** 'not to come', **ġel-eme-mek** 'to be unable to come'; **anla-ma-mak** 'not to understand', **anlı-y-ama-mak** 'to be unable to understand'. The aorist of the impotential is conjugated like the aorist negative, its base being not **-eme+r** but **-emez**. It will be seen that the impotential of any verb-form can be made by inserting **e/a** before the negative suffix: **ġelmedi** 'he did not come'; **ġelemedi** 'he could not come'. **anlamıyor** 'he does not understand'; **anlıyamıyor** 'he cannot understand'. **bilmezler** 'they do not know'; **bilemezler** 'they cannot know'.

(*c*) The positive potential endings may be attached to a negative

[1] For the history of this extension see C. S. Mundy, 'The -e/ü Gerund in Old Ottoman', *BSOAS* xvi/2 (1954), 298–319, and ibid. xvii/1 (1955), 156–9.

or even to an impotential stem: **gel-mi-y-ebil-ir-im** 'I am able not to come', i.e. 'I may not come' or 'I don't have to come if I don't want to'. **gel-emi-y-ebil-ir-im** 'I am able to be unable to come', i.e. 'I may be unable to come'.

56. The order of extensions. The order in which the extensions to the stem are placed is as follows:

    1. reflexive
    2. reciprocal
    3. causative
    4. passive

Examples of verbs containing both reflexive and reciprocal suffixes are hard to find; one such, a product of the language reform, is **dayanış-** 'to practise mutual aid', made up thus:

simple:	**daya-**	to prop up
+reflexive:	**daya-n-**	to prop oneself up
+reciprocal:	**daya-n-ış-**	to engage with other people in propping oneself up

We may continue the extensions to show the full possibilities of the verb:

+causative:	**daya-n-ış-tır-**	to make to practise mutual aid
+passive:	**daya-n-ış-tır-ıl-**	to be made to practise mutual aid

The only departures from this order are apparent rather than real; i.e. the reciprocal suffix can follow the passive or causative suffix but only in the case of pseudo-passive or pseudo-causative verbs, that is, verbs which look like passives or causatives but whose original simple stems have gone out of use, e.g.: **dağıl-** 'to disperse'; **dağıl-ış-** 'to disperse all in different directions'. **seğirt-** 'to hasten'; **seğird-iş-** 'to hasten together'.
Examples of the normal order:

simple	**acı-**	to feel pain
reflexive	**acı-n-**	to feel pain in oneself, to grieve
causative	**acı-n-dır-**	to make grieve
passive	**acı-n-dır-ıl-**	to be made to grieve

simple	tanı-	to know
reciprocal	tanı-ş-	to know one another
causative	tanı-ş-tır-	to make to know one another, introduce
passive	tanı-ş-tır-ıl-	to be introduced to one another
simple	in-	to descend
causative	in-dir-	to bring down
passive	in-dir-il-	to be brought down

To summarize, the 'extended stem' is the simple stem plus any or all of these four extensions, in this order: reflexive, reciprocal, causative, passive. Less commonly, 'reflexive' and 'causative' may be replaced by 'repetitive'. To the extended stem, as to the simple stem, may be added any one of the following options:

   (*a*) negative   **-me-**
   (*b*) potential   **-ebil-**
   (*c*) impotential   **-eme-**
   (*d*) negative+potential   **-miyebil-**
   (*e*) impotential+potential   **-emiyebil-**

Then comes the tense and/or mood characteristic and finally the personal suffix, which, if Type I, may be preceded by the interrogative particle (the Type II endings are followed by the interrogative particle). Using the stem **daya-**, all four extensions, option (*e*), the future characteristic, the interrogative particle, the inferential suffix and the Type I ending of the first-person plural, we arrive at: **dayanıştırılamıyabilecek miymişiz?** 'is it said that we may not be able to be made to practise mutual aid?' This example, though a little contrived, would not strike a Turk as unnatural. The three following examples are all taken from recent writings (the third from an article on anti-aircraft defences): **acındırılmadık** 'we were not made to grieve' (stem **acı-**+reflexive+causative+passive+negative+first-person plural of **di**-past). **tanıştırılamadıysanız** 'if you were not able to be introduced' (stem **tanı-**+reciprocal+causative+passive+impotential+second-person plural of **di**-past conditional). **indirilemiyebilecekler** 'it may be that they will not be able to be brought down' (stem **in-**+causative+passive+impotential+potential+third-person plural of future).

57. Auxiliary verbs.

(*a*) **etmek**. There is a handful of verbal phrases consisting of a Turkish noun and the verb **etmek** 'to do', on the pattern of the English 'to do honour to', e.g. **yardım etmek** 'to help'; **alay etmek** 'to mock'. These served as the model for a vast number of phrases in which the first element was an Arabic verbal noun:

kabul	acceptance	**kabul etmek**	to accept
mukayese	comparison	**mukayese etmek**	to compare
ispat	proof	**ispat etmek**	to prove
teşkil	formation	**teşkil etmek**	to form

The same device is used nowadays to make verbs from foreign words, especially French past participles:

désinfecté	**dezenfekte etmek**	to disinfect
isolé	**izole etmek**	to isolate, insulate
adapté	**adapte etmek**	to adapt
organisé	**organize etmek**	to organize
knock-out	**nakavt etmek**	to knock out (in boxing)

Nouns whose final syllable is subject to any of the changes described in I, 16, 17 (*c*), 19 are usually written as one word with **etmek**, especially if they are monosyllables:

af	forgiveness	**affetmek**	to forgive
fikir	thought	**fikretmek**	to ponder
tehyiç	excitement	**tehyicetmek**	to excite

Exceptional is **haketmek** 'to deserve', with a single **k** although it is from **hak**, acc. **hakkı** (I, 16, last paragraph); **hakketmek** 'to engrave' is regular, < *ḥakk* (A).

(*b*) **eylemek**, formerly an elegant alternative to **etmek**, is now little used except (i) to avoid the constant repetition of **etmek**; (ii) in **Allah rahmet eylesin** and **Mevlâ rahmet eyleye** 'God have mercy ⟨on him/her⟩'; (iii) in the stereotyped expression **ne etseniz ne eyleseniz** 'whatever you do, in spite of all your efforts' (for the syntax see XX, 7).

(*c*) **kılmak** was anciently another alternative to **etmek**, but as an auxiliary verb it now occurs regularly only in **namaz kılmak** 'to perform the rites of Muslim prayer' and **takla kılmak** 'to do a somersault', though in this latter phrase it is often replaced by **atmak** 'to throw'. It is still fairly common in the sense of 'to make someone something', as in **savaş dışi kılmak** 'to render *hors de combat*' ('to make war-outside') and **mecbur kılmak** 'to oblige' ('to make compelled'), with the dative of the -me verbal noun: **beni bunu yapmaya mecbur kıldılar** (or **ettiler**) 'they have obliged me to do this'.

(*d*) **buyurmak** properly means 'to order' and was used in courtly speech as a substitute for other verbs, including **etmek** and **eylemek**, the underlying theory being that exalted persons do not perform any task themselves but simply command; thus 'he forgave me' would be **beni affetti** or **beni affeyledi** if the forgiver were an ordinary man, but **beni af buyurdular** (note the courtly plural) if he were the Sultan.

Nowadays, **ne buyurdunuz** 'what did you say?' is mostly used ironically, but the imperative **buyurun** is regularly used to mean 'deign', in making courteous requests. It may be construed with an accusative when it stands for 'take', or with a dative when it stands for 'enter': **buyurun kahve-niz-i** 'please take your coffee'; **buyurun salon-a** 'please enter the drawing-room'.

(*e*) **yapmak** 'to make, do', unlike **etmek**, does not usually make transitive verbal phrases; one exception, officially sanctioned, though disliked by purists, is **park yapmak** 'to park (a car)'. It replaces **etmek**, however, when a normally transitive verbal phrase is used without an object or when the noun element is defined; i.e. when the noun is really the object of 'to do' and is not just part of a compound verb: **bu iki eser-i mukayese ediyor** 'he is comparing these two works' but **bu iki eser-in mukayese-si-n-i yapıyor** 'he is making a comparison ('doing the comparison') of these two works'; **masraflarımı hesab-ediyorum** 'I am calculating my expenses' but **hesap yapıyorum** 'I am calculating'.

(*f*) The passive and causative forms of **etmek**, but not of **eylemek**, are in full use: **köprüyü tamir ediyorlar** 'they are repairing the bridge'; **köprü tamir ed-il-iyor** 'the bridge is being repaired'; **köprüyü tamir et-tir-iyorlar** 'they are having the bridge repaired'; **köprü tamir et-tir-il-iyor** 'the repair of the bridge is being carried out' ('the bridge is-being-got-repaired').

For the use of **ol-un-mak** instead of **ed-il-mek** see § 54, end.

Some phrases with **etmek** make their passive in **olmak** instead of or as well as in **edilmek** or **olunmak**. This is easily explicable where the first element is not a noun but an adjective, e.g. in **kaybetmek** 'to lose', the **kayb** being a corruption of the Arabic *ghā'ib* 'missing', so the passive **kaybolmak** is literally 'to be missing'. Similarly, from **mahkûm** 'condemned' comes **mahkûm etmek** 'to condemn', passive **mahkûm olmak** or **mahkûm edilmek**. But **olmak** is also used to make the passive of a number of **etmek** compounds whose first element is a noun, e.g.: **tıraş etmek** 'to shave'; **tıraş olmak** 'to be shaved, shave oneself'. **mahvetmek** 'to destroy'; **mahvolmak** 'to be destroyed'. **defetmek** 'to repel'; **defol!** 'buzz off!'

In such phrases, **olmak** seems to mean 'to undergo, be subjected to', as in the following examples too:

öksürük olmak	to catch a cough
tifo olmak	to catch typhoid
ameliyat olmak	to undergo an operation
sünnet olmak	to undergo circumcision
imtihan olmak	to sit an examination
cehennem ol!	get to hell out of it!

(*g*) Any Arabic verbal noun may itself govern an object in the absence of **et-**: **açıklama-y-ı tekrar etmek istemiyorum** 'I do not wish to repeat the explanation'; **açıklama-y-ı tekrardan kaçın-ıl-mış-tır** 'repetition of the explanation has been avoided' ('avoidance-has-been-done from-repeating . . .). **Bursa'-yı ziyaret ettim** 'I visited Bursa'; **Bursa'yı ziyaretim** 'my visiting Bursa, my visit to B.' **saat-i tahmin ettim** 'I guessed the time'; **saat-i tahmin-e çalıştım** 'I tried to guess the time'. **onu teşyi edelim** 'let us see him off ; **onu teşyie hazırlanalım** 'let us prepare to see him off'.

The object may be in the dative; e.g. 'to attend school' is **okul-a devam etmek** ('to do continuance to school'): **her çocuk, ilk okula devam-a mecbur-dur** 'every child is obliged to attend primary school' ('. . . is compelled to-continuance to . . .').

The Italian borrowing seen in the transitive verbal phrase **protesto etmek** 'to protest against' may similarly take a direct object even in the absence of the auxiliary verb: **işçiler, lokavt-ı protesto ettiler** 'the workers protested-against the lockout';

lokavt-ı protesto miting-i için izin alındı 'permission has been obtained for the meeting of protest-against the lockout'. Note that whereas the ı suffixed to **lokavt** is the mark of the accusative, the **i** after **miting** is the suffix of the third person.

The book-title **Bilinmiyen Yönleriyle Atatürk'ü Analiz**,[1] 'Analysis of Atatürk in his unknown aspects', exhibits a similar use with a direct object of the French noun *analyse*.

It is remarkable that the two adjectives meaning 'indebted', **medyun** (A) and its modern equivalent **borçlu,** may take a direct object, as in **hayatımı kendisine medyunum/borçluyum** 'I owe my life to him'.

[1] By Ahmet Eren Doğan (Izmir, 1976).

IX

PARTICIPLES

1. Present. The present participle is formed by adding **-en** to the stem, original or extended: **gelen** 'coming'; **olan** 'being, becoming'; **indirilen** 'being brought down'; **dağılışan** 'dispersing'. The usual **y** is inserted after vowel-stems and narrows the preceding vowel, although this narrowing is not always shown in writing: **anla-** 'to understand', **anlıyan (anla-y-an)** 'understanding'; **bekle** 'to wait', **bekliyen (bekle-y-en)** 'waiting'. The vowel of the negative **-me** is similarly treated: **anlamıyan (anla-ma-y-an)** 'not understanding'; **olmıyan (ol-ma-y-an)** 'not being'; **beklemiyen (bekle-me-y-en)** 'not waiting'; **gelmiyen (gel-me-y-en)** 'not coming'.

These words function as adjectives or nouns: **bekliyen misafirler** 'the guests who-are-waiting'; **bekliyenler** 'those who are waiting'; **oynamıyan çocuklar** 'children who-do-not-play'; **oynamıyanlar** 'those who do not play'.

Participles exercise the same governance as the corresponding finite verb; e.g. **beklemek** is transitive, so its participle governs an accusative: **bizi bekliyenler** 'those who are awaiting us'. But **başlamak** 'to begin' takes a dative, so: **bu iş-e başlıyanlar** 'those who are beginning this job'.

The present participle may overlap the very recent past: **yeni doğ-an çocuk** 'new-born child'; **yeni açıl-an fabrika** 'newly opened factory'; **geç-en hafta** 'last week'. Such idiomatic uses must not be confused with situations where the present participle has to be translated by an English past tense because the main verb of the sentence is in the past: **gül-en adam çıkarıldı** 'the man who laughed (lit. 'the laughing man') was thrown out'.

The participle of the present II is **-mekte ol-an** or **-mekte bulun-an**.

2. Future I. The participle is the tense-base (VIII, 20): **olacak** 'who/which will be'; **olmıyacak** 'who/which will not be'. In formal speech and writing, where the rules of word-order are strictly applied, confusion is unlikely between the future participle used as an attributive adjective and the third person of the future simple tense; an attributive adjective precedes its noun, whereas in formal language the subject precedes its predicate: **iyi haber** 'good news'; **haber iyi** 'the news is good'. **gelecek haber** 'news which will come'; **haber gelecek** 'the news will come'.

In informal speech, however, and in the increasingly popular **devrik cümle** school of writing (XV, 3), the subject may follow its predicate: **gelecek, haber** 'it will come, the news'. To avoid ambiguity the future participle is often used together with the present participle of **ol-** or **bulun-** 'to be'; **gelecek olan haber** 'the news which-is about-to-come'; **bu işi yapacak bulunan amele** 'the workman who-is going-to-do this job'.

The future participle also functions as a noun: **gelecek** 'who/ which will come, the future'; **gelecekler** 'those who will come'; **olacak olur** 'what-is-to-be will-be'. There is a tendency, however, to attach case-endings to **olan** rather than directly to the future participle, especially in the singular: **gelecek olanların** (rather than **geleceklerin**) **çoğu akrabamız** 'of-those-who-are about-to-come most are our relatives'; **bunu okuyacak olana** 'to-the-one-who-is going-to-read this' (**okuyacağa** is theoretically possible but most unlikely).

The habitual use of **-ecek** to translate 'who/which is about to ...' must not hide from us its basic sense of 'pertaining to future ... ing, as in **yiyecek bir şey alalım** 'let us buy something to eat' ('a thing pertaining-to-future-eating'; **okuyacak bir kitap istiyorum** 'I want a book to read'; **softalar saldıracak adam arıyor** 'the bigots are seeking someone to attack'.[1] The future passive participle of intransitive verbs is used impersonally in the same way: **otur-ul-mı-y-acak bir ev** 'a house not to be lived in' ('pertaining-to-future-living's-not-being-done'). This use of the active and the impersonal passive future participle is possible only when the participle is attributive, never when it is predicative,[2] whereas the future passive participle of transitive verbs can be used either way:

Attributive: (*a*) active: **okuyacak bir kitap** 'a book to read';

---

[1] Cf. the English use of the infinitive as an active or quasi-passive future participle; active in 'you are to drink this', quasi-passive in 'something to drink would be a good idea'.

[2] Another way of expressing this is to say that the quasi-passive use is confined to the future participle and is not found with the future finite verb.

(*b*) transitive passive: **oku-n-acak bir kitap** 'a book which will be read'; (*c*) intransitive passive (impersonal): **şaşılacak bir şey** 'a thing at which surprise will be shown'.

Predicative: transitive passive: **bu kitap oku-n-acak** 'this book will be read, is to be read'.

The active participle cannot be used predicatively in the quasi-passive sense: **bu kitap okuyacak** could only mean 'this book will read' and not 'this book is one to read'. Nor can the impersonal passive participle be used predicatively: one can say **bu evde oturulacak** 'living-will-be-done in this house, one will live in this house', where **oturulacak** is a finite verb, but **bu ev** cannot be made the subject of **oturulacak**.

In this sense of 'pertaining-to-future-doing', the future participle may be followed by the postposition **kadar**, e.g. **üretim, ihtiyaçlara yetişmiyecek kadar az-dır** 'production is inadequate to meet requirements', lit. 'small the amount pertaining to future not-sufficing for needs'. Here the future participle might be thought to be an ordinary 'which will not suffice', but such an explanation cannot be applied to **dönemiyecek kadar yürümüştük** 'we had walked too far to turn back'; lit. 'amount pertaining to future inability to turn', not 'the amount which will not be able to turn'. **çocuk, okula gidecek kadar büyüktür** 'the child is big enough to go to school' ('the amount pertaining to future going . . .'). **çocuk, okula gidemiyecek kadar küçüktür** 'the child is too small to go to school' ('the amount pertaining to future inability to go . . .').

Several future participles have become common nouns, e.g.:

giy-	to put on	**giyecek**	garment
çek-	to pull	**çekecek**	shoe-horn
oy-	to drill a hole	**oyacak**	drill
yak-	to burn	**yakacak**	fuel
ye-	to eat	**yiyecek**	food

The future participle of **ol-** often has the ironic sense of 'who is supposed to be, so-called': **damad-ım olacak o kumarbaz** 'that gambler who is supposed to be my son-in-law'; **dişçi olacak o kasap** 'that butcher who calls himself a dentist'. The present participle of 'to be' is not used after the future participle in this use, so **damadım olacak olan kumarbaz** can only mean 'the gambler who is going to be my son-in-law'.

This sense is sometimes found in the future participle of other verbs: **ġüvenlik sağlıyacak insanlar** 'the people who-are-supposed-to-ensure security'. See also XX, 9.

3. Future II. The restricted future base **-esi** may be used as an adjective: **kör olası herif** 'the damned scoundrel' ('the may-he-become-blind scoundrel'); **can-ı çıkası karı** 'the accursed woman' ('the may-her-soul-come-out woman'); **ad-ı batası kâfir** 'the abhorred infidel'. In the colloquial it may occur in non-pejorative contexts, e.g. **şaşılası bir şey** is a sub-standard variant for **şaşılacak bir şey** 'an astonishing thing'.

As a noun: **kör olası bunu yaptı mı?** 'has the damned-one done this?'; **canı çıkası-nın piç-i ġeldi** 'the accursed-one's bastard has come'. The suffix **-ce** may be added without altering the meaning: **kör olasıca, canı çıkasıcanın**.

4. Aorist. The aorist participles, positive and negative, are identical with the respective bases:

Finite verb		*Participle*	
su akar	water flows	**akar su**	flowing water
su akmaz	water does not flow	**akmaz su**	stagnant water
şair ölmez	the poet does not die	**ölmez şair**	the immortal poet

Some aorist participles have become or are becoming common nouns:

yaz-ar	writer
oku-r yaz-ar	literate ('reader-writer')
düşün-ür	thinker
çık-ar	profit, advantage ('what comes out')
çıkmaz	impasse (short for **çıkmaz yol** 'road which does not come out')
ġel-ir	revenue ('what comes')
kes-er	adze ('cutter')

All these can be fully declined: **yazar-ın** 'the writer's'; **okur yazarlar** 'literates'; **çıkmaz-dan** 'from the impasse', etc. But aorist participles which usage has not fixed as nouns are not usually declined. For example, 'he is unreasonable' is **söz anlamaz**, lit. 'he does not understand words'. If we wish to use this expression to translate 'those who are unreasonable' we cannot simply add the plural suffix, since **söz anlamazlar** would naturally be taken as the third-person plural of the finite verb, 'they are unreasonable'. Instead, we either add a noun, e.g. **adam** 'man' or **kişi** 'person', or use the present participle of **anla-ma-**, or of **ol-** following **anlamaz**:

    söz anlamaz adamlar
    söz anlamaz kişiler
    söz anlamıyanlar
    söz anlamaz olanlar

5. miş-past. The participle, though identical in form with the base, has none of the inferential sense of the **miş**-past tense: **plan hazırla-n-mış** 'I gather that the plan has been prepared'; **hazırlanmış plan** 'the plan which has been prepared'. Like the future participle in **-ecek,** it is often used in conjunction with the present participle of 'to be': **gelmiş olan/bulunan arkadaşlar** 'the friends who have come'. So, for example, 'of those who have sat down' may be translated:

    oturmuşların
    oturmuş olanların
    oturmuş bulunanların

The following phrase contains the **miş**-past, present II, and future I participles of **oku-** 'to read, study': **Avrupa'da okumuş, okumakta ve okuyacak olan gençler** 'young people who have studied, are in process of studying, and will study in Europe'.

6. di-past. The participle is identical in form not with the base but with the first-person plural, **-dik**. It appears mostly in frozen forms, of which these are the commonest:

bildik	acquaintance
tanıdık	,,
düşünülmedik	unthought-of
işitilmedik	unheard-of
dedik	said

görülmedik	extraordinary ('unseen')
olmadık	unprecedented ('not-having-happened')
umulmadık	unlooked-for
beklenmedik	unexpected
okumadık	unread
okunmadık	,,
yaratık	creature
yapılmadık	not done

It will be observed that most of these are negative and that **bildik, tanıdık, okumadık, dedik,** and **yaratık**, though active in form, are passive in meaning. This is because the past participle in **-dik** really means not 'having done' but 'characterized by past doing'. So the active **okumadık** 'characterized-by-not-reading' comes to be synonymous with the passive **okunmadık** 'characterized-by-not-being-read'.

The action may be present as well as past, or begun in the past and still continuing; the same ambiguity occurs in the English passive participle: compare 'things done nowadays' with 'things done fifty years ago'.

Apart from the frozen forms listed above, any verb may appear in the negative with **-dik** in sentences of the type of 'I have left no stone unturned': **okumadık gazete kalmadı** 'no newspaper is left unread' ('newspaper characterized-by-not-reading has-not-remained'); **aramadık bir yer komadım** 'I have left no place unsearched'; **sürmedik kara bırakmadılar** 'they have left no evil imputation unmade' ('they-have-not-left black characterized-by-not-smearing'). Care must be taken to distinguish this participle from the first-person plural of the **di-**past, as in **gezmedik memleket bırakmadık** 'we have left no country unvisited'.

7. *The personal participles.*[1] Of paramount syntactic importance are the forms made by adding a personal suffix to the participles in **-dik** and **-ecek**. **bir tanıdık** 'an acquaintance' is literally 'a characterized-by-knowing'; the addition of a personal suffix shows on whose part the knowing was or is, and the resulting word can be used as an adjective or noun: thus **bir tanıdığım** is 'an acquaintance of mine' and **tanıdıklarım** 'my acquaintances';

---

[1] This term is not entirely satisfactory but has been used for want of a better. The term 'relative participle', used by the author in his earlier *Teach Yourself Turkish*, is even less adequate.

tanıdığım bir adam 'a man I know' ('a man characterized-by-my-knowing'); **tanıdığım adamlar** 'the men I know'.

Similarly, as **okuyacak** means 'pertaining to future reading', **okuyacağım** means 'pertaining to my future reading', 'which I shall read', while the plural **okuyacaklarım** means 'things which I shall read'.

When the personal suffix is that of the third person, it may stand in izafet with a qualifier or possessor.

The personal participle is used:

(*a*) As an adjective. **okuyacağım kitap** 'the book which I shall read'; **kardeş-im-in beklediği misafir** 'the guest whom my brother is/was awaiting' ('pertaining to my brother's awaiting'); **Türkiye'ye geldiğiniz uçak** 'the aircraft in which you come/came to Turkey'; **doğduğu şehir** 'the city in which she was born'; **oturacağımız ev** 'the house in which we are going to live'; **konuşacakları meseleler** 'the problems which they are going to discuss'; **sevgi, saygı duyduğumuz bir meslektaş** 'a colleague for whom we feel affection, respect' ('pertaining-to-our-feeling affection . . .'); **kız-ın oynamadığı bebek** 'the doll with which the girl is/was not playing'; **mücevherlerin çalındığı oda** 'the room from which the jewels were stolen'; **bunu aldığım fiata satarım** 'I shall sell this for the price at which I bought ⟨it⟩'.

It will be seen that from the point of view of the English translation these examples fall into two classes: those in which the noun qualified by the personal participle is translated as the direct object of the verb; and those in which the translation requires the insertion of a preposition before the relative pronoun ('in which, for whom, with which, from which, at which').

The device employed to distinguish between past and present time, when the context is insufficient guide, is exemplified in: **dün yaptığım ve bugün yapmakta olduğum işler** 'the jobs which I did yesterday and am doing today' ('pertaining-to-my-doing yesterday and pertaining-to-my-being in-the-act-of-doing today').

(*b*) As a noun meaning 'that which I do', etc. **Avrupa'da gördüklerim** 'the things I saw in Europe'; **size bir diyeceğim yok** 'I have nothing to say to you' ('to-you a thing-of-my-future-saying is-not'); **Beatles'lerin her yaptığı İngiltere'de moda oluyor** 'everything the Beatles do is becoming the fashion in England': **her** 'every' qualifies the noun **yaptığı** 'thing of their doing'; **halk bizim inanmadığımıza inanabilir** 'the people may believe that which we do not believe': **inan-** is construed with a dative; **inanmadığımız** means 'that pertaining to our not believing'. **olduğundan büyük görünür** 'it seems bigger than it is' ('. . . than-that-of-its-being').

In the following phrase the adjectival and nominal uses are exemplified by **baktığı** and **sevdiğini** respectively: **her baktığı kadında sevdiğini gören şair** 'the poet who sees his beloved in every woman he looks at' ('seeing in every woman pertaining-to-his-looking the-one-pertaining-to-his-loving').

(c) As a noun meaning '(the fact of) my doing', etc.: **bugün gideceğim şüpheli** 'it is doubtful whether I shall go today' ('my-future-going is doubtful'); **İstanbul'a geldiğimin dördüncü günüydü** 'it was the fourth day after my arrival in Istanbul' ('. . . of-my-coming to-Istanbul'); **bir parti kurduğunuz, isminin de Türkiye Adalet Partisi olduğu doğru mu?** 'is it true that you have founded a party and that its name is the Justice Party of Turkey?' ('your-founding . . . and its-name's being . . . is true?'); **hazır bulunduğuna göre** 'in view of its being ready' (lit. 'according to-the-fact-of-its-being ready'). Cf. **olduğuna göre** in XXIV, 18.

The third-singular personal participle of the ancient **er-** 'to be' survives in the phrase **ne idüğü belirsiz** 'of doubtful antecedents' ('his-being what, unclear, it being unclear what he is'; for the syntax see XVIII, 1), sometimes modernized to **ne olduğu belirsiz**.

8. **-eceği gel-.** The future personal participle is used with **gel-** 'to come', as in **niçin diyeceğim geliyor** 'I begin to feel like saying "why?"', lit. 'my-future-saying "why?" is-coming'; cf. **uykum geliyor** 'my-sleep is-coming', i.e. 'I feel sleepy'. **İstanbul'u göreceğim geldi** 'I feel like seeing Istanbul' ('my-future-seeing Istanbul has-come').

9. **-esi gel-.** The participle of the future II is colloquially used in the same way: **niçin diyesim geliyor; İstanbul'u göresim geldi**. The third-person suffix, however, is omitted; i.e. **-esi** may stand for **-esi-si: insan-ın niçin diyesi geliyor** 'one feels like saying "why?"' ('man's saying why comes').

10. -eceği tut-. The future personal participle with **tut-** 'to catch hold': **kızını evlendireceği tuttu** 'he was suddenly seized with the idea of getting his daughter married', lit. 'his-future-causing-to-marry caught-hold'; **gitmiyeceğim tuttu** 'I suddenly feel like not going', lit. 'my-future-not-going has-caught-hold').

11. Periphrastic tenses and moods. The verb **ol-** is used with the bases of the present I, the future I, and the **miş**-past, to give a greater suppleness to the tense-system. **geliyor olmalılar** 'they must be coming'; **bakacak olursanız** 'if you will look' ('if-you-are about-to-look'); **gelmiş olacağız** 'we shall have come'; **toplantınıza iştirak edememiş olmaktan müteessirim** 'I regret not having been able to take part in your meeting' ('I-am-regretful from-being having-been-unable-to-do participation'); **unutmuş olmayın** 'I hope you have not forgotten' ('do-not-be having-forgotten').

With the aorist participles, **ol-** gives an inchoative sense: **bunu yapar oldu** 'he started to do this' ('became doing'); **bunu yapmaz oldu** 'he stopped doing this' ('became not-doing'); **bunu yapamaz oldu** 'he became incapable of doing this'; **bu teklifi kabul etmez olur muyum hiç**, lit. 'do I ever become not-accepting this suggestion?', i.e. 'am I ever likely not to accept...?'

The use of the future, aorist, and **miş**-past bases in such periphrases is readily understandable, as these bases are the participles of their respective tenses. The reason for the use of the present base is not so obvious. The fact that in origin it is itself an aorist (VIII, 16) may be the explanation.

The use of **değil** 'not' instead of or as well as the negative verb also makes possible the expression of a number of shades of meaning: **bu tehlike sezilmiyor değil** 'this danger is not unperceived'; **bunu yapacak değilim** 'I do not intend to do this'; **bunu bilmez değilim** 'I am not unaware of this' ('I-am-not not-knowing'); **bir neticeye varmış değilim** 'I don't claim to have reached a conclusion' ('I-am-not having-reached'); **bunu anlamamış değilim** 'I have not failed to understand this' ('I-am-not having-not-understood'); **zor kullanmadı değil, kullandı** 'it's not that he didn't use force; he did' ('force he-did-not-use not, he-used'); **ben, yazınızı okudum değil, görmedim bile** 'I have not *read* your article; I haven't even seen ⟨it⟩'; **ben, yazınızı okudum değil, ezberledim** 'I have not ⟨merely⟩ read your article; I have learned ⟨it⟩ by heart'.

X

VERBAL NOUNS

1. Introductory. The principal suffixes which make verbal nouns are: **-mek, -meklik, -me,** and **-iş**. Primarily, **-mek** denotes pure undefined action, **-meklik** the fact of action, **-me** the action or result of action, **-iş** the manner of action; there is, however, a certain blurring of the boundaries of their functions.

2. -mek. This is usually termed the suffix of the infinitive (cf. VIII, 1). It has two peculiarities of accidence: it never takes the personal suffixes or the suffix of the genitive case (see p. 173, *Addendum*). The other cases are in full use:

Absolute: (*a*) As subject: **bunu bilmek kâfidir** 'to know this is sufficient'; **eskiden kopmak kolay iş değil** 'to break away from the old is no easy task'.

(*b*) As object of **istemek** 'to want' and **bilmek** 'to know': **çalışmak istiyor** 'he wants to work'; **susmak bilmez** 'he does not know ⟨how⟩ to keep quiet'.

(*c*) As qualifier in izafet groups: **yazmak arzu-su** 'the desire to write'; **konuşmak niyet-i** 'the intention to speak'; **eğlenmek ihtiyac-ı** 'the need to amuse oneself'.

(*d*) As object of the postpositions **için** and **üzere**: **dünyayı değiştirmek için ne lâzım?** 'in order to change the world, what is necessary?' This is the usual way of expressing purpose.

üzere 'on, on the basis of' is similarly used: **çarşıya gitmek üzere otobüse bindi** 'she got on the bus to go to the market'. It also translates 'on condition of, on the understanding that': **yarın geri vermek üzere bana on lira verir misiniz?** 'will you give me ten liras on the understanding that I give it back tomorrow?' ('on-the-basis-of to-give back').

olmak üzere, lit. 'on-the-basis-of to-be', may sometimes be translated by 'being' or 'as being' but can often be left untranslated: **bu mektepte on beş ayrı millet-in, ekserisi albay olmak üzere, yüksek rütbeli subay-ı okumaktadır** 'in this school, high-ranking officers of fifteen different nations are studying, the majority of them being colonels'; **altı-sı kız, dördü erkek olmak üzere, on talebe-m var** 'I have ten pupils, six-of-them girls, four-of-them boys'.

-mek followed by **üzere** and part of the verb 'to be' means 'to be on the point of': **tren, hareket etmek üzere-ydi** 'the train was on the point of starting'.

Accusative as object of verbs other than **iste-** and **bil-**: **evlenmeğ-i düşünüyorlar** 'they are contemplating getting-married'; **ay-da iki yüz lira vermeğ-i taahhüt eder** 'he undertakes to pay 200 liras a month' ('in-the-month'); **devletten yardım görmeğ-i umuyoruz** 'we are hoping to receive help from the State'; **atlamağ-ı nasıl becerecekti?** 'how would he manage to jump?'; **ekmek almağ-ı unuttu** 'he forgot to buy bread'.

In front-vowel verbs, the accusatives of **-mek** and the verbal noun in **-me** (§ 7) are phonetically identical: **gitmeğ-i, gitmey-i**. In back-vowel verbs, the accusative and dative of **-mak** are phonetically identical, **almağ-ı** and **almağ-a** both being pronounced almā. Consequently, the accusative of **-mek/mak** is rapidly being supplanted in writing, as it has long been in speech, by the accusative of **-me/ma**; in all the examples in the preceding paragraph **-meği/mağı** can be replaced by **-meyi/mayı**.

Dative: **yürümeğ-e başladık** 'we began to walk'; **istediğini yapmağ-a alışıktır** 'she is accustomed to-doing what-she-wants'; **borcumu ödemeğ-e geldim** 'I have come to-pay my debt'; **sigara almağ-a gitti** 'he has gone to buy cigarettes'. Here too **-meye/maya** is taking the place of **-meğe/mağa**, though for the expression of purpose, as in the last two examples, **-meğe/mağa** seems to be holding its own in the written language for the moment.

Locative: **bunu yapmak-ta beis görmedi** 'he saw no harm in-doing this'. The locative of **-me** can replace that of **-mek** in such a sentence, but not so frequently in the present II tense **gitmekteyim**, etc.

Ablative: **hakikat-i yazmak-tan kendimi alamamıştım** 'I had-not-been-able-to-restrain myself from-writing the-truth'; **polis-e haber vermek-ten başka çaremiz yok** 'we have no remedy other than-to-give information to-the-police'; **âr çekmekten bâr çekmek evlâdır** 'to-bear burdens is-better than-to-bear shame'.

See also **-mektense**, XI, 30.

3. The infinitive with subject. In the older language, the infinitive in -mek could regularly have a subject: **sen böyle za'm ü pindar sahib-i olmak nedendir** 'why are you so puffed up and conceited?' (lit. 'you to-be possessor of such pretension and conceit is-from-what?').[1] Modern usage replaces **sen olmak** 'you to be' by **senin olma-n** 'your being'. The old usage survives, however:

(*a*) In proverbial expressions: **böyle oğul olmaktan olmamak yeğdir** 'better no son than such a son' ('than ⟨for⟩ such a son to be, not-to-be is-better').

(*b*) In dictionary definitions: **bulaşmak: bir nesne, üzerine sürülen bir şey yüzünden kirlenmek** 'to be defiled: ⟨for⟩ a thing to be dirtied because of something smeared on it', where **nesne** is the subject of the infinitive **kirlenmek**.

(*c*) In headlines: **Nurculuk aleyhinde konuşan bir müftü susturulmak istendi** 'it was desired that a mufti who spoke against the Nurcu doctrine should be silenced'; the subject of **istendi** 'was wanted' is **bir müftü susturulmak** '⟨for⟩ a mufti to be silenced'. The text of the story avoids giving the infinitive a subject by making the verb active: **200 kişilik bir grup, müftüyü susturmak istemiştir** 'a 200-person group wanted to silence the mufti'. It may be noted, incidentally, that whereas the text employs the past tense in **-miştir**, the headline has the synonymous but shorter **-di**.

(*d*) Rarely in other contexts, e.g. **parası çalınmak mı kötü, Harpagon olmak mı?** 'is it worse to have one's money stolen or to be a miser?', lit. 'his-money to-be-stolen is bad? to be a miser?'

4. **-mekli.** The adjectival suffix **-li** is occasionally added to the infinitive, as in **insan ağlamaklı oluyor vallahi** 'honestly, one feels like crying' ('man becomes characterized-by-weeping, by-Allah').

On the other hand, **-siz** 'without' is not added to **-mek**, **-sizin** being used instead; see XI, 31.

5. Common nouns in **-mek**. In contrast to the many common nouns which are in origin **-me** verbal nouns, very few **-mek**

---

[1] Kâtip Çelebi, *Mizān al-Haqq* (ed. Ebuzzıyâ), p. 84. See also Deny, §§ 1292–5. For **ü** 'and' see XIII, I.

infinitives have acquired concrete meaning: **yemek** 'food' (as an infinitive, 'to eat'); **çakmak** 'cigarette-lighter' ('to strike'); **tokmak** 'door-knocker' (from the obsolete infinitive **tokımak** 'to knock'). The noun **ekmek** 'bread' and the infinitive **ekmek** 'to sow' are not etymologically connected.

6. -meklik. Unlike the bare **-mek**, **-meklik** can take personal suffixes and all case-endings. It is nothing like so frequent as **-me**, but is rather more precise in its sense of 'the act of doing'; it also has one advantage over **-me**, namely that **-me** with the first-singular personal suffix—**git-me-m** 'my going'—is indistinguishable in spelling and pronunciation from the first-singular negative of the aorist present tense—**git-mem** 'I do not go'; any possible ambiguity can be eliminated by using **gitmekliğim** for the former. Thus **gitmem lâzım mı?** 'is my-going necessary, must I go?' might, if we ignore punctuation, be read as 'I'm not going; must I?', whereas there is no such ambiguity about **gitmekliğim lâzım mı?**

7. -me. Verbal nouns formed with this suffix, unlike those in **-mek**, appear in every case[1] and with the personal suffixes. In the absolute case they are identical in writing with the negative of the second-person singular imperative, but differ in accentuation:

gelmé coming   **gélme** do not come
yapmá doing   **yápma** do not do

In the dative the buffer **y** narrows the preceding vowel: **sormıya ne lüzum vardır?** 'what need is there for-asking?' **hançeresini yırtan hıçkırıkları dindirmiye uğraşıyordu** 'she was striving to still the sobs which tore her throat': **sormıya** < **sor-ma-y-a**; **dindirmiye** < **dindir-me-y-e**, though the unnarrowed spellings are more frequent in writing.

The **-me** and not the **-mek** forms are used in phrases like 'waiting-room, reading-book, working-hours': **bekleme salon-u, okuma kitab-ı, çalışma saatler-i**, since, for example, **beklemek** means the undefined concept of waiting, whereas what goes on in a waiting-room is **bekleme**, the act of waiting. The **-me**

---

[1] **-meden**, however, is not necessarily always the ablative. See XI, 12.

forms are therefore used in indirect commands: **bu yazıyı okuma-m-ı söyledi** 'he told me to read this article' ('he-stated my-action-of-reading'); see XVII, 1.

Forms like **tanımamama** in the following sentence can be confusing when first encountered: **yaptığım hatayı memleketi tanımamama verebilirsiniz** 'you may ascribe the mistake I made to-my-not-knowing the country'. The word is built up thus: stem **tanı-**+negative **-ma**+verbal noun **-ma**+'my' **-m**+ dative **-a**. If the negative is replaced by the impotential **-ama** we get **tanıyamamama** 'to my inability to know'.

The English verbal noun may have active or passive meaning; compare 'the singing of the choir' with 'the singing of the song'. Turkish can make passive verbal nouns by adding **-me** to the passive stem: **bu âlet-i kullan-ma-sı** 'his-using this instrument'; **bu âlet-in kullan-ıl-ma-sı** 'the use ("the being-used") of this instrument'. Although phrases like **bunun yapması kolay** 'the doing of this is easy' do occur, the passive **bunun yapılması kolay** is more usual.

8. Common nouns in **-me**. The sense of 'result of action' appears in the use of a great many **-me** verbal nouns as common nouns, e.g.:

as-	to hang	**asma**	vine
devşir-	to levy	**devşirme**	levy, i.e. recruitment *or* recruit[1]
dol-	to be filled	**dolma**	stuffed vine- or cabbage-leaf, embankment
dondur-	to freeze	**dondurma**	ice-cream
dön-	to turn	**dönme**	convert[2]
ez-	to crush	**ezme**	purée[3]
yaz-	to write	**yazma**	manuscript

[1] Used of the compulsory recruitment of Christian boys to the civil and military service of the Sultans.

[2] Used particularly of the Jewish followers of Sabbatai Zevi the false Messiah, who were converted to Islam after his forced conversion in 1666.

[3] Visitors to Turkey may be surprised to see in confectioners' windows two adjacent trays of marzipan, one labelled **badem ezmesi** 'almond purée', the other **çocuk ezmesi**. This is not made of crushed children but is a mild sort of marzipan considered more to the taste of younger customers.

Many are used as adjectives, e.g.: **asma köprü** 'suspension-bridge'; **asma kat** 'mezzanine floor'; **yazma kitap** 'manuscript book'; **dolma kalem** 'fountain-pen'. **doğ-** 'to be born', **büyü-** 'to grow up': **doğma büyüme bir İstanbullu** 'a born and bred Istanbul man'; **anadan doğma kör bir adam** 'a man blind from birth' ('from-mother birth blind'). **kal-** 'to remain': **babadan kalma emlâk** 'inherited estates' ('from-father remnant'); **Osmanlı İmparatorluğundan kalma birisi** 'someone left over from the Ottoman Empire'. **yap-** 'to make': **yapma çiçekler** 'artificial flowers'.

The passive verbal noun is also possible in this adjectival use: **İngiliz kumaşından yapılma bir ceket** 'a jacket made of English cloth'.

9. **-meli.** For the specialized function of this ending see VIII, 30. Descriptive adjectives of this form are few, e.g. **asmalı** 'having a vine' (**Asmalı Mescit** 'Mosque of the Vine' is the name of a **mahalle**, quarter, of Istanbul); **ağlamalı** 'tearful'.

10. **-masyon.** The ending of such French borrowings as **organizasyon**, **adaptasyon**, and **telekomünikasyon** is jocularly conflated with the **-me** verbal nouns of **uydur-** 'to invent' and **at-** 'to boast', giving **uydurmasyon** 'concoction, fabrication' and **atmasyon** 'line-shooting'.

11. **-iş.** This denotes not only the manner but also the fact of action, e.g. from **yürü-** 'to walk': **bu yürü-y-üş-le kasabaya akşama kadar varmış olacağız** 'with this way of walking, at this rate, we shall have reached the town by evening'. **her gün bir saat yürüyüş yapmalısınız** 'you ought to do an hour's walking every day' (**bir saat** is adverbial). Note that the English verbal noun 'walk' has the same two senses.

When this suffix is added to vowel-stems the buffer **y** does not usually narrow the preceding vowel; thus from **anla-** 'to understand', **anlayış**; from **de-** 'to say', **deyiş**. But from **ye-** 'to eat', **yiyiş** is more common than **yeyiş**, probably because of the cumulative narrowing effect of the two **y**'s: **her yiğid-in bir yoğurt yiyiş-i var** 'every young man has a way-of-eating yoghurt', i.e. everyone has his own way of doing things.

The **-iş** verbal noun can also be made from passive stems: **bu ev-in yap-ıl-ış-ı** 'the structure of this house' ('way-of-being-made').

A limited number of **-iş** verbal nouns form adjectives with **-li**, e.g.:

elver-	to be suitable	**elverişli**	suitable
göster-	to show	**gösterişli**	ostentatious
kullan-	to use	**kullanışlı**	serviceable
yağ-	to rain	**yağışlı**	rainy

12. -mezlik, -memezlik. The addition of **-lik** (IV, 8) to the negative aorist base makes a few abstract nouns such as **anlaşmazlık** 'misunderstanding', **saldırmazlık** 'non-aggression', **doymazlık** 'insatiability'. From this form comes a reduplicated negative in **-me-mez-lik**, denoting persistent non-doing, failure to do: **politikacıların uzağı gör-e-memezliğ-i** 'the short-sightedness of the politicians' ('their-persistent-inability-to-see the-distant'); **gelmemezlik etme, seni bekleriz** 'don't fail to come; we expect you'; **tatillerde çalışmamazlık etme** 'don't fail to work in the vacation'; **ona selâm vermemezlik edemezdim, çünkü o bana 'merhaba' dedi** 'I couldn't not greet him, for he said "hello" to me'.

The ablative, less commonly the dative, of **-memezlik** with **gelmek** 'to come', or the dative of **-mezlik** with **vurmak** 'to strike', means 'to pretend not to':

bilmemezlikten gelemezsiniz
bilmemezliğe gelemezsiniz } you cannot pretend not to know
bilmezliğe vuramazsınız

There is some fluctuation of usage; **-memezliğe vurmak** is used by some speakers of standard Turkish, but **-mezlikten gelmek** is a provincialism.

Addendum to §2. In transliterated Ottoman texts, forms in **-meğin** will be found; this is not the genitive but the instrumental of **-mek**. For example, **olmağın** means 'by being', 'because of being', 'with being', or 'when being'.

XI

GERUNDS

1. Introductory. This chapter deals with the many adverbial forms of the verb. The term 'gerunds' has been chosen from among the several terms in use, which include 'deverbal adverbs', 'adverbials', 'gerundives', 'gerundia', and 'converbs', as it has the merit of brevity. Those who are familiar with the gerund in Latin, however, will find little in common between it and most of the forms here described. Indeed, the only point of approach is that one sense of the adverbial form of the verb **-erek**, 'by doing', is like that of the ablative case of the Latin gerund, e.g. *faciendo*.

The forms treated in §§ 2–12 are made by adding suffixes to verb-stems, with the usual **y** as buffer where necessary and the usual fluctuation of usage about the narrowing or otherwise of a preceding vowel; e.g. **anla-** with the suffix **-erek** may be found spelled as **anlayarak** or **anlıyarak**.

The forms treated in sections §§ 13–33 are formed from participles, tense-bases, or verbal nouns.

2. -e. We have already met this suffix in the formation of the potential verb. It occurs also in a few frozen forms, made from the stems **geç-** 'to pass', **kal-** 'to remain', **rasgel-** 'to meet by chance', **ortaklaş-** 'to enter into partnership', **sap-** 'to deviate', **de-** 'to say', and **çal-** 'to strike, throw'. Cf. **göre**, VII, 4.

geç-e, kal-a are used to indicate the hour of the clock at which something happens: **saat üç-ü yirmi geçe geldi** 'he came at twenty past three' ('twenty passing hour three'); **saat dörd-e beş kala gitti** 'he went at five to four' ('five remaining to hour four'). The **saat** may be omitted: **üçü yirmi geçe; dörde beş kala. kala** is also used in expressions of distance like **eve bir kilometre kala benzin bitmiş** 'a kilometre from home we ran out of petrol' ('to-the-house one kilometre remaining, the petrol finished').

rasgele 'haphazardly': **rasgele bir tanesini aldım** 'I took one of them haphazardly, at random'.

ortaklaşa 'jointly, in common': **bütün bu servisler, Comet 4B jet uçaklarıyla ve Olympic Airways ile ortaklaşa yapılır** 'all these services are carried out with Comet 4B jet aircraft and in conjunction with O.A.'

sapa is an adjective meaning 'out of the way, off the beaten track'.

diye 'saying': **evet diye cevap verdi** 'he answered "yes"' ('he gave answer saying yes'). Its use has been greatly extended:

(*a*) To saying in writing: **yarın gel diye bir telgraf çekti** 'he sent a telegram saying "come tomorrow"'. **GİRİLMEZ diye bir levha** 'a sign saying "NO ADMITTANCE"' (note the impersonal passive **girilmez**, lit. 'entering is not done').

(*b*) To unspoken thoughts: **kim bunu yaptı diye düşünüyordum** 'I was wondering who had done this' ('I was thinking, saying "who has done this?"').

(*c*) To expressions of purpose and intention: **Allah seni dünya boş kalmasın diye yaratmamış** 'God did not create you just to take up room' ('saying "let the world not remain empty"'). Hence the common interrogative **ne diye?** 'with what intention?' ('saying what?'): **ne diye yemek yemiyorsun?** 'what's the idea of not eating?'; **ne diye erken geldin?—sizi bekletmiyeyim diye** 'why have you come early?'—'in order not to keep you waiting' ('saying-what have you come early?'—'saying let me not make you wait').

(*d*) Colloquially it is used for 'named': **Beş Şehir diye bir kitap** 'a book named *Five Cities*'; **Liva diye bir dostum var** 'I have a friend named Liva'. In formal language, the place of **diye** in these two examples would be taken by **adlı, isimli,** or **isminde.**

(*e*) In the sense of **namına** (VII, 7): **ehliyet imtihanı diye bir şey yok Belçika'da** 'there is nothing you could call a driving-test in Belgium'.

çala occurs in some compound adverbs, e.g.: **çalakalem yazmak** 'to write busily' (lit. 'throwing-pen'); **çalakaşık yemek** 'to gobble greedily' ('throwing-spoon'); **çalakürek açılmak** 'to row away at full speed' ('to recede throwing-oar').

These frozen forms apart, the **-e** gerund does not occur singly; either (*a*) the **-e** gerund of one verb is repeated or (*b*) the **-e** gerunds of two verbs are used side by side, indicating repeated action contemporaneous with that of the main verb. The accent falls on the first -e of the pair:

(*a*) **leylek zıplıya zıplıya uzaklaşmıştı** 'the stork had hopped away' (**zıpla-** 'to hop'; lit. 'the stork hopping hopping had receded'); **insan belki döğüle döğüle uslanır** 'perhaps one grows well-behaved with being constantly beaten' (**döğül-** passive of **döğ-** 'to beat'); **her kelime için kalemini dört beş kere hokkasına batıra batıra uzun uzun yazdı** 'dipping his pen into his ink-well four or five times for each word, he wrote at great length' (**batır-** 'to dip', causative of **bat-** 'to sink'). Another example is seen in the formula of farewell: **güle güle!** '⟨go⟩ happily!' (**gül-** 'to laugh').

With phrases formed from a verbal noun and an auxiliary verb, there is no need to repeat the verbal noun; thus from **takib-etmek** 'to follow': **izlerini takibede ede yürüdük** 'we walked, following their tracks'. From **feth-etmek** 'to conquer': **memleketler fethede ede ilerlediler** 'they advanced, conquering country after country'.

(*b*) **yürükler kona göçe yaylaya gittiler** 'the nomads went to the plateau, camping and moving on, camping and moving on' (**kon-** 'to settle', reflexive of **ko-** 'to put'; **göç-** 'to migrate'); **böyle gelmiş böyle gidecek dünyamız: bozula düzele, değişe gelişe, yeni eskiyi, eski yeniyi vura vura** 'thus our world has come ⟨down to us⟩ and thus will it go ⟨on⟩: being-spoilt and put-right, changing and developing, the new constantly-striking the old, the old the new'. Other examples: **hoplıya zıplıya** 'hopping and skipping'; **güle oynıya** 'laughing and dancing'; **ite kaka** 'pushing and shoving'; **düşe kalka** 'falling and rising', i.e. with great difficulty.

The repeated -e gerund has an idiomatic use, exemplified in: **gide gide sinemaya mı gittin?** lit. 'going and going was it to the cinema you went?', i.e. with all that going, after all that, couldn't you find anywhere better to go than the cinema? **gele gele bir küçük paket geldi** 'after all that, one little parcel came'. **bana da kala kala çirkin bir kadının karşısında boş bir yer kalmış** lit. 'and to me, remaining and remaining, an empty place remained opposite an ugly woman', i.e. after all that waiting for a seat, all that was left for me was . . . .

3. -erek. Whereas **-e -e** denotes repeated activity contemporaneous with the main verb, **-erek** denotes a single act or continued activity contemporaneous with or slightly prior to the main verb.[1] The first vowel of this suffix is accented, except (*a*) with negative stems, where, as usual, the syllable before the negative **-me** is accented, thus **bilérek** 'knowingly' but **bílmiyerek** 'unknowingly'; (*b*) in **ólarak** 'being', which has the initial accent usual in adverbs (see § 4). **gülerek cevap verdi** 'laughingly he answered'; **kapıyı açarak sokağa fırladı** 'opening the door, he rushed into the street'; **görmezliğe vurarak geçti** 'pretending not to see, he passed by'.

It often corresponds to the English 'by doing' or 'with doing': **bir okuyucu bu yazıya dayanarak hataya düşebilir** 'a reader, by-relying on this article, may-fall into error'; **geceyi konuşarak geçirdik** 'we passed the night with-talking'.

As repeated actions can merge into continuous action, the senses of **-e -e** and **-erek** overlap to some extent; for 'he came running', **koşa koşa geldi** and **koşarak geldi** are both possible, and 'you did this deliberately' may be **bunu bile bile yaptın** or **bunu bilerek yaptın**. In the next example, the main verb is modified by two **-erek** gerunds, the second of which is itself modified by an **-e -e** gerund: **her yaptığını bana açıklayarak, âletlerini seve seve kullanarak iki saat kadar çalıştı** 'explaining to me everything he did, using his instruments lovingly, he worked for some two hours'.

4. olarak. The **-erek** form of **ol-**, more often than that of other verbs, has a different subject from the main verb: **bu sene ilk defa olarak Amerika'ya gittik** 'this year, for the first time (lit. '⟨it⟩ being the first time'), we went to America'. **yemek olarak bir kilo elma aldım** 'as food ("being food") I bought a kilo of apples'. In the next example, the subject of **olarak** could be 'I' or 'you' (it is in fact 'I'): **size bir dost olarak bunu söylüyorum** 'I am telling you this as a friend'. This word therefore becomes a useful device for creating adverbial phrases: **netice olarak** ('it being the result') 'consequently'; **kat'î olarak** or **kesin olarak** 'definitely'.

5. -ip. Instead of using two verb-stems with identical suffixes side by side or joined by 'and', such as **kalktık gittik** 'we rose,

---

[1] The provincial **-erek-ten** denotes only activity contemporaneous with, never prior to, the main verb.

we went' or **okumaz ve yazamaz** 'he cannot read and he cannot write', **-ip** may be added to the first verb-stem: **kalk-ıp ğittik; oku-y-up yaz-a-maz**. For 'let me go and work in the city', there is no need to say **ğideyim (ve) şehirde çalış-ayım; ğid-ip şehirde çalış-ayım** is sufficient. Instead of **oyna-mak ve şarkı söyle-mek** 'to dance and to sing', **oyna-y-ıp şarkı söyle-mek**.

otur-up konuş-uyorlar 'they are sitting and talking'; **ara-y-ıp bul-madı** 'he did not seek and find'; **iç-ip yi-y-eceğiz** 'we shall drink and eat'; **otur-up dinlen-iniz** 'sit and rest!'; **ğid-ip ğör-meliyiz** 'we ought to go and see'; **seç-ip al-ırsam** 'if I choose and buy'; **ğel-ip ğid-enler** 'those who come and go' (note that this is not synonymous with **ğelenler ve ğidenler** 'those who come and those who go').

If a positive stem with **-ip** is followed by the same verb's negative stem with the suffix of a verbal noun or personal participle, the sense is of a choice between the positive and the negative: **lider'i beğen-ip beğen-me-mek, sanki dâvâya inan-ıp inan-ma-ma ölçüsü olmuştur** 'approval or disapproval of the leader has become as it were the criterion of belief or disbelief in the cause' (lit. 'to-approve-and not-to-approve . . . has become the measure of to-believe-and not-to-believe . . .'); **kendisini sev-ip sev-me-diğimi bilmiyorum** 'I do not know whether I love her or not' ('. . . my-loving-and my-not-loving'); **davetlerinin kabul ed-ip et-mi-y-eceğime karar vermeliyim** 'I ought to decide whether or not to accept their invitation' ('I-ought-to-give decision to-my-future-doing acceptance and my-future-not-doing'); **ğelelim benim bir eleştirmeci ol-up ol-ma-dığıma** 'let us come to the question of whether or not I am a critic' ('let-us-come to-my-being-and my-not-being . . .'); **benim sor-ul-up sor-ul-mı-y-acağından endişe ettiğim sual şuydu** 'the question about which I was anxious whether it would be asked or not was this' ('the question pertaining-to-my-doing anxiety from-its-future-being-asked-and not-asked was-this'; **şuydu = şu+idi**.

The use of **de** 'and' (see XIII, 2) after **-ip** marks a break between the action of the two verbs; this is particularly common when the second verb is negative but the first is not: **Zengo'yu ğör-üp de kork-ma-mak imkânsızdı** 'it was impossible to see Zengo [the name of a bandit] and not be afraid'. Without the

de, the meaning would be '... not to see and not to be afraid'.
bil-ip de söylemek istemiyenlerin tavriyle dedi ki ...
'with the air of those who know and do not want to tell, he said
...'. Here the **de** separates **bil-** from **söylemek** and also from
the negative element in **istemiyenlerin**; i.e. the suffixes which
are replaced by the **-ip** are **-en-ler-in** 'of those who'. Without the
de, the **-ip** would link **bil-** to **söylemek**: '... those who do not
want to know and to tell'. **ne yapacaksın, eve ğidip de?** 'what
will you do when you have gone home?' (in reply to **eve ğidi-
yorum** 'I am going home'). Here the **de** marks a time-lag between the going and the doing. In this example there may be
detected a trace of the original function of **-ip**, which was to create
past participles.[1] There is a clear survival of this use in the
traditional joke about Nasrettin Hoja's planting a tree, making
water on it and saying **görüp göreceğin rahmet bu kadar!**
'that's all the rain you're ever going to see!' ('the rain you-have-
seen-and will-see is this much').

Modern writers tend to avoid using more than one **-ip** in
a sentence, except for some stereotyped phrases: **ne yapıp
yapıp bu işi başarmalı** 'at all costs ("doing-and doing what")
one must make a success of this task'; **düşünüp düşünüp şu
teklifte bulunuyorlar** 'after thinking and thinking they are
making this proposal'.

Other set expressions involving -**ip** include: **durup dururken**
'without provocation', lit. 'while standing and standing': **durup
dururken bana hücum etti** 'he attacked me unprovoked', i.e.
while I was just standing and minding my own business. **olup
bitmek** ('to occur and finish') 'to happen', of which the past
tense **olup bitti** is used as a noun, 'event, *fait accompli*'.

söyüp saymak ('to swear and recount') 'to curse and swear'.

6. **-ince.** This denotes action just prior to that of the main verb.
The accent is on the first syllable of this suffix. **o gelince kal-
karım** 'when he comes I shall get up'. **böyle yaz de-y-ince
yaz-an, sus deyince sus-an haber müessese-si olur mu?**
'can there be an organ of the press like this, which writes when
it is told to write and is silent when told to be silent?' ('thus,
on-⟨someone's⟩-saying "write!" writing, on-⟨someone's⟩-saying

---

[1] This is still its normal function in Azeri, the dialect of Azerbaijan, where,
for example, **gelüpmen** means 'I have come' (**men** = **ben**).

"be-silent!" being-silent news-institution occurs?'). **yanında bıçağ-ı varsa üzerime hücum ed-ince ne yaparım ben?** 'if he has his knife on him, when he attacks me what do I do?' **otobüs gel-me-y-ince bir taksiye bindim** 'as the bus did not come I got into a taxi'. **hoca ol-ma-y-ınca talebe olmaz ya!** 'if there is no teacher there is no pupil, you know!' ('teacher not-being, pupil does-not-occur').

gelince 'on coming' is frequently used after a dative to mean 'as for': **bana gelince** 'as for me'; **paraya gelince, o çok zor bir mesele** 'as for money, that's a very difficult problem'.

In the older language, **-ince** meant 'until'; the sense of 'on doing' was then expressed by the now defunct **-icek**. The older meaning of **-ince** survives only in proverbs: **Arap doy-unca ye-r, Acem çatla-y-ınca** 'the Arab eats until he is satisfied, the Persian until he bursts'.

7. **-inceye kadar, -inceyedek, -inceye değin** 'until'. This is simply the dative of **-ince** with a postposition meaning 'as far as': **rapor neşr-ol-uncaya kadar hiç bir şey yapamayız** 'we cannot do anything until the report is published'; **öl-ünceyedek bekâr kalacak** 'he will remain a bachelor until he dies'.

8. **-ene kadar, -enedek, -ene değin** 'until'. This use of the dative of the present participle with a postposition meaning 'as far as' is no longer[1] confined to the popular language; though more informal than **-inceye kadar**, it has a respectable recent literary past and is especially frequent in newspapers: **hükümet, abluka altına alınan yerlerdeki Türkler açlık tehlikesiyle karşılaş-ana kadar pasif kalmıştır** 'the government has remained passive until the Turks in the blockaded areas ("the places taken under blockade") are faced with the danger of hunger'. **bugün-e gel-ene kadar bunun farkında değildim** 'until [coming to] today I was not aware of this' ('I-was-not in-the-discernment-of this'). **biz gid-ene kadar orada hapis kalacaktır** 'until we go, he will remain imprisoned there'.

9. **-esiye** 'to the point of'. This dative of the participle of the future II occurs mostly in set expressions: **çıldır-asıya sevmek** 'to love to the point of going mad'; **onu öldür-esiye dövdüm**

---

[1] As it was in the nineteen-twenties: Deny, § 1407.

'I beat him to the point of killing'; **bu elbiseyi ver-esiye yaptır-dım** 'I have had this suit made on credit' ('for future giving'; the uninflected **veresi** is also used in this sense); **bayıl-asıya güldüler** 'they laughed to the point of fainting'; **kendilerinden geç-esiye hora teperlerdi** 'they used to dance the *hora* (a cyclic dance) to the point of losing consciousness' ('to-the-point-of-passing from-themselves').

The negative is rare, except for **durmamasıya** 'unceasingly'.

-esiye kadar in the sense of **-inceye kadar** 'until' is a provincialism.

10. -eli, -eli beri, -eliden beri, -dim -eli 'since'. The **-dim** in this last is the first-person singular of the past tense, which changes as appropriate:

biz buraya { **geleli** / **geleli beri** / **geleliden beri** / **geldik geleli** } hiç yağmur yağmadı 'since we came here it has not rained'. **siz geldiniz geleli** 'since you came'; **o geldi geleli** 'since he came'. **biz buraya geleli iki yıl oldu olmadı** 'it is scarcely two years since we came here' ('two years have-been have-not-been').

In the older language, **-eliden** was used without **beri** for 'since'.

-eli suffixed to a negative stem means 'during the time that . . . not', but is best translated 'since': **ben görmiyeli birkaç kilo vermişsiniz** 'since I saw ⟨you⟩, you have lost some weight'.

11. oldum olası or **oldum olasıya.** This expression means 'ever since the beginning of things, for as long as anyone can remember'. It is a corruption of **oldu olalı** 'since it has been'; in spite of appearing to contain the first-person singular **oldum**, it does not mean 'for as long as I can remember'. **bu durum bugün olmuş değildir; oldum olası böyledir** 'this situation has not come into existence today; it has always been thus'.

12. -meden or **-mezden** 'before, without'; **-meden** or **-mezden evvel/önce** 'before'. The forms in **-mezden** are much less common in writing than those in **-meden. ben gelmeden evvel işe başlamayınız** 'don't start work before I come'; **mektubu**

okumadan attı 'he threw the letter away without reading ⟨it⟩';
'**Para bulunmazdan kimse çıkamaz**' **diyor polis** '"Before the money is found no one can go out", says the policeman'.

An adverb may be introduced before the **evvel** or **önce**: **o ölmeden az önce** 'a little before he died'; **biz ayrılmadan iki gün evvel** 'two days before we left'.

çok geçmeden ('before much passes') means 'before long'. To avoid suffixing **-siz** 'without' to **siz** 'you', 'without you' is expressed, in contexts where **sensiz** would be too familiar, by **siz olmadan** 'without your being'.

This ending **-meden** looks like the ablative of the **-me** verbal noun; hence indeed its use with **evvel** and **önce**. Historically, however, it is not so,[1] as is shown by the accentuation, for the ending is enclitic, the accent falling on the verb-stem: **gélmeden**, **okúmadan**, whereas in the ablative of **-me** it is the **-den** that is accented. Because of the risk of ambiguity, the ablative of **-me** is not used as often as it might be; e.g. **kitap okumadan çok eğleniyorum** could be taken to mean 'I get a lot of fun without reading books', so to express 'I get a lot of fun from reading books' one uses the ablative of **-mek** rather than of **-me**: **kitap okumaktan**, etc.

13. -r -mez. The juxtaposition of the positive and negative aorist bases denotes 'as soon as': **ben oturur oturmaz telefon çaldı** 'as soon as I sat down, the telephone rang' (i.e. just as I was on the border-line between not-sitting and sitting); **gerillâcılar, karanlık basar basmaz görev için derhal toplanırlar** 'the guerillas assemble at once for duty as soon as darkness falls'. Colloquially, the positive and negative of the **di**-past are similarly used: **oturdum oturmadım telefon çaldı**. A similar colloquial use is exemplified in **o geldi mi geldi bana haber ver** 'let me know the moment he comes' (lit. 'has he come? he has come; ⟨in that case⟩ tell me').

Three exceptions: **ister istemez** is an adverbial phrase meaning 'whether one wants it or not, willy-nilly'; **olur olmaz** and **bilir bilmez** are adjectival phrases meaning respectively 'ordinary, chosen at random' and 'half-knowing, with a little knowledge'.

---

[1] The oldest recorded form of this suffix was **-meti**, the first syllable being probably the negative **-me**. With the addition of the instrumental **-n** this became **-metin**. The change to **-meden** was due to analogy with the ablative.

14. -dikçe. The suffix -ce (XII, 2) added to the participle in -dik denotes 'so long as, the more': **ben konuş-tukça konuşacağım gelir** 'the more I talk, the more I feel like talking' ('as I talk, my-future-talking comes'). **o gül-dükçe ben de güldüm** 'the more she laughed, the more I laughed'. **sen o mektubu yazmağ-ı geciktir-dikçe başlaması güç olacak** 'the more you delay writing that letter, the harder it will be to begin' ('its beginning will become hard'). **dedem koynunda yat-tıkça benim-sin ey güzel toprak** 'so long as my ancestors lie in your bosom you are mine, o lovely land'. **doğu, insan-ın piş-tikçe sustuğu, sus-tukça piştiği yer** 'the east is the place where the more one matures the more one is silent; the more silent one is the more one matures' ('the place pertaining to his—man's—being silent as-he-matures' and vice versa).

The negative means much the same as **-me-y-ince**: **devlet yardım et-me-dikçe fert ne yapabilir?** 'so long as the State does not help, what can the individual do?'

The use of **her** 'every' before **-dikçe** gives the sense of 'whenever': **bize her geldikçe çiçek getiriyor** 'every time he comes to us he brings flowers'.

Particularly common are **gittikçe** 'gradually' (lit. 'as it goes'); **gün geçtikçe** 'as the days pass'; **oldukça** 'rather': **o akşam oldukça az yedi** 'that evening he ate rather little'.

15. -diğince. The suffix -ce can also be added to the personal participle; this was an old Anatolian equivalent of **-dikçe** but is sometimes used nowadays in the sense of 'in the measure of his doing' or 'inasmuch as he does': **herkes, elden geldiğince, fakirlere yardım etmeli** 'everyone ought to help the poor as much as he can afford' ('in-the-measure-of-its-coming from-hand'). **uzağı göremediğimizce, manzara hakkında bir fikrimiz yoktu** 'inasmuch-as-we-could-not-see far, we had no idea about the landscape'.

16. -dikte 'on doing, at the moment of doing'. This is not very frequent. It may be reinforced by **her**: **her mektup aldıkta yüzü güler** 'every time he gets a letter his face lights up'; **her geldikte kavga eder** 'every time he comes he quarrels'.

17. -dikten sonra 'after doing'. This is the regular converse of

-meden evvel/önce: roman-ı oku-duktan sonra fikrimi söyliyeyim 'let me state my opinion after reading the novel'.

Expressions of time may be inserted: yeni vazife-m-e başladıktan bir hafta sonra tuhaf bir şey oldu 'a week after ⟨my⟩ beginning my new duty, a queer thing happened'.

18. -dikten başka 'apart from doing, in addition to doing': o, bu şiiri oku-duktan başka ezberlemiş de 'apart from reading this poem he has learned it by heart too'; ingilizce konuş-ama-dıktan başka, türkçe de bilmez-sin 'apart from not being able to speak English, you don't know Turkish either'.

19. -diğinden başka. The personal participles are similarly used in the ablative with başka: konuşamadığından başka 'apart from the fact that you cannot speak'; ev parasını ödiyemiyeceğinden (öde- 'to pay') başka 'apart from the fact that he is not going to be able to pay the rent'.

20. Gerund-equivalents. The personal participles form the basis of a number of gerund-equivalents (i.e. phrases equivalent in meaning to a gerund) about which one important general observation must be made:[1] if they have a third-person subject expressed, it is in the absolute case. Consider first these three sentences:

(*a*) o gün yağmur yağdı 'that day, it rained'.

(*b*) o günlerde çok yağmur yağdı 'in those days it rained a lot'.

(*c*) o gün yağmurlu idi 'that day was rainy'.

In (*a*), o gün is an adverb modifying yağdı; in (*c*) it is a noun-phrase, subject of idi. o günlerde in (*b*) is, in Turkish terminology, a locative complement to the verb (in English we should call it an adverbial phrase of time). Now if o is replaced by a personal participle whose subject is a noun, that noun stays in the absolute form in (*a*) but goes into the genitive in the other two sentences:

(*a*) **Orhan geldiği gün yağmur yağdı** 'the day Orhan came, it rained'.

(*b*) **Orhan'ın Rize'de kaldığı günlerde çok yağmur yağdı** 'in the days Orhan stayed at Rize, it rained a lot'.

---

[1] The point here explained is instinctively known to every Turkish-speaker but seems not to have been mentioned by any grammarian, native or foreign, before C. S. Mundy (*BSOAS* (1955), xvii/2, pp. 293–4).

(*c*) **Orhan'ın geldiği gün yağmurlu idi** 'the day Orhan came was rainy'.

The words **Orhan geldiği gün** in (*a*) are a gerund-equivalent; i.e. they could be replaced by **Orhan gelince** or **Orhan geldikte**, and so **Orhan** remains in the absolute case. But **Orhan'ın kaldığı** in (*b*) and **Orhan'ın geldiği** in (*c*) are adjectival phrases, qualifying **günlerde** and **gün** respectively, and as their subject **Orhan** is definite it is put in the genitive. The same distinction is seen in these examples: **soyadı kanunu çıkacağı zaman ben Avrupaya gitmiştim** 'when the surname-law was about to be promulgated, I had gone to Europe'; **soyadı kanununun çıkacağı günlerde bir akşam yemeğinde Atatürk'ün sofrasında idim** 'in the days when the surname-law was about to be promulgated, at one evening meal I was at Atatürk's table'. In the first example, the words **soyadı . . . zaman** are a gerund-equivalent, i.e. an adverbial clause; in the second, **günlerde** is a noun and **soyadı . . . çıkacağı** an adjectival phrase qualifying it.

insan vasiyetnamesini yazacağı zaman avukatını çağırmalı 'when a man is about to write his will he should call his lawyer'; here again **insan . . . zaman** is a gerund-equivalent, i.e. an adverbial clause. But in **insanın vasiyetnamesini yazacağı zaman olur** 'the time occurs when a man is about to write his will', **zaman** is the subject, qualified by **insanın . . . yazacağı**.

Such gerund-equivalents, containing the words **zaman** or **vakit** 'time', **gün** 'day', **an** 'instant', **sıra** 'moment', used adverbially in the absolute form and qualified by a personal participle, correspond to English adverbial clauses of time.

gittiğimiz zaman o kaldı	when we went, he stayed
gittiğimiz zaman o kalır	when we go, he stays
gittiğimiz zaman o kalacak	when we go, he will stay
gideceğimiz zaman o geldi	when we were about to go, he came
gideceğimiz zaman o gelir	when we are about to go, he comes
gideceğimiz zaman o gelecek	when we are about to go, he will come

21. -diği müddetçe 'as long as': **Osman bahçede çalıştığı müddetçe şarkı söyler** 'Osman sings all the time he works in the garden'. In this gerund-equivalent, **müddetçe** may be replaced by the neologism **sürece**.

22. -diği halde. The most frequent sense of this gerund-equivalent is 'although': **'ve' edat-ı o kadar yaygın kullanışlı olduğu halde dilimizdeki yer-i sanıldığından çok daha az önemlidir** 'although the particle ve is so widely used, its place in our language is much less important than is thought'. **bağırdığım halde kimse yardıma gelmedi** 'although I shouted, no one came to help'.

It is, however, also found in its literal sense of 'in a state of ...ing': **bacağı alçıda olduğu halde eve döndü** 'she returned home with her leg in plaster' ('in a state of being her-leg in-plaster'; for the construction, see XVIII, 1).

23. -diği takdirde lit. 'in-the-assumption pertaining-to-his-...ing' is still used to mean 'in the event of his ...ing', though disapproved by modernists, who prefer a simple 'if'-clause: **hazır bulunamadığınız takdirde toplantıyı tehir ederiz** 'in the event of your being unable to be present, we shall postpone the meeting'.

24. -diği için or **-diğinden.** The personal participle followed by **için** or in the ablative case means 'because of his ...ing': **bir müslüman-la evlendiği için** (or **evlendiğinden**) **kiliseden atılmıştır** 'because she married [with] a Muslim she has been expelled from the church'; **bu akşam tiyatroya gideceğimiz için** (or **gideceğimizden**) **yemek erken yiyeceğiz** 'because we are going to go to the theatre this evening, we shall dine early'.

25. -diği nispette 'in proportion to his ...ing'. This is a little antiquated, and most writers nowadays would prefer **-dikçe**: **annem, kendini müdafaaya çalıştığı nispette fazla ezilirdi** 'my mother used to be all the more bullied as she tried to defend herself' ('. . . used to be more crushed in the proportion of her trying . . .').

26. -diği kadar. The personal participle followed by **kadar** 'amount' means 'as much as': **istediğiniz kadar kalabilirsiniz** 'you can stay as long as you want'; **doyacağımız kadar**

yiyeceğiz 'we shall eat as much as will fill us' ('. . . the amount pertaining-to-our-future-being-satiated').

27. -diği gibi 'as soon as he does/did' or 'as he does/did'. In the first of these two uses the **-diği gibi** is a gerund-equivalent. The two uses are paralleled in English; cf. 'I left as he came in' and 'I think as he thinks'.

-eceği gibi, however, is not a gerund-equivalent but is used only in the literal sense of 'as he will . . .': **öğretmen, dersi öğrenci-nin anlıyacağı gibi anlatmalı** 'the teacher should explain the lesson in a way the pupil will understand' ('like what-he-will-understand').

28. -eceğine or **-ecek yerde** 'instead of ...ing': **herif benden özür dileyeceğine** (or **dileyecek yerde**) **küfüre, hakarete başladı** 'the scoundrel, instead of begging pardon from me, began swearing and insulting'. **başkası yüz vereceğine** (or **verecek yerde**) **siz yirmi verin** 'instead of someone else giving a hundred, you give twenty' (the speaker is offering a bargain). **kışın burada kal-ıp yük olacağıma** (or **olacak yerde**) **şehre gid-ip ekmeğimi ararım** 'instead of staying here in winter and being a burden, I shall go to the city and seek my bread'.

29. -mekle. This and the next two gerunds are based on the infinitive.

-mekle is frequent in its literal sense of 'with/by ...ing': **gitmekle aptallık ettim** 'I did a silly thing by going'; **günümü hep çalışmakla geçirdim** 'I spent my day entirely with working'.

As the infinitive could regularly have a subject in the older language (see X, 3), **-mekle** used to occur with a subject different from that of the main verb, as in **İstanbul'da büyük bir yangın zuhur etmekle, Sultan Selim Edirne'ye gitti** 'with a great fire occurring in Istanbul, Sultan Selim went to Edirne'. This now sounds highly archaic; the modern idiom would be **zuhur ettiği için** or **çıktığı için** 'because ... occurred'.

-mekle beraber or **-mekle birlikte** 'together with ...ing' is common in the sense of 'although': **pek zengin olmamakla beraber her zaman nikbindir** 'although he is not very rich he is always optimistic'. In this construction the subject of **-mekle** can still be different from that of the main verb.

30. **-mektense** or **-mekten ise** 'rather than': **şehrin yakıcı havası altında bunalmaktansa Sariyer'e gitmiye karar verdim** 'rather than be suffocated under the scorching air of the city, I decided to go to Sariyer' (**gitmiye** is the dative of **gitme**, the verbal noun of **git-**). **onu bu halde görmektense ölümü bin defa tercih ederim** 'rather than see him in this state I would a thousand times prefer death'.

The explanation of this form is that the **-ten** is comparative, so, for example, **görmektense** means 'if it is by comparison with seeing, if it is from the starting-point of seeing'. An alternative explanation[1] would make it a corruption of **görmekten eyisi** (the modern **iyi-si** 'its good'), so that the original sense would have been 'that which is better than seeing'. This is unlikely, as the **eyisi** would have been too obviously redundant in such proverbial expressions as **nâdân ile konuşmaktan ise ehl-i irfan ile taş taşımak yeğdir** 'rather than converse with the ignorant it is better to carry stones with the learned' (**ehl-i irfan** is a Persian izafet group).

An older alternative is **-medense**: **ağlamadansa ağlatmak ilâhî bir kanundur** 'to-make-weep rather-than-to-weep is a divine law'.

This use of the ablative followed by **-se** is not confined to the verbal nouns; cf. **hiç yoktansa ona da razı olduk** 'we agreed to that as being better than nothing', lit. 'if it is by comparison with nothing'.

31. **-meksizin** 'without ...ing'. This is less frequent than **-meden**. **insan çalışmaksızın para kazanmaz** 'one cannot earn money without working'; **demindenberi sebebini bilmeksizin rahatsız oluyordum** 'recently I have been getting unwell without knowing the-cause-of-it'.

32. Equivalents of 'as if'.

(*a*) **-cesine**. This suffix is used with nouns (see XII, 2 (*h*)) and with the base of the aorist and **miş**-past and with the third-person singular inferential of the aorist, present I, future I, and **miş**-past: **yağmur, bardaktan boşan-ır-casına yağıyordu** 'the rain was falling as-if-being-emptied out-of-glasses'; **makinenin bir**

---

[1] By Elöve, p. 1030, footnote 1.

parçası ol-muş-çasına, sağa sola bakmadan çalışıyorlardı 'they were working without-looking to right ⟨or⟩ to left as-if-having-become a part of the machine'; **pek eskiden tanış-ır-mış-çasına el sıkıştık** 'we shook hands as-if-being acquainted long-since', i.e. as if we had known each other for years; **kendi kendine söylü-yor-muş-çasına mırıldandı** 'he murmured as-if-talking to himself'; **öl-ecek-miş-çesine sık sık soluyordu** 'he was breathing rapidly as-if-about-to-die'; **iyi bir söz söyle-miş-miş-çesine böbürleniyordu** 'he was showing off as if he had said something worth saying' (lit. 'as-if-having-said a good saying').

An example of the suspended affixation of this suffix: **hiç bir şey görmüyor ve duymuyormuşçasına gözünü kapadı** 'he closed his eyes as if seeing and hearing nothing', i.e. **gör-mü-yor-muş-çasına ve duy-mu-yor-muş-çasına**.

(b) The same sense is conveyed by **gibi** 'like': **boşanır gibi, olmuş gibi, tanışırmış gibi**, etc.; the whole phrase can be introduced by **sanki** (see XIII, 30).

(c) The dative of the base of the **miş**-past is often used with **benzemek** 'to resemble', as in **tımarhaneden kaç-mış-a benziyorsun** 'you look as if you had escaped from the asylum' (lit. 'you resemble having-escaped . . .'). The same construction may be found with other bases, even that of the present ('even', because this base is not a participle): **uyuyor'a benziyor** 'he looks as if he is sleeping'. This is not a normal literary construction; hence the apostrophe to separate the verb from the dative suffix. The closest English equivalent is 'he looks like he's sleeping'.

33. -mecesine. The addition of **-cesine** to the verbal noun in **-me** makes a form meaning 'on condition of', used colloquially to express the terms of a wager: **salı gününe kadar işi bitir-mecesine bahse girdik** 'we made a bet ("we entered wager") to finish the job by Tuesday'. This form cannot be made to refer to a particular person, nor, in this meaning, can it be made negative. For example, for 'we bet that you would not be able to finish the job by Tuesday', the dative of the future personal participle is used (with the impotential **-eme-**): **salı gününe kadar işi bitiremiyeceğine bahse girdik**.

The negative appears in **durmamacasına** 'ceaselessly'. Synonymous are **durmamasına** and **durmamasıya**.

34. iken. Originally a participle of the obsolete **er-** 'to be', this now means 'while being'. Like the other surviving forms of that stem, it may be used as an independent word or a suffix. The **e** being invariable, the suffixed forms are **-ken** after consonants and **-yken** after vowels: **bu kelime aslında isimken, edat olarak da kullanılır** 'while this word is originally a noun, it is also used as a particle' ('this word in-its-origin noun-while-being, it is also used being a particle'); **kendisi çocukken babası ölmüştü** 'while he was a child ("himself child-while-being"), his father had died'; **ben oradayken (orada iken) öyle bir niyeti yoktu** 'while I was there he had no such intention'.

It may be suffixed to any tense-base, positive or negative, singular or plural, except the **di**-past, but is most frequent with the aorist: **o giderken muhakkak beni uyandır** 'when he is going, be sure to wake me'; **çocuklar parkta oynarlarken biz çarşıya gittik** 'while the children were playing in the park, we went to the market'; **bizde kanunların çoğu tatbik edilmezken acaba bu kanun neden kemal-i ciddiyetle tatbik edilir?** 'amongst us while most of the laws are not applied, why, I wonder, is this law applied in real earnest?' (**kemal-i ciddiyetle** is a Persian izafet: 'with perfection-of seriousness').

With bases other than the aorist: **kapıcı bana anahtarı veriyor-ken zıl öttü** 'while the janitor was giving me the key, the bell rang'; **biz hana girmekte-yken güneş battı** 'while we were entering the inn, the sun set'; **gürültüden neşemiz kaçmış-ken hepimiz gene güldük** 'while our pleasure had fled ("our-pleasure while-being-having-fled") because of the noise, we all laughed again'; **sırası gelmiş-ken şunu da söyliyeyim** 'as the time for it has come ("its-time while-being-having-come"), let me say this too'; **kâğıdı alacak-ken durdu** 'while-about-to-take the paper, he stopped'; **tam gömecek-ler-ken ölmekten cayıyor** 'just as-they-are-about-to-bury ⟨him⟩ he changes his mind about ("swerves away from") dying'.

With the future participle in the depreciatory sense (IX, 2, end): **marangoz olacak-ken güzelim dolabımızı berbat etti** 'while he is supposed to be a carpenter, he has ruined our lovely cupboard'.

Like the English 'while', **iken** may be used concessively: **o delikanlı, çok iri yapılı iken (or yapılıyken) gürbüz sayılamaz** 'that young man, while of very large build, cannot be considered robust'.

In the next example, **iken** is followed by the ablative suffix: **senin kadarkendenberi pul topluyorum** 'I have been collecting stamps since I was your age' (-ken-den-beri 'since while being'; **senin kadar** 'as big as you').

35. Compound verbs.

(a) The addition of **durmak** to the -e or -ip gerunds denotes continuous action: **söylenedurmak** or **söylenip durmak** 'to keep grumbling'. Less commonly, a finite tense may be followed by the same tense of **durmak**: **söylenir durur** 'he keeps grumbling'; **söylendi durdu** 'he kept grumbling'. A colloquial alternative is -e **komak** or -e **koymak**: **çalışako** or **çalışakoy** 'keep on working!'

(b) The addition of **gelmek** to the -e gerund has the same effect: **neler çek-e-geldi** 'what things he-has-always-suffered!'; **böyle işler ol-a-geldi** 'such things have always happened'; **kullan-a-geldiğimiz Arapça ve Farsça kelimeler** 'the Arabic and Persian words which we have always used'; **okun-a-gelen kitaplar** 'books which are always being read'.

The **gel-** is sometimes written separately: **bu hata yine yapıla gelmektedir** 'this mistake is still being constantly made'.

Exceptional is **çık-a-gelmek**, meaning not 'to keep coming up' but 'to come up suddenly'.

(c) **-e-kalmak** 'to remain, be left ...ing': **don-a-kaldım** 'I was left freezing, was petrified'; **bak-a-kaldılar** 'they remained staring'.

(d) The imperative of **görmek** placed after the negative of the -e gerund of other verbs means 'mind you don't', or, if the imperative be that of the third person, 'beware, lest . . .': **oraya git-mi-y-e-gör** 'mind you don't go there'; **harp ateş-i yan-mı-y-a gör-sün** 'beware lest the fire of war be kindled'.

(e) Until the nineteenth century, the **miş-** or **di**-past of **yazmak**, added to an -e gerund, indicated that the action of the first verb was narrowly averted. Even as an archaism, the only example one might see now is **düş-e-yazdı** 'he well-nigh fell'. The modern expression for this is **az kaldı düşüyordu**, lit. 'little remained he-was-falling'.

(f) Rapid or sudden action is conveyed by suffixing to a verb-stem an **i** (or, after a vowel, **yi**), which changes with vowel-harmony, and **vermek**: **onu kaldır-ı-verelim** 'let us quickly remove it'; **Abdurrahman'ın içinden şu herifi denize uçuru-vermek geçti** 'Abdurrahman felt like chucking that fellow into the sea' ('to-suddenly-make-to-fly that fellow to-the-sea passed through-his-inside'); **köprü çök-ü-verdi** 'the bridge suddenly collapsed'. Colloquially, the first element may be repeated: **köprü çökü çöküverdi**.

The passive suffix is attached to the **ver-**, not to the first verb-stem: **kapıdan dışarıya koy-u-ver-il-di** 'he-was-rapidly-put outside the door'.

There are alternative forms of the negative. If the **ver-** is negated, it means that the rapid or sudden action was not done: **gid-i-vermek** 'to go quickly, to dash'; **çarşıya kadar gidiverdi** 'he dashed as far as the market'; **çarşıya kadar niçin gidiverme-di?** 'why did he not dash as far as the market?' If the main stem is negated, it means 'to stop abruptly': **çarşıya kadar niçin git-me-y-i-verdi?** 'why has he suddenly stopped going as far as the market?'

(*g*) The addition of **gitmek** to the **-e** gerund denotes continuity or finality according to context: **iyiliğe kemlik ol-a-gelmiş ol-a-gider** (proverb) 'kindness has always been requited with evil and always will be' ('for-good, bad has-always-happened, always-will-go-on-happening'). **kadınlarımızın yüzünden atılan peçe bütün gerçeklerimizin yüzünden at-ıl-a-gidecek-tir** 'the veil cast away from the face of our women will-be-cast-away-and-done-with from the face of all our realities'.

See also the use of **-dir gider** in VIII, 43.

(*h*) Colloquially, finality is expressed by the use of the third-person past **gitti** after any person of the **di**-past of another verb: **bir türlü ısınamadım gitti şu koltuğa** 'I just haven't been able to get used to this ministerial post, and that's all there is to it' (said in false modesty by a 'man of the people'). **böyle kapandı gitti hırsızlıktan çok daha büyük bir suç, insanları yok yere suçlandırma suçu** 'thus was a crime much graver than theft covered up and done with, the crime of accusing people falsely'.

Alternatively, the base of the **miş**-past may be followed by **gitmiş** with the appropriate personal ending: **ölmüş gitmişler** 'they're dead and gone'; **unutmuş gitmişim** 'I've totally forgotten'.

XII

ADVERBS

1. General observations. Almost any 'adjective' may modify a verb:

iyi	good	iyi çalışır	he works well
doğru	straight	yol doğru gider	the road goes straight
heyecanlı	excited	heyecanlı konuşuyor	he is talking excitedly
açık	open	açık konuşalım	let us speak openly
ağır	heavy	ağır bastı	it pressed heavily
yavaş	slow	yavaş git	go slowly

A repeated adjective or noun may serve as an adverb:

yavaş yavaş yürüyorduk we were walking slowly
hikâyeyi güzel güzel anlattı he told the story beautifully
kapı kapı dolaştım I wandered ⟨from⟩ door ⟨to⟩ door
ev ev aradılar they conducted a house-to-house search
efendi efendi davrandı he behaved in a gentlemanly way
Cf. **hanım hanımcık oturdu** 'she sat like a proper little lady'.

There are a host of reduplicated adverbial expressions, including onomatopoeic words like **horul horul**; see XIV, 29. Some verbs have reduplicated cognate adverbs ending in **m: sürüm sürüm sürünmek** 'to grovel grovellingly', i.e. to drag out a wretched existence; **burum burum burulmak** 'to be contorted gripingly'; **kıvrım kıvrım kıvrılmak** 'to writhe convulsively'. The adverb **için için** 'inwardly' must not be confused with the postposition **için** 'for'.

The distributive numerals when repeated serve as adverbs: **ikişer ikişer girdiler** 'they entered two by two'; **dörder dörder** 'in fours'; **ikişer üçer** 'in twos and threes'. For 'one by one', however, **bir bir** or **teker teker** is preferred to **birer birer**. 'Little by little' is **azar azar**.

Adjectives can be made into adverbs with the help of **suret** (A) 'shape' and **hal** (A) 'condition': **hafif surette** 'lightly' ('in light shape'); **ağır surette** 'heavily'; **fena halde** 'badly, unpleasantly'. The same is done for adjectives and nouns with the help of **olarak** (XI, 4): **azamî olarak** 'at most' ('being maximal'); **şaka olarak** 'jokingly' ('it being a joke').

2. -ce. This enclitic suffix makes adverbs from substantives.

(*a*) From adjectives it makes adverbs of manner: **iyi** 'good', **iyice** 'well'; **güzel** 'beautiful', **güzelce** 'beautifully, properly'; **doğru** 'straight', **doğruca** 'directly'.

(*b*) The pronominal **n** which appears before the case-endings of third-person pronouns appears also before this suffix: **bu** 'this', **bunca** 'this much, so much'; **o** or **kendisi** 'he', **onca** or **kendisince** 'according to him'; **bazıları** 'some people', **bazılarınca** 'on the part of some people'.

(*c*) The translation of the adverbs it makes from nouns depends on the context, as with the pronouns in the preceding paragraph. In the next two examples we have plain adverbs of manner: **çocukça konuşuyorsun** 'you are talking childishly'; **ordumuz, düşmana aslanca saldırdı** 'our army attacked the enemy like lions'.

Sometimes it means 'in respect of' and corresponds to the American use of '-wise': **adanın arazisi toprakça zayıftır** 'the island's land is weak in respect of soil'; **karısı kendisinden yaşça büyük ve zekâca üstündür** 'his wife is older than he ("bigger age-wise") and intellectually superior'; **karınca karar-ı-n-ca** lit. 'the ant according to its assessment', i.e. 'one contributes to the extent of one's modest ability'.

(*d*) In the sense of 'on the part of', it is rapidly superseding **tarafından** as an indicator of the agent of a passive verb: **Millî Savunma Bakanlığı-n-ca hazırlanan teklif, Maliye-ce reddedilmiştir** 'the proposal prepared by the Ministry of National Defence has been rejected by the Finance Department'. **dünyaca meşhur** 'world-famous' ('famed on-the-part-of-the-world').

(*e*) The adverbs it makes when attached to names of peoples come to be used as names of their languages and then as adjectives:

türkçe konuşmak 'to speak like-the-Turks, to speak Turkish'; akıcı bir türkçe ile dedi ki ... 'in a fluent Turkish, he said ...'; türkçe sözlük 'Turkish dictionary'; ingilizce konuşmak 'to speak English'; ingilizceniz nasıl? 'how is your English?'; ingilizce bir kelime 'an English word'. Hence, with the interrogative ne: nece? '(in) what language?'

(f) In numerical expressions: kilometrelerce uzak 'kilometres away' ('distant kilometres-wise'); haftalarca önce 'weeks before'; uçakların miktarı 10.000 lercedir (read on binlercedir) 'the number of aircraft is in the tens of thousands'; bu millet, benim gibi daha binlerce Mustafa Kemal çıkarır 'this nation will produce thousands more Mustafa Kemals like me'.

(g) Added to the demonstratives böyle, şöyle, and öyle. böylece and the far rarer öylece have the sense of 'therefore' as well as 'thus': iş böyle tamamlandı 'the job was completed in this way'; böylece iş tamamlandı 'that's how the job came to be completed'. şöylece is synonymous with şöyle 'thus'.

(h) Extensions of -ce. (i) -cene: böylecene, iyicene are colloquial alternatives for böylece, iyice.

(ii) -cesine. This is occasionally used to make adverbs from nouns: domuz-casına 'piggishly'; canavar-casına 'like a monster'; eşek-çesine 'like a donkey'. See XI, 32.

3. Nouns used adverbially.

(a) In the absolute case: sabah akşam çalışıyorum 'I am working morning ⟨and⟩ evening'; hava alanı, şehirden on beş kilometre uzaktır 'the airfield is 15 km. distant from the city'; sizden bir baş uzundur 'he is a head taller than you'; iki hafta evvel 'two weeks ago'. Note, however, that bir an evvel does not mean 'a moment before' but 'as soon as possible'.

(b) In the dative and ablative cases (cf. § 13 (e)): doğrudan doğruya 'directly, without intermediary'; ince-den ince-ye 'in fine detail'; dar-a dar 'narrowly, only just'; baş-a baş 'on equal terms'; baş baş-a '*tête à tête*'; baş-tan baş-a 'entirely'; bir-den-bir-e (written as one word) 'immediately'; gün-den gün-e 'from day to day'; günü günü-n-e 'to the very day', 'by return of post', (lit. 'its-day to-its-day').

(c) In the old instrumental case. See § 13 (f) and I, 39 (d), and note the widely used neologism örneğin 'for example', this being the instrumental of örnek 'pattern'.

4. Foreign adverbs. Arabic substantives with the Arabic accusative ending *-an* (unaccented) are used as adverbs, e.g.:

iktisat	economics	iktisaden	economically
kaza	accident	kazaen	accidentally
muvakkat	temporary	muvakkaten	temporarily
nispet	proportion	nispeten	relatively
siyaset	politics	siyaseten	politically
şeri	religious law	şer'an	canonically

Possibly because the two adjacent vowels of **kazaen** (originally separated by a glottal stop) are hard to pronounce, the Persian synonym **kazara** is more usual.

It is not unknown for people of limited education to coin analogous adverbs from non-Arabic words, such as **kültüren** for 'culturally', properly **kültürce** or **kültür bakımından** 'from the point of view of culture'.

In a number of commonly used adverbs of this formation the original *-an* came to be pronounced as long **a**; e.g. **evvelâ** 'first of all'; **asla** 'never'; **acaba** 'I wonder' (lit. 'wonderingly'); **mutlaka** 'absolutely'; **faraza** 'hypothetically, for argument's sake'; **meselâ** 'for example'; **hâlâ** 'still' (which exists side by side with **hâlen** 'at present', both from *ḥālan*). The final **a** of **acaba** is now pronounced short.

From the Arabic *ān* 'moment' and *sā'at-* 'hour' were formed, with the Persian preposition *ba* 'by', **anbean** 'from moment to moment', and **saatbesaat** 'from hour to hour'. Analogous formations with the Turkish **yıl** and **gün** are **yılbeyıl** 'year by year' and **günbegün** 'day by day'. These are disapproved by purists (just as English purists disapprove 'per day'), as is **özbeöz** 'one hundred per cent genuine', similarly formed from the Turkish **öz** 'self, essence', which has a firm place in the colloquial.

French has recently contributed **otomatikman** 'automatically', though purists prefer **otomatik olarak**.

From the dialectal Italian *giaba* comes **caba** 'gratis, into the bargain'.

5. Comparison of adverbs. This follows the pattern of comparison of adjectives (III, 4): **sen benden iyi bilirsin** 'you know better than I'; **uçak, sesten süratli gidiyor** 'the aircraft is going faster than sound'.

ADVERBS

çok 'much' when following an ablative translates 'more, rather than': **annesinden çok babasına benziyor** 'he resembles his father more than his mother'. The Arabic **ziyade** ('increase') is similarly used: **speleoloji, bir spordan ziyade bir ilimdir** 'speleology is a science rather than a sport'. In negative sentences 'much' is translated by **pek** (which as an adjective means 'strong'): **pek sevmiyorum** 'I don't much like'; **pek gelmez** 'he doesn't come much'.

'Most' is **en çok: en çok teyzesine benziyor** 'he most resembles his aunt'. In the presence of another adverb, however, **çok** is unnecessary: **en süratli giden uçak, jet uçağıdır** 'the aircraft which goes most quickly is the jet aircraft'.

6. bir. Besides meaning 'one' and 'a', **bir** is used adverbially to mean 'once' and 'only': **her hafta bir geliyor** 'he comes once every week'; **bir görürse ne yapabiliriz?** 'if once he sees, what can we do?'; **bir ona, bir bana baktı** 'he gave a look at him, a look at me'; **her şey bitti, bir bu kaldı** 'everything is finished, only this is left'; **bir ben, bir de Allah bilir** 'only God and I know' (said when hinting at a dark secret); **bunu bir sen yapabilirsin, bir de o** 'only you and he can do this'.

bir de 'and another thing, moreover' (for **de** see XIII, 2): **sen gel, bir de arkadaşın gelsin** 'come, and let your friend come too'. In the locution **bir de ne göreyim** it conveys surprise; 'and all of a sudden': **pencereden baktım bir de ne göreyim, bir polis kapıya doğru yürüyor** 'I looked through the window and all of a sudden what should I see—a policeman is walking towards the door'.

7. bir türlü 'by no means' (in negative sentences). As a noun, **türlü** means 'kind, category'; as an adjective, 'various'. **bir türlü onu kandıramadım** 'I just could not convince him'.

8. . . . bile or **hattâ** . . . 'even'. **sırrını benden bile sakladı** 'he concealed his secret even from me'.

hattâ (A), though less common than **bile**, is standing its ground: **camiyi kaç defa gezdim hattâ minarelerine çıktım** 'how many times have I gone round the mosque; I have even been up its minarets'.

Colloquially, it may be repeated, or preceded by **daha**, for emphasis: **iyi futbolcular, klâs adamlar, hattâ hattâ yıldızlar** vardı aralarında 'there were good footballers among them, men of class, even stars'. **bugünün insanı lüks aramıyor, hattâ istemiyor, daha hattâ, lüks'den kaçıyor** 'the man of today is not looking for luxury, he does not even want it; he even runs away from luxury'.

It may reinforce **bile**: **hattâ onu sevmiyenler bile cesaretini inkâr edemiyorlar** 'even those who do not like him cannot deny his courage'.

9. âdeta 'virtually, as it were'. The *OTD* defines the word thus: 'as usual; simply; merely; sort of; nearly; as good as: walk! (*riding command*).' This, though all true, obscures the fact that ninety-nine times out of a hundred **âdeta** is used to tone down an exaggeration or to apologize for a metaphor: **bu kitap, âdeta sizin için yazılmış ĝibidir** 'this book seems virtually to have been written for you'. **bu dar ve uzun vâdi, âdeta bir korkulu rüyaya benziyordu** 'this narrow and long valley resembled, as it were, a dreadful dream'.

10. Adverbs of place:

içeri	inside	dışarı	outside
yukarı	up	aşağı	down
ileri	forward	ĝeri	backward
öte	yonder	beri	hither
karşı	opposite		

(*a*) All these can be used as nouns: **ev-in yukarı-sı kiralık** 'the upper part of the house is to let'; **kuyu-nun aşağı-sı karanlıktı** 'the bottom of the well was dark'; **iş-in öte-sini bana bırak** 'leave the rest of the business to me'.[1]

(*b*) Or as adjectives: **ileri fikirler** 'progressive ideas'; **saatim beş dakika ĝeri** 'my watch is five minutes slow'; **karşı yaka** 'the opposite shore'. For 'inside' and 'outside' as adjectives, however, **iç** and **dış** (VII, 6) are commoner than **içeri** and **dışarı**.

(*c*) All but **öte** and **karşı** can be used as postpositions with the ablative.

(*d*) As adverbs, they indicate motion towards, either in the absolute form or in the dative, except that **öte-ye** and **karşı-ya**

---

[1] An interesting recent coinage is **fizikötesi** ('the beyond of physics') for 'metaphysics'.

are used in preference to **öte** and **karşı**: **içeri** or **içeriye gitti** 'he went inside'; **dışarı** or **dışarıya gitti** 'he went outside'; **öteye gitti** 'he went further on'; **karşıya gitti** 'he went to the opposite side'. The earthy expression for being on the horns of a dilemma is: **aşağı tükürsem sakalım, yukarı tükürsem bıyığım** 'if I spit down, my beard; if I spit up, my moustache'.

To indicate rest in or motion from, they are put in the locative or ablative respectively; in these cases **içeri, dışarı, yukarı**, and **ileri** generally lose their final vowel: **içerden** 'from inside'; **dışarda** 'on the outside'; **ilerde** 'in front, in the forefront, in future'; **yukardan** 'from above, from upstairs'; **aşağıda** 'down below, downstairs'.[1]

11. aşırı. As an adjective or adverb this word means 'excessive(ly)'. With a noun preceding, it means 'beyond, at an interval of': **deniz aşırı bir memleket** 'an overseas country'; **gün aşırı** 'every other day'; **bizden bir ev aşırı oturuyorlar** 'they are living one house [beyond] from us, next door but one'. It is not a postposition; the construction is as in **köyden bir kilometre uzak** 'one kilometre distant from the village'.

12. -re. The addition of this unaccented suffix to **bu, şu, o**, and **ne** turns them into nouns of place, of which the dative, locative, and ablative cases make the equivalents of English adverbs of place:

buraya hither	**şuraya, oraya** thither	**nereye** whither?
burada here	**şurada, orada** there	**nerede** where?
buradan hence	**şuradan, oradan** thence	**nereden** whence?

The locative and ablative forms may lose their first **a** or **e**: **burda, nerden**, etc.

The absolute forms **bura, şura**, etc., can theoretically occur as subject of a sentence but seem never to do so in standard Turkish; instead, they take the third-person suffix (II, 22 (*d*)): **orası güzel** 'that place is beautiful'; **burası neresi?** 'what place is this?'

It must be emphasized that when 'here' means 'this place' and not 'in this place' **burası** and not **burada** must be used: **burası**

---

[1] Turkish schoolchildren are taught to parse, for example, **içeri** as an adverb but **içeriye, içerde**, and **içerden** as nouns. The distinction hardly seems worth making, at least for the foreign student.

Ankara Radyo'su 'Here is Ankara Radio'. So with **şurası** and **orası**; for example, in this extract from a description of darkest Africa: **Vahşî hayvanlar orada idi. Yamyamlar, pigmeler orada idi. Siyah ırkın anavatanı orası idi** 'Wild animals were there. Cannibals and pygmies were there. The motherland of the black race was there'.

bura, etc., can be the first or second element of izafet groups: **bura halk-ı kuzu gibi** 'the people of this place are like lambs'; **ora-nın hava-sı güzel** 'the weather of that place is beautiful'; **şehr-in ora-sı çok pahalı** 'that part of the city is very expensive'; **şehr-in nere-si-n-de oturuyorsun?** 'in what part of the city are you living?'; **bura-sı-n-da** 'in this part of it'.

orası and **şurası** may mean 'that/this aspect of the matter under discussion': **ne istiyorlar senden?—orasını bilmiyorum** 'what do they want of you?'—'I don't know that part of it'; **şurası da var** 'there is the following point about it too'; **şurasını unutmıyalım** 'let us not forget the following fact'.

buraca, oraca (§ 2) mean 'on the part of this/that place, institution', etc.: **buraca verilecek bilgi yok** 'there is no information to be given by-this-department'; **Senato'ya arzolunan fakülte kararı, oraca kabul edilmiştir** 'the faculty decision submitted to the Senate has been accepted by-that-body'.

The diminutive of **şurada** is quite common: **şuracıkta** 'just over there'; **buracıkta** and **oracıkta** are rarer.

In the plural: **oraları gezdiniz mi?** 'have you toured those parts?'; **buraların yabancısıyım** 'I am a stranger in these parts' (lit. 'the stranger of these parts'); **buralarda otel var mı?** 'is there an hotel hereabouts?'

With **-li**: **buralı değilim** 'I am not a native of this place'; **nerelisiniz?** 'of what place are you a native?'

neredeyse or **nerdeyse** (i.e. **nerede ise**, lit. 'wherever it is'), means 'soon' or 'almost'.

Ahmet nerede, Mehmet nerede (lit. 'where is A., where is M.?') means 'how can you compare Ahmet and Mehmet?' Cf. the colloquial use of a single **nerede** for 'far from it!', 'not likely!'

13. Adverbs of time.

(*a*) Nouns used as adverbs of time usually appear in the absolute case: **ne zaman oldu?** 'when did it happen?' ('what time?' in the broad sense, not 'at what o'clock?', for which see the next section); **o zaman oldu** 'it happened then'; **bugün geldi** 'he came today'; **yarın akşam gelecek** 'he will come tomorrow evening'; **geçen yıl** 'last year'; **dün sabah** 'yesterday morning'; **bir gün** 'one day' (but **günün birinde** 'some day'); **sabah sabah** 'early in the morning'; **sabah akşam** 'all day long' (lit. 'morning evening'); **o saat** 'straight away' (see I, 16).

Izafet groups in the absolute case: **akşamüstü**, less commonly **akşamüzeri**, means 'at sunset', lit. 'evening-top', i.e. 'on evening'; cf. **yemeküstü** 'at dinner-time', **suçüstü** 'redhanded' ('on guilt'). Names of days of the week are mostly used in izafet with **gün** 'day': **çarşamba günü geldi** 'he came on Wednesday'.

(b) Some adverbs consist in plural nouns with the third-person suffix, which has the defining function mentioned in II, 22: **akşamları** 'in the evenings, of an evening'; **geceleri** 'by night'; **sabahları** 'of a morning'. The adverbs **sonra** 'after' and **önce** 'before' are similarly treated: **sonraları** 'afterwards'; **önceleri** 'previously'.

(c) The locative occurs in, for example, **ilkbaharda** 'in spring', **sonbaharda** 'in autumn'; **bu esnada, o esnada** 'meanwhile'; **bu/o sırada** 'at this/that time'; **geçenlerde** 'recently'; **simdilerde** 'round about now'; **bayramda** 'at the festival'. With names of months: **haziranda** 'in June'; **haziran ayında** 'in the month of June'; **yirmi altı temmuzda** 'on 26 July'. With years: **1453 te** (**bin dört yüz elli üçte**) 'in 1453'; **1453 yılında** or **senesinde** 'in the year 1453'.

(d) The dative: **haftaya görüşürüz** 'we'll meet next week'; **akşama gelir** 'he'll come this evening'.

(e) **çoktan** ('from much') means 'for a long time, long since', but the meaning of other adverbs formed in the ablative case is not so readily apparent: **bugünden yarını düşünmeli** means 'one should think about tomorrow today', not 'from today onward' but 'from the standpoint of today'. Similarly, **şimdiden** means 'already now', not 'from now'; **eskiden** 'in the old days'; **önceden** 'at first'; **sonradan** 'subsequently'.

(f) The old instrumental case appears in: **yazın** 'in summer'; **kışın** 'in winter'; **güzün** 'in autumn' (for which **sonbaharda** is much commoner); **gündüzün** 'in the daytime'; **ilkin** 'first of all'; **dönüşün** 'on the return journey'. It is also the last element in the invariable suffix **-leyin**: **sabahleyin** 'in the morning', **akşamleyin** 'at evening' (see p. 205, *Addendum*).

14. Telling the time.

(a) saat kaç?	what is the time?
saat bir	one o'clock
saat biri beş geçiyor	five past one ('five is passing hour one')
saat biri çeyrek geçiyor	quarter past one
saat biri yirmi beş geçiyor	five-and-twenty past one
saat bir buçuk	half past one ('hour one and a half')
saat ikiye yirmi beş var	five-and-twenty to two ('there are twenty-five to hour two')
saat ikiye çeyrek var	quarter to two
(b) saat kaçta?	at what time?
saat birde	at one o'clock
saat biri beş geçe (XI, 2)	at five past one
saat biri çeyrek geçe	at a quarter past one
saat bir buçukta	at half past one
saat ikiye yirmi beş kala	at five-and-twenty to two
saat ikiye çeyrek kala	at a quarter to two

saat may be omitted in the answers though not in the questions: **bir buçuk** 'half past one'; **bir buçukta** 'at half past one'.

15. ertesi. The bare **erte** 'the morrow' is no longer used, but the form with the third-person suffix appears in izafet in, for example, **bayram ertesi** 'the day after the festival', **savaş ertesi** 'the day after the battle' and, somewhat abraded, in **cumartesi** 'Saturday' (< **cuma ertesi** 'the morrow of Friday') and **pazartesi** 'Monday' (**pazar** 'Sunday'). As an adjective, **ertesi** means 'the following' and occurs in such adverbial phrases as **ertesi gün** '(on) the following day', **ertesi ay** '(in) the following month'. **ertesi ve daha ertesi günler** '(on) the two following days'.

16. evvelsi, evvelki. evvelsi 'previous' is for an earlier **evvel--i-si** (for the doubled suffix cf. V, 7). It is far commoner than the synonymous **evvelki**. Despite its third-person suffixes, it is used only adjectivally: **evvelsi gün** 'the previous day, the day before yesterday', **evvelsi yıl** 'the previous year, the year before last'. The spellings **evelsi, evelki**, are common but not recommended.

17. evvel and **sonra**, 'before' and 'after', both used as postpositions with the ablative, are also adverbs: **bir hafta evvel oldu** 'it happened a week ago'; **üç gün sonra gelecek** 'he will come three days later, three days from now'; **on seneden az bir zaman evvel** '[a time] less than ten years ago'. Note **bir hafta evvel-i-n-e kadar**, 'until a week ago'; **on sene evveline kadar** 'until ten years ago'.

18. şimdi 'now'. Colloquially it may take the diminutive suffix: **şimdicik**, also **şimcik** and **şimdik**, 'just now, right away'. In the dative: **şimdiye kadar** 'until now'. In the ablative: **şimdiden sonra** 'after now, henceforth'. Colloquial in the same sense is **şimdengeri**. For **şimdiden** see also § 13 (*e*); a colloquial variant is **şimden**.

19. artık, bundan böyle, gayrı. artık means 'at last' or 'henceforth', i.e. it marks a turning-point: **artık gidelim** 'that's enough of that; let's go'; **artık yaz geldi** 'summer has come at last'; **bıktım artık** '⟨I've stood it long enough and⟩ now I'm fed up'; **artık bunu yapmaz** 'he won't do this any more'.

Synonymous in the sense of 'henceforth' are **bundan böyle** and the provincial **gayrı**: **bundan böyle oraya gitmem** 'I'm not going there any more'; **insaf gayrı!** (lit. 'justice henceforth') 'it's time we had a bit of fair play!'

20. daha 'still, more, (not) yet' partly overlaps the senses of **hâlâ** (A) 'still, (not) yet' and **henüz** (P) 'just, (not) yet': **daha** (or **hâlâ**) **burada mısın** 'are you still here?'; **hâlâ daha burda mısın** (coll.) 'are you *still* here?'; **daha** (or **hâlâ** or **henüz**) **gitmedi mi** 'hasn't he gone yet?'; **daha okuyacak bir şey kalmadı** 'there is nothing more left to read'; **daha dün geldi** 'he came only yesterday'; **mektubu henüz aldım, daha** (or **hâlâ**) **okuyamadım** 'I have just received the letter; I haven't yet been able to read it'; **daha bir hafta bekledik** 'we have already waited a week'; **daha bir hafta beklemeliyiz** 'we must still wait a week'; **bir hafta daha beklemeliyiz** 'we must wait one more week'.

bir daha in negative sentences means 'no more, not again': **oraya bir daha gitme** 'don't go there any more'.

21. **hemen** (P) 'at once, just about': **güneş doğunca hemen yola çıktılar** 'the sun having risen, they at once set out'; **hemen o sıralarda tanışmıştık** 'we had become acquainted just about that time'.
hemencek and **hemencecik** (both coll.) 'instantly': **beni görünce hemencecik duvardan atladı** 'seeing me, he instantly leaped over the wall'.
hemen hemen 'almost, very soon': **hemen hemen iki yıl oldu** 'it has been almost two years' (lit. 'almost two years have happened'); **muamma hemen hemen halledilecek** 'the riddle will very soon be solved'. Cf. **neredeyse**, § 12.

22. **gene, yine** 'again, still': **gene o adam** 'it's that man again'; **hiç telefon etmedi, gene iyi; konuşacak vaktim yok** 'he has not telephoned at all; still, that's all right; I have no time to talk'.

23. The verb 'to be' in temporal expressions: **elli seneden fazladır şu evde oturuyor** 'she has been living in that house for over fifty years' ('it is more than fifty years she is living . . .'); **bir buçuk saattir seni arıyorum** 'I have been looking for you for an hour and a half'; **kaç zamandır konuşmadık** 'we haven't talked for quite some time' ('how much time it is we have not talked'); **iki yıl oluyor Paris'e gitti** 'getting on for two years ago he went to Paris' ('two years are coming into being he went . . .'); **Ahmet köyden ayrılalı** (XI, 10) **altı ay oldu** 'it has been six months since Ahmet left the village' ('six months have happened since . . .').
The adverb **bıldır** 'a year ago, a year before' (which is not very common in writing) appears to be a phonetic simplification of **bir yıl-dır** 'it is a year'.[1]

24. **derken** (i.e. the aorist of **demek**+**-ken**, lit. 'while saying') is used for:

(*a*) 'while everyone is saying . . .': **yeni yol bitti bitiyor derken hâlâ bitmedi** 'while everyone is saying the new road is just about finished (lit. "has finished is finishing"), it still is not finished'.

---

[1] For an alternative etymology, see Elöve, p. 89.

(b) 'at that precise moment': **sokağa çıkıyordu, derken telefon çaldı** 'he was going out and at that precise moment the telephone rang'.

(c) 'while attempting to', with a first-person singular subjunctive: **onu kurtarayım derken ben de düştüm** 'while attempting to save him, I fell too' (lit. 'while saying "let me save him" ...').

Addendum to § *13.* The same suffix is seen in **-cileyin**, meaning '-like'. Though otherwise obsolete, it survives in **bencileyin** and **bizcileyin**, sometimes used (especially by Yakup Kadri Karaosmanoğlu) in mock humility for 'like poor old me' and 'like us ordinary mortals' respectively.

XIII

CONJUNCTIONS AND PARTICLES

1. ve (A) 'and' is little used in speech and many Turks contrive to dispense with it entirely in writing, employing the native resources of the language instead: **ile, -ip, de** (see next section), or simple juxtaposition.[1] That is to say, co-ordinate words and clauses may be put one after the other with no conjunction at all, on the pattern of Caesar's *veni vidi vici*. Thus 'he came into the room and sat down on a chair' may be **odaya girdi ve bir sandalyeye oturdu** or **odaya girip bir sandalyeye oturdu** or **odaya girdi, bir sandalyeye oturdu**. 'You and I' may be **sen ve ben** or **seninle ben**. 'You, I, and your brother' may be **sen, ben, ve kardeşin** or **sen, ben, kardeşin de** or just **sen, ben, kardeşin**.

The Persian for 'and', *ō*, came into Ottoman as **u** or **ü** after consonants, **vu** or **vü** after vowels, forms which survive in some compound nouns: **abuhava** 'climate' (*āb* (P) 'water', *hawā* (A) 'air'); **hercümerc** 'turmoil, Armageddon' (*harj wa-marj* (A), with Persian *ō* replacing Arabic *wa*). Similar Ottoman expressions which have not attained the status of Turkish words are now written separately: **yar ü ağyar** 'friend and foe'; **kaza vü kader** 'fate and destiny'.

2. de 'and, also, too' never begins a sentence and, though written as a separate word, is enclitic and changes to **da** after back-vowels. It also changes to **te** or **ta** after unvoiced consonants, although the modern tendency, supported by YİK, is not to show this change in writing.

(*a*) When it means 'too', it follows the word it modifies: **oraya ben de gittim** 'I too went there'; **ben oraya da gittim** 'I went there too' (as well as elsewhere); **ben oraya gittim de** 'I *went* there too' (I did not only read about it); **şapka-n-ı, palto-n-u da giy** 'wear your hat and your coat too'.

[1] Cf. Ediskun, p. 315: 'Ve bağlacının görevi, virgülün görevine benzer', 'the function of the conjunction **ve** resembles that of the comma'.

(b) Repeated, it equals 'both ... and ...': **şapkanı da paltonu da giy** 'wear both your hat and your coat'; **ben de sen de kardeşin de** 'both I and you and your brother'.

(c) When it comes between two words which it connects, it can usually be translated 'and', but often it has an overtone: 'and then', 'and so', or even 'but': **çalışmış da kazanmış** 'he worked and he won'; **evime kadar gideyim de size geleyim** 'let me go as far as my house and then come to you'; **bizi gördü de selâm vermedi** 'he saw us but did not give us greeting'; **nasıl oldu da seçilmediniz?** 'how was it that you were not chosen?' ('how did it happen and so you were not chosen?'); **ne yaptı da kurtuldu?** 'how did he manage to escape?' ('what did he do and so was saved?'); **ne iyi ettin de geldin** 'how well you did to come!' ('... and came'); **söyle de gelsin** 'tell him to come' ('say, and so let him come'); **ölür de söylemez** 'he will die rather than tell' ('he will die and will not tell').

(d) A common elliptical use is seen in: **niçin sormadın?— utandım da ...** 'why didn't you ask?'—'I was ashamed, that's why', lit. 'I was ashamed and ⟨therefore did not ask⟩'. A fuller form is: **utandım da ondan** 'I was ashamed and therefore ...' (lit. 'and from-that').

(e) After a repeated verb, **de** indicates sudden action after a long delay: **misafir gelmez gelmez de, birden gelir** 'the guest does not come, does not come, and suddenly he comes'; **durdu durdu da, turnayı gözünden vurdu** 'he stood, stood, then shot the crane in the eye' (i.e. after a long spell of apparent indecision he acted with great speed and efficiency).

(f) It has an emphasizing function after pronouns and adverbs: **ikisi de** 'both of them'; **üçümüz de** 'all three of us'; **sus, sen de!** 'quiet, you!'; **ne de güzel şey!** '*what* a pretty thing!'; **o filim çok da güzel imiş** 'that film is said to be *very* good'; **bu söylenti hiç de doğru değil** 'this rumour is not at all true'; **durum, daha da ağırlaşmıştır** 'the position has become even more serious'.

(g) **dahi** 'too, also', from which **de** is derived, is seldom used by the younger generation of writers.

3. ne ... ne ... or ne ... ne de ... 'neither ... nor ...'. The number of **ne**s is not restricted to two. Whether to use a positive or a negative verb with **ne** is to some extent a matter of taste.[1] The following rules sum up the general literary usage.

The verb is positive:

(*a*) When each **ne** introduces a separate verb or separate clause: **kitabı ne aldım ne de okudum** 'I neither bought the book nor read it'; **ne tiyatroya gider ne radyoyu dinler** 'he neither goes to the theatre nor listens to the radio'.

(*b*) When one verb, in the non-initial position, covers both or all clauses: **bu sabah ne çay ne kahve içtim** 'this morning I drank neither tea nor coffee'; **ne şiş yansın ne kebap** 'let neither the spit burn nor the meat' (i.e. I hope no harm comes to either party); **ne Türkçe, ne Arapça, ne Farsça biliyor** 'he knows neither Turkish nor Arabic nor Persian'.

The verb is negative:

(*a*) When one verb covers and precedes both or all the elements introduced by **ne**: **gelmez ne dost ne düşman** 'there does not come either friend or foe'; **bugün çıkmadım ne bahçeye ne sokağa** 'today I have not gone out either to the garden or to the street'.

(*b*) When it is conditional: **ne sen, ne ben bu işe karışmasaydık böyle olmazdı** 'if neither you nor I had interfered in this business, it would not be like this'. For the first-person plural verb see XVI, 3 (*d*).

(*c*) If the negative nature of the sentence is emphasized by an adverb or particle: **ne tütüne, ne içkiye sakın alışmayın** 'mind you don't become accustomed to tobacco or drink' (lit. 'beware do not').

(*d*) If the subjects or complements introduced by **ne** are resumed by another subject or complement before the verb: **ne İstanbul'a ne Konya'ya, bir yere gitmiyor** 'neither to Istanbul nor Konya, he does not go anywhere'; **ne sen, ne o, ikiniz de bilmediniz** 'neither you nor he, both of you did not know'.

(*e*) If the verbal element is a gerund other than **-ip, -erek**, or **iken**: **ne memlekette konuşulan dili, ne oranın âdetlerini bilmediğinden çok zahmet çekti** 'because he did not know either the language spoken in the country or the customs of that place, he had a lot of trouble'; **ondan ne bir selâm ne bir mektup almadıkça adını bile anmıyacağım** 'so long as I do

---

[1] For a full discussion see Elöve, pp. 645–55, footnote.

not receive from him either a greeting or a letter, I shall not mention even his name'.

(*f*) If a number of words or a pause intervene between the second **ne** and the verb, so that the negative nature of the sentence needs to be reasserted: **ne şapka almak, ne de şapkasız gezmek—bilhassa kış aylarında—istemiyorum** lit. 'neither to buy a hat nor to go about hatless—especially in the winter months—I do not want'; **bu sabah ne çay, ne kahve . . . içmedim** 'this morning neither tea nor coffee . . . I did not drink'.

4. **gerek . . . gerek . . .** or **gerek . . . gerekse . . .** 'both . . . and . . .': **bu haber, gerek Ankara'da gerekse Vaşington'da fena bir hava yaratmıştır** 'this news has created a bad atmosphere both in Ankara and in Washington'. In some contexts the translation 'whether . . . or . . .' is possible but may be misleading, as the words introduced by **gerek** are not mutually exclusive, e.g. **gerek ben gideyim, gerek siz gidin, gerek o gitsin, işin sonu değişmez** 'whether I go or you go or he goes, the end of the affair will not change' (lit. 'both let me go and you go and let him go . . .').

5. **hem . . . hem . . .** or **hem . . . hem de . . .** 'both . . . and . . .': **hem ziyaret hem ticaret** 'both pilgrimage and trade' (a proverbial expression, cf. our 'combining business and pleasure').

A single **hem** or **hem de** means 'and indeed, moreover': **sıcak, hem ne sıcak** or **hem de ne sıcak** 'it's hot, and how hot!'

6. **ha . . . ha . . .** 'both . . . and . . .': **ha bağ, ha bahçe, ha tarla** 'both orchard and garden and field'.

7. **ister . . . ister . . .** '(either . . .) or . . .'. In origin, **ister** is the aorist participle of **iste-** 'to want'; its use is not confined to the third person: **ister git, ister kal, bana ne?** 'go or stay; what is it to me?'; **ister gitsin, ister kalsın, bana ne?** 'let him go or let him stay; what is it to me?'

8. **. . . olsun . . . olsun** 'both . . . and . . .', 'whether . . . or . . .'. This is the third-person imperative of **ol-**, lit. 'let it be', repeated: **lokanta olsun, otel olsun, her şey var orada** 'let it be restaurants, let it be hotels, there's everything there'.

A single **olsun** means 'if only', as in: **yüzünü bir kere olsun görmek istiyorum** 'I want to see his face, if only once' (lit. 'let it be one time'); **bir dakika olsun istirahat edelim** 'let us rest, if only for a minute'.

9. **ya . . . ya . . . veya . . .** 'either . . . or . . . or . . .': **ya ben, ya sen, veya Mehmet** 'either you, or I, or Mehmet'. The third choice can be introduced by **ya da** (occasionally written **yada**) or **yahut** (P) instead of **veya**.

veya, yahut, and **veyahut** all mean 'or': **elma veya** (or **yahut** or **veyahut**) **şeftali, ne istersen al** 'apples or peaches, buy whatever you want'. **yahut** is decreasingly used in this sense, but is current in the sense of 'or indeed', offering a total change of plan: **bu akşam bize gelin yahut biz size gelelim** 'come to us this evening—or let us come to you'; **bu mektubu postaya ver, yahut dursun, ben kendim götürürüm** 'post this letter—or let it stay, I shall take it myself'. Cf. the use of **yoksa** in § 34. **veyahut** is distinctly old-fashioned.

10. **ama, fakat, lâkin** 'but'. All three words are Arabic in origin, but **ama**, being the least alien in shape and having many idiomatic uses, is the most assured of survival: **sen de gel, ama gel** 'you come, but come!' (i.e. be sure to come). **güzel ama!** is a slightly surprised 'it's good, mind you!'

At the end of a sentence it may convey a slight reproof: **bu söz söylenmez ama!** 'one does not say this, though!'

For emphasis it reverts to its original form **amma**, with the second a long: **amma** (or **amma da**) **yaptın ha!** 'now you've done it!'

It is sometimes preceded, sometimes followed, by a comma; the latter if it introduces a change of subject or if for any other reason there is a slight pause after it: **kız güzel, ama bencil** 'the girl is beautiful but selfish'; **yarın gelmek istiyor ama, ben evde bulunmıyacağım** 'he wants to come tomorrow but I shall not be at home'; **ben de ona yardım ettim ama, pişman oldum** 'I too helped him—but I repented ⟨of it⟩'.

Between simple adversative clauses, 'but' need not be expressed at all (cf. **ve**): **yalancının evi yanmış, kimse inanmamış** (proverb) 'the liar's house burned, ⟨but⟩ no one believed'.

The purist expression for 'but' is **ne var ki**.

11. ancak, yalnız 'only'. Both, like English 'only', are originally adverbs but also have an adversative use: **ancak iki buçuk liram var** 'I have only two and a half liras'; **yalnız şunu demek istiyorum** 'I want to say only this'; **kütüphanede çalışmam lâzım, ancak bugün gidemem** 'I have to work in the library, only I cannot go today'; **geldi, yalnız biraz geç kaldı** 'he came, only he was a bit late'.

12. mamafih (maamafih), bununla beraber, bununla birlikte 'however, nevertheless'. The first is from *maʻa mā fīh* (A) 'with what is in it' and the first two syllables are long, despite the modern spelling with a single **a**; the **i**, however, is short and the final **h** is often omitted in pronunciation. The other two equivalents mean literally 'together with this'.

13. madem, mademki, değil mi, değil mi ki 'since'. The **a** of **madem(ki)** is long < *mā dām* (A) 'as long as'; for the **ki** see § 15. **madem(ki) anlamıyorsun, niçin karışıyorsun?** 'since you do not understand, why do you interfere?'

değil mi (ki) is a provincialism now being groomed to succeed **madem(ki)**: **değil mi** or **değil mi ki dediğimi yapmadın, yüzüme bakma** 'since you have not done what I said, do not look at my face'.

14. meğer, meğerse 'it seems that, apparently'. This introduces inferences and is consequently used with an inferential verb: **ben de seni arkadaş sanırdım; meğerse aldanmışım** 'I thought you a friend; it seems I have been deceived'; **meğer ne kadar seviyormuşum bu kızı ... yanımda iken neye anlamamışım?** 'I realize how much I love this girl; why did I not understand when she was by my side?'

It may end a sentence: **çarşıya çıkmış meğer** 'he has gone to the market, apparently'.

A colloquial alternative is **meğerleyim**.

15. ki 'that'. The importation of this Persian conjunction opened the door to the Indo-European pattern of sentence, which is in many respects the reverse of the native Turkish literary pattern:

yarın geleceğine eminim }   'I am sure he will come to-
eminim ki yarın gelecek }   morrow'

geleceği şüphesiz şüphesiz ki gelecek	'it is indubitable that he will come'
beklemesini istiyorum istiyorum ki beklesin	'I want him to wait'
kapıyı kapamıyan bir çocuk bir çocuk ki kapıyı kapamaz	'a child who does not shut the door'

In this last example, the only one from which the **ki** cannot be omitted, even in the roughest colloquial, **ki** looks like a relative pronoun. It is possible that such uses may have been helped to gain currency by the resemblance between **ki** and the Turkish interrogative pronoun **kim**; indeed, **kim** is a very ancient alternative for **ki**, still occasionally heard and, in the written language, surviving in **nitekim** (§ 17). Grammatically, however, **ki** is purely a conjunction. This is not a pedantic question of nomenclature but is of practical importance in translation, particularly of sentences like the following: **kirazı yedim** 'I ate the cherry' **ki şeker gibi** *not* 'which was like sugar' *but* 'and found it was like sugar'. This will be clearer if we consider such uses as **baktım ki**, lit. 'I looked that', but to be translated 'I looked and saw that', 'I looked and behold!': **baktım ki, kapı açık** 'I looked and saw that the door was open'; **geldim ki, kimseler yok** 'I came and found there was no one there'; **çantamı açtım ki, bomboş** 'I opened my bag and found it absolutely empty'.

Although the attachment of **ki** has become habitual in such sentences, it is not essential. Ediskun[1] gives an example of the use of **ki**—**biberi dilime değdirdim ki zehir gibi** 'I let the pepper touch my tongue and found it was like poison'—which he glosses by simply putting a comma in place of the **ki**.

ki cannot be omitted:

(*a*) When it introduces a relative clause of the Indo-European type, with a finite verb, following the qualified word instead of the Turkish type with a participle preceding it.

(*b*) When it introduces a noun clause which is the subject of a preceding verb: **görülüyor ki, bu karar haksızdır** 'it is evident that this decision is wrong'; **anlaşılıyor ki, yeni yol haziran ayında açılacaktır** 'it is understood that the new road will be opened in June'.

---

[1] p. 386.

This situation often occurs when a sentence begins with an introductory adverbial clause such as **bu sebeptendir ki** 'it is for this reason that . . .'. Many such expressions contain a postposition: **bundan dolayıdır ki, plân tatbik edilemedi** 'it is on account of this that it has not been possible to apply the plan'; **İslâmiyetin kabul-i-yle-dir ki, Türk dili üzerine bir taraftan Arapça, öbür taraftan Farsça etki yapmaya başlamıştır** 'it is with the acceptance of Islam that Arabic on the one hand, Persian on the other, began to influence the Turkish language'.

(*c*) When it links two sentences of which the first indicates the time at which the action of the second occurs: **güneş batmıştı ki köye vardık** 'the sun had set when we reached the village'; **telefonu kapayarak yerime henüz dönmüştüm ki, kapı zilinin üstüste birkaç kere çalındığını duydum** 'putting down the telephone, I had just returned to my seat when I heard the doorbell ring several times in quick succession'.

The link between **ki** and the preceding word is very close (in accent it is enclitic and some writers habitually put a comma after it), whereas it may be separated by a word or phrase from the clause it introduces:[1] **bu dil kalkmalı ki ortadan, başlıyabilelim türkçe düşünmeğe** 'this language must disappear, so that we may be able to start to think in Turkish'. The writer has chosen to reverse the normal phrase for 'to disappear', **ortadan kalkmak**, but has not displaced **ki** from immediately after the verb. The sentence would be complete without it, but it serves as a warning that a consequence is coming; cf. the premonitory use of **eğer** in XX, 8. **bir yazar diyormuş ki, bir dergide geçen ay,** . . . 'it seems a writer was saying, in a journal last month, that . . .'. **daha ileri gidip diyeceğim ki size** . . . 'I shall go further and say to you that . . .'.

Parenthetic remarks are introduced by **ki**. **eğer bu mektubu okuduysan—ki okuduğuna eminim—onun ne yaptığını biliyorsundur** 'if you have read this letter—as I am sure you have—you certainly know what he is doing'. **hal öyle olsa bile—ki değil elbette—sana ne?** 'even if the situation were thus—as it certainly isn't—what is it to you?'

---

[1] Sabahattin Eyüboğlu is particularly fond of this practice; all three examples here are from his *Mavi ve Kara* (pp. 39, 78). Cf. the example beginning **o kadar sevdim ki** in the second paragraph on the next page, from Orhan Hançerlioğlu, *İnsansız Şehir*, p. 68.

Clauses expressing consequence are introduced by **ki**. **öyle zayıfladı ki bir deri bir kemik kaldı** 'he grew so thin that he remained only skin and bone'. **bir bağırdı, bir bağırdı ki yer yerinden oynadı** 'he gave such a shout, such a shout that the earth started from its place'.

The consequence, however, is frequently left to the imagination, as it is in English, except that the **ki** is retained whereas we omit the 'that': **öyle zayıfladı ki!** 'he grew so thin!' **o kadar güldük ki!** 'we laughed so much!' **bir bağırdı ki!** 'he gave such a shout!' An intermediate stage is seen in the next example, where the three dots indicate that the expression of the thought is not going to be completed and then the writer completes it after all: **o kadar sevdim ki bu şehri . . . oturup ağlayasım geldi** (IX, 9) 'I loved this city so much . . . I felt like sitting down and crying'. The vestigial **ki** is very frequent in the colloquial: **Deveye 'Neden boynun eğri?' demişler. 'Nerem doğru ki?' demiş** 'They said to the camel, "Why is your neck crooked?" Said he, "What part of me is straight, that ⟨you should single out my neck⟩?"' **'Hangi partidensiniz?' 'Memlekette birkaç parti var mı ki?'** 'Of which party are you?' 'Are there several parties in the country that ⟨you need to ask⟩?' The **ki** in both these last examples may be translated 'then'.

There are several other colloquial uses of **ki**. At the end of a question it indicates anxiety: **bu borcun altından kalkabilir mi ki?** 'can he ever rise from-under this debt?' Between repeated words it shows admiration or surprise: **vapur ki vapur!** 'such a fine ship!' **okudu ki okudu!** 'my goodness how he studied!' **bilmem ki** means 'I wonder' (not 'I don't know that . . .'): **bilmem ki ne yapsam?** 'I wonder what I should do?' **bilmem ki kime şikâyet etsem?** 'I wonder who to complain to?'

For **ki** with the subjunctive, see Chapter XIX.

16. meğerki 'unless' is followed by the subjunctive: **ümidimiz yok, meğerki hükümet müdahale etsin** 'we have no hope, unless the government intervene'; **vapura yetişmiyeceksin, meğerki koşasın** 'you will not catch the steamer unless you run'.

17. nitekim, netekim 'just so, just as' introduces the second clause of a comparison: **ben hata yaptım, nitekim siz de hata yaptınız** 'I made a mistake, just as you made a mistake'.

Even when it begins a sentence, it refers not forward but back to the preceding sentence: **Dünkü toplantıda bulunmadım. Nitekim yarınki toplantıda bulunmak niyetinde değilim** 'I was not at yesterday's meeting. In just the same way, I do not intend to be present at tomorrow's meeting'.

The synonymous[1] **nasıl ki** is used in the same way, but may also introduce the first clause of a comparison, often with an **öyle** 'thus' in the second clause: **nasıl ki ben hata yaptım, siz de öyle hata yaptınız** 'just as I made a mistake, so did you too make a mistake'.

18. **halbuki, oysa(ki)** (accented on the **u** and **o** respectively) 'whereas', 'though'. These represent the backward- not the forward-looking 'whereas' or 'though', even when they begin a sentence, in which case the correct translation is 'Yet' or 'But': **bana gücenmiş, halbuki aramızda bir şey geçtiğini hatırlamıyorum** 'I gather he is vexed with me, though I do not recall that anything has passed between us'. **Halk, çok defa softayı idealistle karıştırır. Oysaki softa, idealistin tam tersidir** 'The people often confuse the bigot with the idealist. Yet the bigot is the exact opposite of the idealist'.

19. **çünkü, zira** 'for'. These Persian borrowings almost always begin a sentence but, like the English 'for', always explain the preceding statement (cf. **nitekim, halbuki**): **Dillerin doğuşu demek, kelimenin doğuşu demektir. Çünkü, her dilin en küçük birliği kelime'dir**[2] 'The origin of languages means the origin of the word ("to say 'the origin of languages' is to say . . ."). For the smallest unit of every language is the word'. **'lâkin' bağlacı, Eski Türkçede yoktur; çünkü Arapça asıllıdır** 'the conjunction **lâkin** does not exist in Old Turkish, because it is of Arabic origin'.

zira could replace **çünkü** in these examples, but is not much used. The spelling **çünki** for **çünkü** is not recommended.

20. **demek** 'it means': **Sene yıl demektir. Senevi de yıllık demek olacak** '**Sene** means "year". So **senevi** will mean ("will be to say") "annual"' (**senevi** < *sanawi* (A)).

[1] **nite** being an old word for 'how?', now replaced by **nasıl** except in **nitekim** and the neologism **nitelik** 'quality'. The spelling **netekim**, though less common, is preferred by *YİK*.

[2] For the function of the apostrophe here, see I, 2 (*b*).

A fuller form of the expression is seen in the first example of the preceding section and in: **demokrasi demek adalet demektir** 'to say "democracy" is to say "justice"'.

At the beginning of a clause **demek, demek ki,** or **demek oluyor ki** ('it becomes to say') signifies 'that is to say': **düşünüyorum, demek ki varım** 'I think, that is to say, I exist'.

demek alone can be used when seeking corroboration of an impression: **paranız yok demek?** 'so you have no money?' **demek o da geliyor?** 'that means he is coming too?'

21. **diğer taraftan** (P, A) and **öte yandan**, lit. 'from the other side', are not as adversative as they look; 'moreover' or 'at the same time' is usually the best rendering, not 'on the other hand'.

22. **gerçi** (P) 'it is true that': **gerçi pek sevimli değil, ama gayet iyi bir arkadaştır** 'it is true he is not very attractive, but he is a very good friend'.

23. **gûya** (P) 'allegedly, forsooth'; usually pronounced and sometimes written **göya**. For an example see XXIV, 12.

24. **hani**, an old word for 'where?', has several idiomatic uses. It may be reinforced by **ya**, written together with it or as a separate word.

(*a*) It asks the whereabouts of someone or something expected but not in evidence, or lost and unattainable: **haniya arkadaşınız, gelmedi mi?** 'where is your friend; hasn't he come?' **hani o günler!** 'where are those ⟨good old⟩ days!'

(*b*) It draws attention to a failure to carry out a promise: **hani ya bana bir hediye getirecektin?** 'I thought you were going to bring me a present?'

(*c*) 'You know' covers most other uses: **hani büronuzda esmer çocuk var ya, işte o sizi arıyor** 'you know there's the dark boy in your office; well it's he who is looking for you' (for **işte** see § 28); **hani yanlış da değil** 'and it's not wrong, you know'.

(*d*) A parenthetic **hani yok mu** lends weight to the following words: **bu problemi çözmek için, hani yok mu, tam üç gün çalıştım** 'to solve this problem, would you believe it, I worked exactly three days'.

(e) **hanidir** (for the syntax see XII, 23) 'for a long time now': **hanidir onu görmüyorum** 'I haven't seen her for ages'.

25. **hele** 'above all, at any rate'; with an imperative, 'just': **hele siz bunu söylememeli idiniz** 'you above all should not have said this'; **hele insan kaybı yokmuş** 'at any rate there is no loss of life reported' ('there-is-said-to-be-no human-loss'); **şuna bak hele** 'just look at that!'; **çocuk sınıfını geçmesin hele, döğerim onu** 'just let the child not pass up ("let-him-not-pass his-class"), I'll give him a good hiding'.

A doubled **hele hele** is hortatory: **hele hele söyle daha neler olmuş** 'come on then, tell what else happened!'

26. **herhalde** 'certainly, surely'. The literal translation is 'in every case', but this gives a misleading idea of the use. In the first two examples it indicates a strong supposition: **herhalde biliyorsunuz** 'you surely know'; **gazetede okumuşsunuzdur herhalde** 'you have certainly read ⟨it⟩ in the newspaper'.

ister darılsın, ister darılmasın, herhalde hakkımı istiyeceğim 'let him get cross or not, as he chooses, I shall certainly ask for my due'.

27. **ise** ('if it is') and its suffixed forms (VIII, 8) draw attention to the preceding word and may be translated 'as for', 'whereas', 'however' according to context: **ben ise (bense) patates hiç yemiyorum** 'as for me, I don't eat potatoes at all'; lit. 'if it is I ⟨about whom you are asking⟩...'; **babası İngiliz, annesi ise (annesiyse) Amerikalı** 'his father is English, whereas his mother is American'.

The word before **ise** may be in whatever case the syntax of the rest of the sentence demands: **kendisi pek hoş, sesini ise hiç sevmem** 'he himself is very pleasant; his voice, however, I don't like at all'. Here **sesini** is accusative, object of **sevmem**. For **ise** following the genitive, see the fifth example in XVI, 6.

28. **işte** 'behold!' 'there!' 'precisely': **hani benim kalem?— işte!** 'where is my pen?'—'there!'; **işte otobüs geldi** 'there, the bus has arrived'; **işte bu sebepten dolayı ona oy vermedim** 'precisely for this reason I did not vote for him'.

29. sakın. In origin it is the imperative of **sakınmak** 'to be cautious'. As an interjection, **sakın** or **sakın ha** means 'beware! don't do it!' It is also used with a negative imperative: **sakın düşme!** 'mind you don't fall!'

With the periphrastic perfect tense of the negative imperative it expresses anxiety: **sakın unutmuş olmayın** 'I do hope you have not forgotten' ('do-not-be having-forgotten'). A question-mark may emphasize the doubt in the speaker's mind: **söylediklerimi sakın unutmuş olmasın?** 'I do hope he hasn't forgotten what I said?'

30. sanki 'as if' (lit. 'suppose that') is usually construed with **gibi** following an inferential verb or a tense-base, or with an inferential verb alone: **sanki dünyada başka bir kadın yokmuş gibi hep Ayşe'yi düşünüyor** 'as if there were no other woman in the world, he thinks entirely of Ayesha'; **sanki bilmiyor gibisiniz!** 'as if you didn't know!'; **sanki kabahat benimmiş!** 'as if the fault were mine!'

In a question it conveys an argumentative or scornful 'do you think?': **ablanıza çok üzüntü verdiniz de iyi mi ettiniz sanki?** 'you have greatly upset your sister and have you ⟨thereby⟩ done well, do you think?'

Otherwise it is to be translated 'it is as if': **makine sabah akşam çocuğa elinin ve kafasının gücünü hesaplatıyor; sanki ona "Benimle yapacağın her işten sen sorumlusun, ben karışmam" diyor** 'the machine all day long makes the child take account of the power of his hand and head; it is as if it says to him "For every job which you are going to do with me, you are responsible; I do not interfere"'.

31. şöyle dursun (lit. 'let it stand thus'), with an infinitive as subject, means 'let alone . . ., never mind about . . .'.

radyoda dinlediğimiz şarkıların çoğu millî olmak şöyle dursun musiki bile değildir 'most of the songs we listen to on the radio are not even music, let alone national'.

tercümanlık yapmak şöyle dursun, kendi lisanını bile bilmez 'never mind about acting as interpreter, he doesn't even know his own language'.

32. ya has a wide variety of functions in the colloquial, e.g. at the end of rhetorical questions: **her akşam sinemaya gidilir mi ya?** 'does one go to the cinema every evening?'

In the sense of 'you know', 'isn't that so?': **bugün niye okula gitmedin?—pazar ya!** 'why haven't you gone to school today?' —'it's Sunday, you know!' **köşede küçük bir dükkân var ya, işte orada aldım** 'you know there's a little shop on the corner; well I bought ⟨it⟩ there'.

In the sense of 'yes indeed' the **a** is pronounced long: **size çay vereyim mi?—ya, ver** 'may I give you ⟨some⟩ tea?— 'yes, do'.

At the beginning of a clause it means 'and what about . . . ?'; this is the use most likely to be found in the written language: **bu kadar yetişir, diyorsun, ya yetişmezse?** 'this much will be enough, you say; and what if it isn't enough?' A versified slogan of the 1960 revolution ran: **Gerekirse / Ölürüz / Biz. / Ya siz?** 'If necessary / We are ready to die, / We. / And you?'

33. **yok** is used colloquially for 'no' in reply not only to questions containing **var mı** or **yok mu** (VIII, 45): **gidiyor musun?— yok** 'are you going?'—'no'. In this sense it is often pronounced without the **k** and with the vowel lengthened, and may be phonetically spelled **yooo!**

When retailing an unlikely story, a sarcastic **yok** may preface each clause: **yok kâğıdı kalmamış, yok mürekkebi iyi değilmiş, hasılı bir alay bahaneler** 'oh no he had no paper left, oh no his ink was no good; in short, a host of excuses'.

34. **yoksa** 'if not, otherwise': **uslu durursun, yoksa seni bir daha buraya getirmem** 'you'll stand nice-and-quiet, otherwise I shan't bring you here again'.

It then comes to be used to introduce the second half of double questions: **bu mümkün mü yoksa değil mi?** 'is this possible, or is it not?'; **bugün mü yoksa yarın mı gidiyorsunuz?** 'is it today or tomorrow you are going?'

When two possibilities are considered but only one is expressed, **yoksa** may begin the sentence: **yoksa gitsem mi?** 'or should I go?'

At the end of a sentence it means 'it suffices': **Allah insanın aklını başından almasın yoksa!** (Sabahattin Ali), 'Let God not take a man's wits out of his head, that's all we ask.'

XIV

WORD-FORMATION

1. Deverbal substantives. In sections 2–18 are discussed the principal suffixes which are added to verb-stems to make nouns and adjectives, excluding those dealt with under the headings of participles and verbal nouns in Chapters IX and X. These suffixes have been and are the chief weapons of the language-reformers in their campaign to substitute words from Turkish roots for Arabic and Persian borrowings.

2. -ici. Like the related **-ci** (IV, 4), this indicates regular activity:

ak- to flow	**akıcı** fluent
oku- to read	**oku-y-ucu** reader
gül-dür- to make to laugh	**güldürücü** amusing
uyuş-tur- to benumb	**uyuşturucu** narcotic
et- to make	**sarhoş edici** intoxicant ('drunk making')
öl-dür- to kill	**haşarat öldürücü ilâç** insecticide ('insect killer medicament')
dön-dür- to turn	**baş döndürücü hız** vertiginous ('head turning') speed
ver- to give	**hayat verici** life-giving

The first vowel of the suffix has been lost in **dilenci** 'beggar' from **dilen-** 'to beg'. **öğrenci** 'student, pupil' was manufactured analogously from **öğren-** 'to learn'.

3. -men. A suffix **-man** occurs in a number of time-honoured words, apparently with intensive significance; e.g. from **koca** 'big', **kocaman** 'huge'; from **şiş** 'swollen', **şişman** 'fat'.[1]

---

[1] As for **Türkmen** 'Turcoman', the author inclines to the view of Vámbéry and Ligeti, who considered the **-men** in this word to be a collective suffix. See İbrahim Kafesoğlu, '*Türkmen* adı, manası ve mahiyeti' in *Jean Deny Armağanı* (TDK 1958), pp. 121–33.

The language-reformers have used **-men** to make nouns of occupation:

öğret-	to teach	öğretmen	teacher
oku-t-	to make to read	okutman	lector
say-	to count	sayman	accountant
seç-	to choose	seçmen	elector
yaz-	to write	yazman	secretary

In this use the suffix is a hybrid, deriving, on the one hand, from the Turkish **-man** and, on the other, from the English *-man*, familiar to the Turks in four borrowings from French: **vatman** 'tram-driver', **sportmen** 'sportsman', **barmen** 'barman' and **rekortmen** 'record-holder'. The last three may have come via Russian.

eğemen 'sovereign' purports to be derived from **eğe** or **eğe** 'guardian'. In fact it is a distortion of the Greek ἡγεμών 'leader'; the French *hégémonie* was borrowed in the form **heğemonya** by Ziya Gökalp (d. 1924).

4. **-ik** makes adjectives, mostly with passive meaning, and nouns, mostly denoting the result of action:

birleş-	to unite	birleşik	united
boz-	to destroy	bozuk	unserviceable, spoilt
çık-	to come out	çıkık	dislocated
değiş-	to change	değişik	varied
öksür-	to cough	öksürük	cough
sök-	to undo	sökük	unravelled
tükür-	to spit	tükürük	saliva

5. **-i** denotes action or result of action. It occurs (neologisms aside) only with monosyllabic consonant-stems:

dol-	to be filled	dolu	full
kork-	to fear	korku	fear
koş-	to run	koşu	race
öl-	to die	ölü	dead, corpse
yap-	to make	yapı	construction

6. -ti, -inti.

(*a*) **-ti** denotes action or result of action:

belir-	to appear	belirti	symptom
buyur-ul-	to be ordered	buy(u)rultu	command
bula-n-	to be nauseated	bulantı	nausea
çalka-n-	to be agitated	çalkantı	agitation

(*b*) Analogously with the last two examples, there are a number of nouns ending in **-in-ti** from verbs with no reflexive in use, e.g.:

ak-	to flow	akıntı	stream
bur-	to twist	buruntu	colic
çık-	to come out	çıkıntı	projection
çök-	to collapse	çöküntü	debris
kur-	to brood	kuruntu	melancholy fancy

7. -ġi denotes action or its result or its instrument:

sev-	to love	sevġi[1]	affection
iç-	to drink	içki	(alcoholic) drink
çal-	to play	çalġı	musical instrument
as-	to hang	askı	pendant, braces (U.K.), suspenders (U.S.A.)
bur-	to twist	burġu	gimlet
ör-	to interlace	örġü	plait
yar-	to split	yarġı	decision

8. -ç makes adjectives and abstract nouns, primarily from reflexive stems and other stems in **n**:

iğren-	to be disgusted	iğrenç	loathsome, loathing
inan-	to believe	inanç	belief
kazan-	to win	kazanç	gain
kıskan-	to envy	kıskanç	jealousy
usan-	to be bored	usanç	boredom

By analogy, **korkunç** 'terrible' is formed from **kork-** 'to fear', although this verb has no reflexive.

[1] A cautionary word may not be out of place about **sevda** 'passionate desire', since at least two Western writers on Turkish grammar have described it as from **sev-** 'to love' with an otherwise unknown invariable suffix **-da**. It is in fact the Arabic *sawdā* 'blackness > black bile > melancholy > longing'.

9. -ek, -k (the latter after vowel-stems) makes nouns of place and instrument and nouns or adjectives with active meaning:

dur-	to stop	durak	stopping-place
yala-	to lick	yalak	trough
bat-	to sink	batak	marsh
ele-	to sift	elek	sieve
tara-	to comb	tarak	comb
ölç-	to measure	ölçek	scale of a map
yed-	to tow	yedek	tow-rope > led animal > spare
aksa-	to limp	aksak	lame
büyü-	to become great	büyük	great
soğu-	to become cold	soğuk	cold
ürk-	to shy	ürkek	timid

From **at-** 'to throw' comes **atak** 'bold, daring', not to be confused with the identical-looking noun meaning 'attack', a French borrowing used by sports-writers and military experts.

10. -gen makes adjectives with intensive meaning:

çekin-	to withdraw	çekingen	retiring
döğüş-	to fight	döğüşken	bellicose
sokul-	to push oneself in	sokulgan	ingratiating
unut-	to forget	unutkan	forgetful

The neologism for 'planet' is **gezegen** from **gez-** 'to wander'.

A rare by-form is **-eğen**:

ol-	to happen	olağan	normal
piş-	to cook (intr.)	pişeğen	easily cooking

There is another **-gen**, a made-up invariable suffix inspired by the suffix seen in *pentagon* and *hexagon*. Added to numerals it makes the names of plane figures: **üçgen** 'triangle', **altıgen** 'hexagon', **çokgen** 'polygon' and so on.

11. -gin makes nouns and adjectives with active or passive meaning:

er-	to mature	ergin	adult
kız-	to become heated	kızgın	fevered
sol-	to fade	solgun	faded
sür-	to exile	sürgün	exile
bit-	to end	bitkin	exhausted

küs-	to sulk	küskün	sulky
şaş-	to go astray	şaşkın	bewildered
coş-	to overflow	coşkun	exuberant

geçkin 'past' and aşkın 'exceeding' may govern an object: **elli yaşını geçkin bir adam** 'a man past his fiftieth year'; **boyu, iki metreyi aşkındı** 'his height was over two metres'.

12. -it, -t (the latter after vowel-stems). This noun-suffix, though not very productive in former times, is a favourite of the neologizers; witness the last five examples:

ayır-	to distinguish	ayırt	distinction
geç-	to pass	geçit	passage, ford
yoğur-	to knead	yoğurt	yoghurt
an-	to call to mind	anıt	memorial
soy-	to strip	soyut	abstract
taşı-	to carry	taşıt	vehicle
yak-	to burn	yakıt	fuel
yaz-	to write	yazıt	inscription

It occurs also in a few adjectives deliberately derived from substantives:

yaş	age	yaşıt	coeval
eş	mate	eşit	equal
karşı	opposite	karşıt	contrary

An analogous recent coinage is **somut**, from **som** 'solid', for 'concrete', as opposed to **soyut** 'abstract'.

13. -im makes nouns, many of them denoting a single action.[1] This too is an abundant source of neologisms, e.g. **basım** and the three following examples on the next page.

---

[1] Elöve (p. 507, footnote 4) criticizes Deny's calling these 'noms d'unité', on the ground that the idea of unity is not intrinsic in, for example, **içim** but emerges only when one puts **bir** before it. Deny, however, is right and his translator is wrong, not having realized that one can have more than one unité; e.g. 'two draughts of medicine', **iki içim ilâç**. 'Noun of unity' is an accepted term of Arabic grammar for, for example, *ḍarba* 'blow'; the fact that this can be put in the plural, as in 'he struck him ten blows', does not make it any the less a noun of unity.

iç-	to drink	içim	draught
yut-	to swallow	yudum	swallow, mouthful
dil-	to slice	dilim	slice, strip
doğ-	to be born	doğum	birth
öl-	to die	ölüm	death
tut-	to hold	tutum	thrift, behaviour

The noun of unity of **tut-** is irregular: **tutam** 'handful'.

bas-	to press, print	basım	printing
de-	to say	deyim	expression
dur-	to stand	durum	situation
yat-ır-	to lay, deposit	yatırım	investment

See also XII, 1; **sürüm sürüm**, etc.

Examples of deverbal nouns in **-m** from vowel-stems are few, but the reformers have created some on the analogy of **anlam**, alleged to be used in Konya in the sense of 'meaning'; cf. **anla-** 'to understand'.

gözle-	to observe	gözlem	observation
kavra-	to grasp	kavram	concept

The same dubious suffix is seen in the neologism **gündem** 'agenda' from **günde** 'in the day'.

14. -in makes nouns:

ak-	to flow	akın	stream, rush, raid
ek-	to sow	ekin	crop
tüt-	to smoke (intr.)	tütün	tobacco
yığ-	to pile up	yığın	heap
bas-	to press, print	basın	the Press

15. -ğeç, -ğıç. These two related suffixes make a few nouns, mostly denoting agent or instrument:

dal-	to plunge	dalğıç	diver
süz-	to filter	süzğeç	filter, strainer
yüz-	to swim	yüzğeç	swimmer, float
başla-	to begin	başla-n-ğıç	beginning
patla-	to explode	patla-n-ğaç, patla-n-ğıç	
		pop-gun, fire-cracker	

16. -tay. This neologism seems to have been extracted from the Mongol *quriltai* 'assembly of the nobles', as if the word were derived from the Turkish **kur-ul-** 'to be established' +*-tay. **kurultay** is the name given by the Turkish Language Society to its annual congress. The 'suffix' has been used (with substantives as well as with verb-stems) to create a number of administrative terms, all of which, except the first, are often used in official language. The older terms are given in brackets.

kamu (an archaic Persian borrowing for 'all', now a noun meaning 'the public')
Kamutay Grand National Assembly (**Büyük Millet Meclisi**)

danış- to consult
Danıştay Council of State (**Şura-yı Devlet, Devlet Şurası**)

sayış- to settle accounts
Sayıştay Exchequer and Audit (**Divan-ı Muhasebat, Muhasebat Divanı**)

yargı decision
Yargıtay Supreme Court of Appeal (**Temyiz Mahkemesi**)

17. -ev, -v. This was borrowed by the neologizers from the Tatar dialect.

gör-	to see, perform	**görev**	duty
sayla-	to choose	**saylav**	deputy, M.P.
sına-	to test	**sınav**	examination
söyle-	to tell	**söylev**	speech

18. -ey, -y. This suffix, of Chaghatai origin, is also beloved of the neologizers.

dene-	to try	**deney**	experiment
dik-	to set up	**dikey**	perpendicular
ol-	to happen	**olay**	event
yat-	to lie	**yatay**	horizontal

It is rarely added to nouns:

yüz-	face	**yüzey**	surface

19. Denominal verbs. Relatively few substantives are also verb-stems; among the commonest are:

acı	grief	acı-	to grieve (intr.)
boya	paint	boya-	to paint
ekşi	sour	ekşi-	to become sour
eski	old	eski-	to become worn out
gerek	necessary	gerek-	to be necessary
göç	migration	göç-	to migrate
kuru	dry	kuru-	to dry (intr.)

On the other hand, many verbs are formed by adding suffixes to substantives. These suffixes are discussed in §§ 20–29.

20. -e-. This is now unproductive.

boş	empty	boşa-	to divorce
harç	expenditure	harca-	to spend
kan	blood	kana-	to bleed
oyun	game	oyna-	to play
yaş	age	yaşa-	to live

21. -le-. This, with its derivatives (§§ 22–24), is the most productive of all verbal suffixes. The precise relationship between the meanings of the basic substantive and the derived verb is not always guessable; compare the last two examples in list (*a*). It is added to:

(*a*) Nouns:

balta	axe	baltala-	to sabotage
göz	eye	gözle-	to keep an eye on
kilit	lock	kilitle-	to lock
kir	dirt	kirle-	to dirty
su	water	sula-	to irrigate
yumurta	egg	yumurtala-	to lay eggs
kuzu	lamb	kuzula-	to lamb
köpek	dog	köpekle-	to cringe

(*b*) Adjectives:

küçük	small	küçükle-	to slight
serin	cool	serinle-	to become cool
temiz	clean	temizle-	to clean

üç	three	üçle-	to increase to three, to let a farm in exchange for one-third of the crop

(c) Onomatopoeic words:

hav hav	bow-wow	**havla-**	to bark
miyav	miaow	**miyavla-**	to mew
püf	puff	**püfle-**	to puff, blow out

22. -len-. In origin the reflexive and passive of **-le-**, it also makes some verbs synonymous with those in **-le-**, and some of which there is no **-le-** form in use:

temizle-	to clean	**temizlen-**	to be cleaned
kirle-	to dirty	**kirlen-**	to become dirty
serinle-	to become cool	**serinlen-**	to become cool
can	soul, life	**canlan-**	to come to life
ev	house	**evlen-**	to marry

Its causative is **-len-dir-**: **canlandır-** 'to vivify', **evlendir-** 'to give in marriage'.

23. -let-. The causative of **-le-**. Some verbs formed with it are synonymous with the forms in **-le-**:

temizle-	to clean	**temizlet-**	to get cleaned
kilitle-	to lock	**kilitlet-**	to get locked
kirle-	to dirty	**kirlet-**	to dirty

24. -leş-. Originally the reciprocal of **-le-**, it is also freely used to make verbs meaning 'to become ...':

karşıla-	to meet	**karşılaş-**	to meet one another
serin	cool	**serinleş-**	to become cool
dert	pain, trouble	**dertleş-**	to tell each other your troubles
mektup	letter	**mektuplaş-**	to correspond

Tanrı	God	tanrılaş-	to become divine
ölmez	(IX, 4) immortal	ölmezleş-	to become immortal
Amerika-lı	American	amerikalılaş-	to be Americanized
ġarp-lı	westerner	ġarplılaş-	to be westernized
bir	one	birleş-	to become united

Its causative is **-leş-tir-**: **ölmezleştir-** 'to immortalize', **ġarplılaştır-** 'to westernize', **birleştir-** 'to unite'.

25. -el-, -l-. Added to a number of adjectives, but very few nouns, it conveys 'to become . . .'. Disyllables in final **k** lose it before this suffix:

az	little	azal-	to diminish (intr.)
çok	much	çoğal-	to increase (intr.)
boş	empty	boşal-	to be emptied
sivri	sharp-pointed	sivril-	to become prominent
alçak	low	alçal-	to condescend
ufak	tiny	ufal-	to diminish (intr.)
yüksek	high	yüksel-	to rise
yön	direction	yönel-	to direct oneself

The causative is **-elt-, -lt-**: **azalt-** 'to diminish' (tr.), **yükselt-** 'to raise'.

26. -er-. No longer productive, with adjectives of colour it conveys 'to become . . .'; added to other words it usually has an active sense. Disyllabic adjectives of colour lose their final syllable before it; more accurately, **-er-** is added to the monosyllabic stem from which the disyllabic adjective is derived:

ak	white	ağar-	to become white
boz	grey	bozar-	to become grey

gök	blue, green	göğer- *or* göver- (I, 10)	to become blue, green
yeşil	green	yeşer-	to become green
kızıl	red	kızar-	to become red, be roasted
sarı	yellow	sarar-	to become yellow
ev	house	ever-	to marry off
ot	grass	otar-	to pasture
su	water	suvar- (p. 29)	to water an animal
yaş	moisture	yaşar-	to become wet

27. **-se-**. This was once not uncommon in the sense of 'to want . . .', e.g. in **tütünse-** 'to crave tobacco'. The only surviving example in common use is:

su	water	susa-	to thirst

In a few words it has the sense of 'to regard as . . .':

benim	of me	benimse-	to regard as one's own
mühim	important	mühimse-	to think important
çirkin	ugly	çirkinse-	to think ugly
ğarip	stranger	ğaripse-	to consider strange, to feel lonely

28. **-imse-**. A suffix **-imse-** with the same meaning appears in:

az	little	azımsa-	to consider inadequate
çok	much	çoğumsa-	to consider excessive

In the postvocalic form **-mse-** it is used to make two neologisms:

iyi	good	iyimse-	to be optimistic
kötü	bad	kötümse-	to be pessimistic

The aorist participles **iyimser** and **kötümser** are commonly used for 'optimistic' and 'pessimistic' respectively.

Different in sense and in being formed from a verb-stem is **gülümse-** 'to smile'; cf. **gül-** 'to laugh'. This **-imse-** may have been formed on the analogy of **benim-se-** and **mühim-se-** but is more likely related to the adjectival suffix **-imsi** (IV, 3).

29. **-de-** is added to a number of onomatopoeic words ending in **r** or **l** which, when repeated, are used as adverbs. For example, **cızır** imitates the sound of sizzling; 'sizzlingly' is **cızır cızır**, while **cızırdamak** is to make this noise. Such verbs in **-de-** have a corresponding noun in **-di** or **-ti**, thus **cızırtı** 'sizzling'. There is also a verb **cız-la-mak** 'to sizzle'. A series like this exists for many onomatopoeic words, though in some the verb in **-le-** is wanting. Where there is a verb in **-le-**, it is used with the doubled adverb in preference to the verb in **-de-**. Thus 'to snore snortingly' is **horul horul horlamak** rather than **horuldamak**.

Imitative Word	*Represents*	*Verb*	*Noun*	*Verb*
gıcır	creaking	**gıcırda-**	**gıcırtı**	—
hırıl	growling	**hırılda-**	**hırıltı**	**hırla-**
horul	snoring	**horulda-**	**horultu**	**horla-**
kütür	crunching	**kütürde-**	**kütürdü**	—
patır	footsteps	**patırda-**	**patırdı**	—[1]
parıl	glittering	**parılda-**	**parıltı**	**parla-**
pırıl	,,	**pırılda-**	**pırıltı**	—
takır	tapping	**takırda-**	**takırtı**	—

30. Compound nouns and adjectives. The various ways in which these may be formed are dealt with in this and the following sections.

Two nouns juxtaposed:

baba father   **anne** mother   **babaanne** paternal grandmother

anneanne maternal grandmother

kayın brother-in-law   **ata** father   **kaynata** father-in-law

ana mother   **kaynana** mother-in-law

---
[1] There is a verb **patlamak** but it means 'to explode, to go *bang!*'

baş head	bakan minister	başbakan prime minister
	çavuş sergeant	başçavuş sergeant-major
	parmak finger	başparmak thumb
iç interior	yüz face	içyüz 'the inside story'
	kale fort	içkale citadel
orta middle	çağ epoch	ortaçağ the Middle Ages
	okul school	ortaokul intermediate school
göz eye	kulak ear	gözkulak alert, interested
ağa chief	bey lord	ağabey elder brother
hanım lady	efendi master	hanımefendi Madam
		beyefendi Sir
yüz face	göz eye	yüzgöz over-familiar
ön front	ayak foot	önayak pioneer, ringleader
	söz word	önsöz foreword
	yargı judgement	önyargı prejudice[1]

31. Abbreviated nouns. Some military terms have been formed on the pattern of Russian officialese, from abbreviated nouns:

tümen division	general general	tümgeneral major-general
ordu army		orgeneral general (4-star)
	donatım equipment	ordonat equipment branch[2]
üst top	teğmen 2nd lieutenant	üsteğmen 1st lieutenant

32. Izafet groups:

yıl year	baş head	yılbaşı New Year
el hand	birlik oneness	elbirliği co-operation
iç interior	acı pain	içler acısı heart-rending
hanım lady	el hand	hanımeli honeysuckle
saman straw	yol way	samanyolu the Milky Way
cuma Friday	erte morrow	cumartesi Saturday
tarih history	önce before	tarih öncesi prehistoric

[1] A verb **öngörmek** 'to foresee, envisage' has been coined from **ön** and **görmek** 'to see' and is popular with journalists and politicians.

[2] The English *ordnance* contributed to this coinage.

milletler nations	**ara** interval	**milletlerarası** international
harp war	**sonra** after	**harp sonrası** post-war
su water	**üst** top	**suüstü** surface (adj.)[1]

33. Frozen izafet groups. The following words, though izafet groups in origin, are treated like simple vowel-stems (cf. II, 24). Thus **denizaltılar** 'submarines', **binbaşıya** 'to the major', **ayakkabıcı** 'shoemaker'.

deniz sea	**alt** underside	**denizaltı** submarine (noun and adj.)
bin thousand	**baş** head	**binbaşı** major
yüz hundred		**yüzbaşı** captain
on ten		**onbaşı** corporal
ayak foot	**kap** cover	**ayakkabı** footwear, shoes

34. Proper names consisting in izafet groups. Patronymics in **-oğlu** are strictly declined as izafet groups: **Osmanoğulları** 'the Ottoman dynasty', **Karamanoğluna** 'to the Karamanid'. Colloquially, however, surnames of this type are sometimes treated as simple vowel-stems; thus 'to Azizoğlu', strictly **Azizoğluna** or **Azizoğlu'na**, may occur as **Azizoğlu'ya**.

İnönü as a place-name is declined as an izafet group; as a surname (derived from the place-name) it is declined as a simple noun. Either way it may be written with or without an apostrophe before the case-endings:

	Place-name	Surname
abs.	İnönü	
acc.	İnönü'nü	İnönü'yü
gen.	İnönü'nün	
dat.	İnönü'ne	İnönü'ye
loc.	İnönü'nde	İnönü'de
abl.	İnönü'nden	İnönü'den

35. Adjective+noun:

büyük great	**anne** mother	**büyükanne** grandmother
kara black	**yel** wind	**karayel** north-wester
top-lu knobbed	**iğne** needle	**topluiğne** pin
kırk forty	**ayak** foot	**kırkayak** centipede

[1] **denizaltının suüstü sürati** 'the submarine's surface-speed'.

36. Noun+noun+-li:

cin demon	fikir thought	cinfikirli shrewd
koyun sheep	baş head	koyunbaşlı mutton-headed
orta middle	boy stature	ortaboylu of medium height

37. Adjective+noun+-li:

aç hungry	göz eye	açgözlü avaricious
alçak low	gönül soul	alçakgönüllü humble
deli mad	kan blood	delikanlı young man
iki two	can soul	ikicanlı pregnant

38. Noun+adjective:

süt milk	beyaz white	sütbeyaz milk-white
kömür coal	siyah black	kömürsiyah coal-black

39. Noun+third-person suffix+adjective:

din-i his religion	bütün whole	dinibütün devout
el-i his hand	açık open	eliaçık generous

See XVIII, 1.

40. Noun+verb:

kül ash	bastı it pressed	külbastı grilled meat
dal branch		dalbastı fine and large (of cherries)
unutma do not forget	beni me	unutmabeni forget-me-not

To this class belong the names of two dishes made with aubergines: **hünkârbeğendi** 'the Sovereign approved' and **imambayıldı** 'the Imam swooned'.

The verb may be a participle:

kervan caravan	kıran breaking	kervankıran the morning star
oyun game	bozan spoiling	oyunbozan spoilsport
yurt land	sever loving	yurtsever patriot

41. Onomatopoeic word+verb:

çıt crack!	kırıldım I have been broken	çıtkırıldım fragile, effeminate
şıp plop!	sevdi he has fallen in love	şıpsevdi susceptible, impressionable

42. Verb+verb:

çakar it strikes fire	almaz it does not take	çakaralmaz useless, not working; facetious term for gun, 'blunderbuss'
kaptı it snatched	kaçtı it fled	kaptıkaçtı small, privately owned omnibus, 'pirate bus'
vurdum I hit	duymaz he does not feel	vurdumduymaz thick-skinned

43. Hyphenated compounds. The hyphen is hardly ever used in compound words, except in one or two modernisms like **okuryazar** 'literate' ('reader–writer') and **aktör-rejisör** 'actor–producer', phrases like **Ankara-Konya yolu** 'the Ankara–Konya road' and **öğretmen-öğrenci oranı** 'teacher–pupil ratio', and names of commercial firms like **İpek-İş** 'Silk-Work'.

Some of the compounds shown above as one word may be spelt as two, and vice versa.

44. Repetitions. As in English, a verb may be repeated to indicate the duration of activity: **çalışacaksın, çalışacaksın ve muvaffak olacaksın** 'you will work, you will work, and you will be successful'.

Similar locutions are frequent in the colloquial: **kalalım kalalım akşama kadar kalalım, sonra?** 'all right, suppose we stay till evening; what do we do then?' (lit. 'let us stay, let us stay, till evening let us stay; after?'). **arabayı sürmüş, sürmüş, nihayet yetişti** 'he drove and drove the car and finally arrived'. If the object of the repeated verb is also repeated, inversion is automatic: **sürmüş arabayı, sürmüş arabayı, nihayet yetişti**. **yapacağım da yapacağım** 'I'll certainly do it' ('I shall do and I shall do'). **gitti mi gitti!** 'I'll say he went!' ('did he go? he went!').

This last construction is found with adjectives too: **ihtiyar**

zengin mi zengin! 'the old man is certainly rich' ('... rich? rich!').

When repeated adjectives qualify nouns in the plural, the sense is more than simply intensive; e.g. **güzel güzel kızlar** means not just 'very beautiful girls' but 'girls each more beautiful than the last'; **yeni yeni ümitler** are not 'very new hopes' but 'ever-new hopes'.

Repeated nouns: **avuç avuç paralar** 'coins by the handful'; **araba araba odun** 'cartload after cartload of wood'; **sıra sıra dağlar** 'range on range of mountains'; **demet demet otlar** 'bundles and bundles of grass'; **küme küme evler** 'masses and masses of houses'. A once-popular song begins: **Ey miralay, miralay!** / **Askerin alay alay** 'O Colonel, Colonel! / Your soldiers are regiment on regiment'.

45. Doublets. On almost every page of the dictionary will be found nouns and adjectives consisting in pairs of assonant words. Such doublets are of three kinds.

(*a*) Each element is a regular word:

iş güç	employment	('work toil')
kanlı canlı	robust	('having blood and life')
yorgun argun	dead-tired	('tired exhausted')
derme çatma	jerrybuilt	('collection fitting-together')

(*b*) Only one element is a regular word, the other exists only in this doublet:

çocuk	child	**çoluk çocuk**	wife and family
sıkı	close	**sıkı fıkı**	intimate
çarpık	crooked	**çarpık çurpuk**	crooked and twisted
alaca	motley	**alaca bulaca**	garish and discordant of colour

(*c*) Neither element has independent existence:

allak bullak	topsy-turvy
abuk sabuk	nonsensical
ıvır zıvır	miscellaneous rubbish

46. m-doublets. The largest class of doublet is that in which a word is followed by an echo of itself but with **m** replacing its initial consonant or preceding its initial vowel. The meaning of this form beginning with **m** is 'and so on, and suchlike'. **dergi okumuyor** 'he doesn't read journals'; **dergi mergi okumuyor** 'he doesn't read journals or periodicals or magazines'. **bahçede ağaç yok** 'there are no trees in the garden'; **bahçede ağaç mağaç yok** 'there are no trees or shrubs or bushes in the garden'. It must be emphasized that it is useless to seek such words as **mergi** and **mağaç** in the dictionary; they are manufactured *ad hoc*.[1] **partiler, martiler, hep reform meform diye bağırıp çağırıyorlar** 'the political parties and suchlike are always shouting and screaming about reform and all that'. **işin sonunu anlat—sonu monu yok** 'tell ⟨me⟩ the end of the business'—'it has no end or anything resembling an end'. A rough equivalent of 'but me no buts' is **fakatı makatı yok**, lit. 'it has no "but" (**fakat**) or anything like a "but" (**makat**)'.

The late Prime Minister Menderes, on hearing that Harold Stassen was retiring as administrator of United States foreign aid, remarked: **Stassen giderse, yerine Mtassen gelir. Yardımı ondan alırız** 'If Stassen goes, some close facsimile of Stassen will take his place. We'll get the aid from him.' A Turkish workman who had married a German girl, when asked how they managed to communicate, replied: **Tarzanca marzanca idare ediyoruz işte** 'We manage in the manner (or 'language'; see XII, 2 (*e*)) of Tarzan and his mate Jane, that's how it is'.

It will be seen that this is essentially a colloquialism; for a neat literary use of it see XXIV, 14. An ingenious political use was made of it before the 1960 revolution, when the opposition weekly *Kim* was suspended but immediately reappeared under the title of *Mim*. Besides meaning, in this context, 'something closely resembling *Kim*', this is the name of the Arabic letter *m* which was used by the Ottoman bureaucracy as a 'black mark', to put against the names of those politically suspect. Hence **mimli**, still current for 'on the black list'.

This device is possible only because of the lack of native

---

[1] A similar device, of Yiddish origin, exists in New York colloquial, e.g. 'mink, schmink', meaning approximately 'who cares about mink or any such expensive furs?' It differs from the Turkish locution (*a*) in being used as an exclamation and not as a syntactic member of a sentence, and (*b*) in being intrinsically depreciatory.

Turkish words with initial **m**. With words which do begin with **m**, **m**-doublets cannot be made and **falan** or **filân** (V, 21) is used instead:

müfettişler falan ġeliyor 'the inspectors and all that lot are coming'.

XV

THE ORDER OF ELEMENTS IN THE SENTENCE

1. Nominal sentences and verbal sentences. Turkish grammarians classify all sentences as either nominal or verbal, the former being those in which the verbal element, if any, is one of the parts of 'to be' not formed from the stem **ol-**; the latter, those in which the verbal element is from **ol-** or any other normal stem.

Thus these are nominal sentences:

hava güzel(dir)	the weather is fine
yorgun değilim	I am not tired
yüzü temiz idi	his face was clean
ev çok ucuz imiş	the house is said to be very cheap
bekliyordur	he is sure to be waiting

These are verbal sentences:

hava güzel oldu	the weather has become fine
yorulmadım	I have not become fatigued
yüzünü yıkadı	he has washed his face
ev satılmış	the house appears to have been sold
bekliyor	he is waiting

The distinction, however, has no practical value; the weight given to it by Turkish schoolteachers is a relic of the days when Turkish grammar was taught with the technical terms of Arabic, in which the distinction is of fundamental importance.

2. The principles of word-order. The cardinal rule is that the qualifier precedes the qualified; i.e. the adjective, participle, or qualifying noun precedes the noun; the adverb or complement precedes the verb; the modifying phrase or adverb precedes the adjective:

büyük ev	the big house
dönen tekerlek	the turning wheel

çiftçinin evi	the farmer's house
çabuk geldi	he came quickly
çabuk ol	be quick!
buraya geldi	he came here
buradan uzak	far from here
Hindistan kadar büyük bir memleket	a land as big as India
Hindistan'dan büyük bir memleket	a land bigger than India
pek küçük bir bahçe	a very small garden

To leave aside, for the moment, the flexibility given to the written word by writers of the **devrik cümle** school (see § 3), the typical order of the elements in a literary sentence is: (1) subject, (2) expression of time, (3) expression of place, (4) indirect object, (5) direct object, (6) modifier of the verb, (7) verb. If any of these elements is qualified, the qualifier precedes it. The definite precedes the indefinite, so elements (4) and (5) will change places if the indirect object is indefinite and the direct object is definite. Thus **çocuğa hikâyeyi anlattı** 'she told the child the story', but **hikâyeyi bir çocuğa anlattı** 'she told the story to a child'.

By 'modifier of the verb' in position (6) is meant what Turkish grammarians call **tümleç** 'complement'. This may be:

(*a*) a noun in the dative, locative, or ablative case: **vesikaları Ankara'ya yolluyorum** 'I am sending the documents to Ankara'; **misafir otelde bekliyor** 'the guest is waiting in the hotel'; **onu penceremden gördüm** 'I saw him from my window'.

(*b*) an adverb or the equivalent: **dışarı çıkalım** 'let us go outside'; **bizim kadar çalışmıyor** 'he is not working as much as we are'.

(*c*) a particle: **evet, gelirim** 'yes, I am coming'; **hayır, gelmem** 'no, I am not coming'.

An example of the typical word-order: (1) **ressam** (2) **geçen hafta** (3) **Bebek'te** (4) **bize** (5) **resimlerini** (6) **ikinci defa olarak** (7) **gösterdi**; i.e. (1) the artist (7) showed (4) us (5) his pictures (6) for the second time (2) last week (3) at Bebek. The definite precedes the indefinite, so, if he showed his pictures not to us but to a journalist, elements (4) and (5) will change places: **resimlerini bir gazeteciye gösterdi**. Any element which is to be emphasized may be placed immediately before the verb:

geçen hafta Bebek'te bize resimlerini ressam gösterdi 'it was the artist who showed us . . .'; **ressam Bebek'te bize resimlerini geçen hafta gösterdi** 'it was last week that the artist showed us . . .'; **ressam geçen hafta bize resimlerini Bebek'te gösterdi** 'it was at Bebek that the artist showed us . . .'.

If the verb is intransitive, elements (4) and (5) are replaced by the complement or modifier of the verb: (1) **kızkardeşim** (2) **şimdi** (3) **Paris'te** (4) **trenden** (5) **iniyordur**; (1) my sister (5) must be alighting (2) now (4) from the train (3) in Paris. (1) **iki sarhoş** (2) **dün akşam** (3) **Taksim'de** (4) **karakolluk** (5) **oldular**; (1) two drunkards (5) became (4) police-station-material (2) yesterday evening (3) at Taksim (i.e. they were locked up).

The subject of **var** and **yok** adjoins them as a rule: **dağda bir ayı var** or **bir ayı var dağda** 'there is a bear on the mountain'; **evde kimse yok** or **kimse yok evde** 'there is no one in the house'.

It will not escape the reader's attention that such 'typical' sentences are relatively infrequent among the enormous variety that can occur in human speech, especially in its written form. Nevertheless, although not every sentence will have all these elements, the order given above will be found to fit not only most sentences but also most clauses within the sentence.

3. The inverted sentence, **devrik cümle**. In English, which has discarded most of its inflexions, the rules of word-order must be obeyed or the syntactic relationships of the various parts of the sentence will be upset. In an inflected language like Turkish or Latin one can take liberties with the conventional word-order and still be intelligible.

> **Romalılar, barbarları yendiler**
> Romani barbaros superaverunt
> The Romans defeated the barbarians

Let the Romans and the barbarians change places in the Turkish or Latin sentences and the basic implication remains the same, though there is a shift of emphasis ('it was the *Romans* who defeated the barbarians'). If they change places in the English, the sense is totally reversed.

In the past, Turkish prose-writers, like Classical Latin authors, have in the main denied themselves the freedom of word-order which the structure of their language offers. The qualifier in a definite izafet had to precede its noun, just like the attributive adjective; the verb had to come at the end of the sentence.[1] Any departure from these laws could be dismissed as colloquial. Under the Republic, however, new factors have altered the situation: the 'Anatolian' school of novelists and short-story writers have made peasant speech a familiar element of literature; the language-reformers have largely succeeded in establishing the principle that the gap between the written and spoken languages must be narrowed if not eliminated; the writers of the **devrik cümle** ('inverted sentence') school have deliberately departed from the conventional word-order even in formal writing. As they are widely admired and imitated by the younger generation, it seems likely that their style will one day impose itself on all but the most formal and solemn prose. To ignore the 'inverted sentence' in the hope that it will go away, as some conservative Turkish grammarians do, is to confuse the duties of grammarian and literary critic.

In fact, the **devrik cümle** school's deviations from conventional word-order can all be paralleled in the works of the most venerated writers of all periods. Where the more inept adherents of the school go wrong is that they do not use such deviations occasionally, so that by their novelty they may be the more telling, but make them into a new norm.

In the 'inverted sentence', the qualifier in a definite izafet may follow the word it qualifies. The rule that attributive adjectives (in which we may include participial qualifiers, the equivalent of English relative clauses; see XVIII, 2) must precede their nouns is unbreakable, simply because an adjective which is placed after its noun automatically becomes predicative. **mavi deniz** 'the blue sea' reversed becomes **deniz mavi** 'the sea is blue'. **ilk aklıma gelen cevap** means 'the answer which comes first to my mind'. If **ilk . . . gelen** is put after **cevap**, the meaning becomes 'the answer is that which first comes to my mind'. But if we take a definite izafet group such as **çiftçi-nin ev-i** 'the farmer's house' and invert it, **evi çiftçinin**, the grammatical suffixes still show the relationship between the two words and the meaning is unaltered. True, they might also mean 'his house is the farmer's', but in context there could be no ambiguity. In informal speech

[1] See Deny, §§ 1062-5.

the answer to a question like 'what's that place over there?' may well be in the form **evi çiftçinin**. The inverted order is even more likely if the phrase is part of a longer sentence, e.g. **evi büyük, çiftçinin** 'his house is big, the farmer's'. This may look as if the qualifier **çiftçinin** is added as an afterthought, but in fact this is at least as common a form of sentence in the spoken language as the formal **çiftçinin evi büyük**. In poetry this type of inversion is frequent; there are two instances of it in two consecutive lines in Yahya Kemal's *Açık Deniz* ('The Open Sea'): **Gittim o son diyâra ki serhaddidir yerin. / Hâlâ dilimdedir tuzu engin denizlerin!** 'I went to that last land which is earth's boundary. / Still on my tongue is the salt of the limitless seas!' In formal prose, **serhadd-i-dir yer-in** would be **yerin serhaddidir**, while **tuz-u engin deniz-ler-in** would be **engin denizlerin tuzu**. Prose examples are not so frequent: **giyinişi ... hayli acayipti bu adamın** (Yakup Kadri) 'this man's mode of dress was very peculiar'; here the effect is racy and conversational. **halkın konuştuğu dille, bilim, felsefe ve edebiyatın dilini birleştirmek, başka bir deyimle düşündüğünü konuşur gibi yazmak, ilk işi olmuştur Avrupa'da aydın kişilerin** (Eyüboğlu) 'to unify the language spoken by the people and the language of science, philosophy, and literature, in other words, to write one's thoughts as if speaking them, has become the first task of the intellectuals in Europe'. The inversion in the last six words is probably due to the desire to avoid the ugly assonance of **kişilerin ilk işi**.

The commonest manifestation of the **devrik cümle** and the one which most infuriates conservative critics is that the verb does not come at the end of the sentence. In the colloquial, an imperative often begins a sentence, because someone with urgent instructions to give will naturally put the operative word first: **çık oradan!** 'get out of there!' **yakma lâmbayı!** 'don't light the lamp!' Similarly with an urgent question; the bus-conductor in the rush-hour will shout, with his finger poised over the bell, **var mı inecek?** (or **var m'inecek?**) 'is there anyone about to get off?' although at quieter moments he may ask **inecek var mı?** In a statement, however, the verb tends not to come first. The use of **şey** 'thing' for 'what-d'ye-call-it?' is an indication of the strength of this tendency; if one wants to say 'I saw the exhibition' and momentarily forgets the word, one is more likely to say **şeyi**

gördüm—sergiyi than **gördüm—sergiyi**. Similarly for 'I am going to the what-d'ye-call-it—the exhibition': **şeye gidiyorum —sergiye** rather than **gidiyorum—sergiye**.[1] Consequently, even in the writings of the **devrik cümle** school, it is rare to find a sentence beginning with a verb other than an imperative or a question, except for introductory formulas which are part of the standard written language, such as **görülüyor ki** 'it seems that'. But the verb frequently precedes its subject, object, or modifier. **beni burada bulursa Abdi Ağa, öldürür** (Y. Kemal) 'if Abdi Agha finds me here, he'll kill me'. **Nasıl der Yunus Emre: Bir ben var bende benden içeri. Fiili sona koyun da, bakın ne oluyor cümle** (Eyüboğlu) 'What is it Yunus Emre says? "There is an 'I' in me, within the 'I'"'. Put the verb at the end and see what becomes of the sentence' ('what the sentence becomes'). **üç beş kişiyiz böyle söyliyen, biliyoruz çoğunluğa bunu anlatamıyacağımızı** (N. Ataç) 'we are a handful of people who talk like this; we know that we could not make the majority understand it'.

4. The sentence-plus. For one type of **devrik cümle** there is a useful term invented by C. S. Mundy;[2] he applies the name 'sentence-plus' to the sort of statement in which qualifiers or modifiers are added to the end of a sentence which is already grammatically complete in itself. Mundy gives the example **Kayseri'de bir damadı var** 'he has a son-in-law at Kayseri'. If this is expanded into 'he has a son-in-law who is a doctor at Kayseri', it becomes, in the formal written language, **Kayseri'de doktor olan bir damadı var**, but in speech **Kayseri'de bir damadı var, doktor**. Besides being the regular mode of expressing such meanings in speech, the sentence-plus occurs frequently in Old Ottoman texts, so that the outsider's sympathies are with those modernist writers who make full use of it, rather than with the pedants who condemn it as un-Turkish. The first of the three following examples is from the fifteenth-century historian Aşık Paşazade: **hem iki yıldız doğdu ol zamanda kuyruklu** 'moreover two stars rose at that time, tailed' (i.e.

---

[1] **şey** is also used when there is no lapse of memory, as a device to draw attention to the object or complement which is then placed after the verb: **şeye gidiyorum—Paris'e** 'I'll tell you where I'm going—Paris'.

[2] *BSOAS* (1955), xvii/2, pp. 303-4.

comets). **kapılar da gönülleri gibi hep yarı açılır misafire; görülmeden önce görmek, görmekten de çok gözetlemek ister gibi** (Eyüboğlu) 'the doors too, like their hearts, are always ⟨only⟩ half-opened to the guest, as if wanting to see before being seen and to spy rather than to see'. **güreşçiler, bir avuç tuz alıp yere atarlar, şans getirsin diye** 'wrestlers take a handful of salt and throw it on the ground for luck' (lit. 'saying let it bring luck').

XVI

NUMBER, CASE, AND APPOSITION

1. Concordance of subject and verb. It used to be stated as a rule of grammar that inanimate plural subjects took a singular verb, plural verbs being used with animate subjects or with inanimates personified or considered as individuals, e.g. **ağaçlar, yüzümüze konfeti atıyorlar** 'the trees are throwing confetti into our faces'. Conversely, an animate plural subject could take a singular verb if it represented a number of people acting as one. This rule needs to be modified in one respect: another factor nowadays seems to be the distance between subject and verb; i.e. if an inanimate plural subject takes a plural verb for no obvious reason, it will be because subject and verb are widely separated: **F-84 jet uçakları, tâyin edilen hedefleri roket atarak tahrip etmişlerdir** 'the F-84 jets destroyed the assigned targets by firing rockets'.

2. Singular and plural in izafet groups. In an izafet group whose qualifier is a plural, the qualified word, if singular, has the singular possessive suffix: **çarklar dönüyor, küçük çark büyüğünü döndürüyor** (Sait Faik) 'the gears turn, the little gear turns the big one'. If the penultimate word were **büyüklerini** it would mean 'the big ones'. The singular suffix of **büyüğ-ü-nü** refers to the plural **çarklar. Bu yüzükler çok pahalı. Daha ucuzu yok mu?** 'These rings are very expensive. Have you no cheaper ones?' Here the singular suffix of **ucuz-u** refers to the class **yüzük** of which **bu yüzükler** are individual members. This rule may be broken to avoid ambiguity: **türlü ailelere—bilhassa Hind-Avrupa—mensup olan dil ve lehçelerin yayılma tarzı mekanizması, ait oldukları aile çerçevesi içinde incelenerek tespit edilmiş bulunuyor** 'the mechanism of the manner of spreading of languages and dialects belonging to the various families—especially the Indo-European—has been established by being studied within the framework of the family to which they belong'. As 'to which they belong', **ait oldukları**, refers to the inanimate plural 'languages and dialects', it could have been singular, **ait olduğu**, except that this might have been taken as referring to one of the preceding singulars 'spreading', 'manner', or 'mechanism'.

3. Idiomatic uses of the plural.

(*a*) A plural noun is sometimes employed where English prefers the singular: **soğuklar sebebiyle şehrin su boruları patlamıştır** 'because of the cold, the city's water-pipes have burst'. What the city has had to face is not just **soğuk**, cold in general, but **soğuklar**, some specific instances of cold. **bu haber, bizi hayretlere düşürdü** 'this news reduced us to astonishment'. **iyi geceler!** 'good night!' **Allah akıllar versin!** 'God give ⟨you⟩ sense!' **Fethi Bey, o ğece hasımları üzerine saldırarak birer birer yerlere seriyordu** (Ağaoğlu Ahmet) 'Fethi Bey that night, attacking his opponents, was strewing them one by one on the ground'. **ğelin, beyazlar ğiymişti** 'the bride wore white'; **dul kadın, karalar ğiymişti** 'the widow wore black'. In English, although brides wear white, cricketers wear whites.

Note also: **ğerilere ğitmek** 'to go back'; **uzaklarda** 'in the far distance'; **yakınlarda** 'in the vicinity'; **ğeçenlerde** 'in recent times'; **buralarda** 'in these parts'. See further III, 3.

(*b*) The use of the plural for a single second or third person is a mark of respect: **orada yalnız başınıza mı oturuyorsunuz?** 'do you live there all alone?' **eşiniz daha ğelmediler mi?** 'has your wife not yet arrived?'

(*c*) On the other hand, the use of the first plural for the first singular is modest: **boynumuz kıldan ince** 'our neck is finer than a hair'. This is a jocular expression meaning 'I'll have my head cut off if I don't do as I'm told'.

(*d*) The first person plural verb is used when the speaker and another person are joint subjects: **onunla tiyatroya ğittik** 'he and I went to the theatre'.

If the speaker is regarded as the prime mover while the other plays a subordinate part, the verb may be singular: **onunla beraber tiyatroya ğittim** 'I went to the theatre, together with him'.

In the next example, from a newspaper headline, the first plural of the reciprocal verb means not 'I and another' but 'we and others': **Bulğaristan'la yenişemedik** 'we and Bulgaria were unable to defeat each other'; i.e. the Turkish and Bulgarian football-teams drew.

(e) The fact that the plural suffix **-ler** never occurs more than once in the same word[1] makes possible a useful distinction: **tanıştığımız adamlar mühendistiler** (= **mühendis idiler**) 'the men whom we met were engineers'; **tanıştığımız adamlar mühendislerdi** (= **mühendisler idi**) 'the men whom we met were the engineers'. **misafirseler** (= **misafir iseler**) 'if they are guests'; **misafirlerse** (= **misafirler ise**) 'if they are the guests'. **mahkemeye ġirenler yarġıç deġiller** 'those entering the court are not judges'; **mahkemeye ġirenler yarġıçlar deġil** 'those entering the court are not the judges'.

4. **The accusative with bir.** Although the accusative suffix shows that the word to which it is attached is definite, the use of it is not precluded by the presence of **bir**, since this, as well as being the 'indefinite article', is the numeral 'one'. Nevertheless, even in such contexts, 'a' and not 'one' may often be the better translation. Compare **her ġün bir ġazete okuyorum** with **her ġün bir ġazeteyi okuyorum**. Both may be translated 'every day I read a newspaper', but the second, unlike the first, implies that I always read one particular newspaper. **Türk hükümeti, anlaşmaların aynen uygulanmasını istiyen bir notayı Bulgar hükümetine vermişti** 'the Turkish government had given the Bulgarian government a note asking that the agreements should be given strict effect'. Here the **nota** is defined by the preceding participial clause ending in **istiyen**, and therefore has the definite accusative ending. **ne açıkları, ne açları, ne beni kızına münasip görmeyen zengin tüccarı hiç bir şeyi düşünmiyeceğim** (Sait Faik) 'I am going to think neither of the naked nor the hungry, nor the rich businessman who does not regard me ⟨as⟩ suitable for his daughter; not anything at all'. Because **hiç bir şey** 'not any thing' resumes and expands the definite objects about which he is not going to think, it too goes into the accusative.

5. **Two idiomatic uses of the dative case.**

(a) The absolute case of the present participle, followed by its

---

[1] Unless the word contains **-ki**, as in V, 3, last example.

dative case, conveys the idea of a multitude of people doing something in excessive haste: **kumsalı görseydin yıkanmak için gelen gelene** 'if you had only seen the beach; people coming in droves to bathe', lit. 'the-one-coming in order to bathe ⟨is added⟩ to-the-one-coming'. **bizde hükümetten kaçan kaçana** 'amongst us, people fall over each other to avoid being in the government', lit. 'in us, the-one-fleeing from-government ⟨is added⟩ to-the-one-fleeing'.

(b) **kardeşim bize gelmiyor diye merak etme; geliyor gelmesine** 'don't worry about my brother not coming to us; he does come, for what it's worth', lit. 'he comes for its coming'. The implication is perhaps that he does not come very often and certainly that when he does come the visit is never a great success. **gerçi, bulunduğum yer denizi görmüyor değil; görüyor görmesine, ama en aşağı bir, bir buçuk saatlik bir yerden** (Orhan Veli) 'It is true that the place where I am is not without a view of the sea; it has a view all right, but from a distance of at least an hour to an hour and a half'; lit. 'not it-does-not-see the-sea, it-sees for-its-seeing'. **para ödenmesine ödenirdi ama aradan aylar geçerdi** 'the money would be paid all right, but months would intervene'. **Rustaveli, batı dillerine çevrilmiştir çevrilmesine** 'Rustaveli [a Georgian poet] has been translated into the western languages, for what it is worth' (i.e. but nobody reads the translations).

This construction, with the third-person suffix of the **-me** verbal noun, is used even when a second person is addressed; the antecedent of the third-person suffix is vague:[1] **güzel olmasına güzelsin ama bir de kusurun var** 'you are beautiful, as far as that goes, but you also have a fault'. An old-fashioned English equivalent is 'you're beautiful, to say beautiful', i.e. but not to say anything complimentary beyond that. Cf. 'Oh she's *beautiful* enough, if that were all!'

In the first of the two following examples, which are from Aziz Nesin, the **-me** verbal noun has no personal suffix; in the second, there is no verbal noun at all, the abstract noun being used instead: **şair olmaya ben şairim ama okuyamam** 'I am a poet all right, but I can't recite'. **güzelliğine güzelmişsin**

---

[1] *OTD* mentions this locution under **olmasına** but adds 'in reply to a question', which seems to limit its function unduly.

'you are said to be beautiful, as far as that goes'. **güzelliğine** here is ambiguous, as its suffix might be that of the second- or third-person singular; it is in fact third-person singular, 'for the beauty of it', and does not vary with the person, so 'I am said to be beautiful, as far as that goes' would be **güzelliğine güzelmişim**.

6. The genitive as logical subject. Although as a rule the subject comes first in the sentence, we not infrequently find a sentence beginning with a word or phrase in the genitive case. The reason is that if the logical subject, the topic-word of the sentence, does not coincide with the grammatical subject, it is the logical subject which comes at the beginning. **bu gelişmelerin, doğulu vatandaşlarımızın hayatlarında ne gibi etkiler yaratacağı ortadadır** 'what sort of effects these developments will create in the lives of our eastern fellow-citizens is manifest'. The predicate is **ortada-dır** 'is in the middle', that is, *in medio*, in full view. The subject is all the rest of the sentence, **bu . . . yaratacağı**, these words being the substantivized form of the question **bu gelişmeler . . . ne gibi etkiler yaratacak?** 'what sort of effects will these developments produce . . .?' The process of turning this question into a noun-clause has put **bu gelişmeler** into the genitive—'these developments' creating what sort of effects'—but these two words are still the logical subject and are marked as such by their position at the beginning of the sentence and by the comma.

bu kazanın, hayatına mal olmasına ramak kaldı 'this accident all but cost him his life', lit. 'of this accident, a last breath remained to its being the cost for his life'.

kocasının, tıpkı dışarda olduğu gibi, evde de az konuşmak âdetiydi (İlhan Tarus) 'her husband was in the habit of speaking little at home too, just as he was outside', lit. 'of her husband, just as it was outside, to speak little at home too was his custom'.

Hamdi beyin çocuklarına tek bir fiske bile vurduğu görülmemişti. Oysa ki ikisi de oldum olası haşarıydılar (idem) 'Hamdi Bey had never been seen to strike his children even a single flick of the fingers. Yet both of them were pests and always had been.' Lit. 'Hamdi Bey's striking . . . had not been seen'. The lack of a comma after **beyin** must be due to an oversight, as one is needed not merely to mark the logical subject but to prevent the reader from taking the first three words as an izafet group ('to Hamdi Bey's children').

bu insanlarınsa içine bir kurt düşmüştür (Eyüboğlu) 'as for these people, they are full of misgivings', lit. 'as for of these people, a worm has fallen into the inside of them', with the logical subject **insanlar-ın** emphasized by **-sa**; see XIII, 27.

iktisadî buhranın, bu güne kadar bir türlü önü alınamadı 'the economic crisis has been not at all preventable up to now'. 'To prevent something' is **bir şey-in ön-ü-nü almak** 'to take the front of a thing'. Here the phrase is in the passive: 'its front has not been able to be taken'.

bu kuvvetin önüne durulmaz 'this force is irresistible'.

The largest class of sentence with the logical subject in the genitive is that denoting possession or the lack of it: **Mehmed'in parası var** 'Mehmet has money'; **Mehmed'in parası yok** 'Mehmet has no money'. Such expressions must not be thought of as consisting in an izafet group+**var** or **yok**.[1] The syntactical grouping is not **Mehmed'in parası / var** 'Mehmet's-money exists' but **Mehmed'in / parası var** 'Mehmet has-money'. The proof is as follows.

An izafet group cannot be split by an adverb unless the qualified element is a verbal noun; see p. 43, footnote. Consider the group **cemiyet-in toplantı-sı** 'the society's meeting'. If the meeting occurred yesterday and we wish to include that information in the izafet group, we cannot insert the adverb **dün** but must make it into an adjective: **cemiyetin dünkü toplantısı** 'the society's hesternal meeting'. **sakallı ihtiyar-ın durum-u hoşuma giderdi** 'I liked the bearded old man's attitude' ('his attitude used-to-go to-my-pleasure, used to appeal to me'). If we wish to insert in the izafet group the adverbial clause 'especially when he was accepting a tip', **bilhassa bahşiş alır-ken**, this clause must be made adjectival by the addition of **-ki**: **sakallı ihtiyarın bilhassa bahşiş alırkenki durumu**. ... Similarly, if we wish to say 'Mehmet's money in that bank is over a million', the adverbial phrase of place **o banka-da** must be made into an adjective: **Mehmed'in o bankadaki parası** (or **Mehmed'in o bankada bulunan parası**) **bir milyondan fazladır**. But the Turkish for 'Mehmet has money in that bank' is **Mehmed'in o**

---

[1] The author is indebted to C. S. Mundy for bringing this point home to him; loc. cit., pp. 294–5.

bankada parası var.[1] It follows that what we have here is not an izafet group plus **var** but a statement, **o bankada parası var** 'he has money in that bank', to which **Mehmed'in** is the subject.

7. Apposition. Although the normal method of linking two nouns is by izafet, considerable use is also made of apposition. The usual way of saying 'a shepherd-girl' is **bir çoban kızı**, i.e. a girl belonging to the category of shepherd. Also possible, however, is **bir çoban kız** or **bir kız çoban**, the emphasis in the latter being on **kız**, 'a *girl* shepherd'. So with **bir kadın doktor** 'a *woman* doctor', as against **bir doktor kadın** 'a woman doctor'. **bir kadın doktoru**, however, with izafet, is 'a gynaecologist'. A lady gynaecologist might be referred to as **bir kadın kadın doktoru**, but **bir kadın jinekolog** would be more usual in sophisticated circles.

Izafet groups may be used as qualifiers in apposition to nouns: **ev sahibi** is 'householder' and **ev sahibi takım** is 'the home team'; **söz konusu** 'subject of discussion' and **söz konusu kanunlar** 'the laws under discussion'. **tarih öncesi**, literally 'the before of history', and **harp sonrası**, literally 'the after of war', are nouns of this class, although we translate them as adjectives: **tarih öncesi dünya** 'the prehistoric world'; **harp sonrası Avrupa** 'post-war Europe'.

A curious use of apposition is seen in such expressions as **siz yaşta** 'of your age', literally 'in you-age'; **siz yaştakiler** 'people of your age'; **ben yaşta yahut daha büyük çocuklar** 'children of my age or older'.

Apposition is the rule with titles: **Osman Gazi, Balaban Bey, Zenbilli Ali Efendi, Halide Hanım, Mareşal Fevzi Çakmak, Doktor Adnan, Profesör Mansuroğlu**. It will be noticed that the ancient titles follow the name,[2] whereas the modern Marshal, Doctor, and Professor precede it, in the western fashion. **Sultan** is an interesting exception; it preceded the names of sovereigns—**Sultan Mehmet, Sultan Süleyman**— but followed the names of non-regnant members of the dynasty,

---

[1] **Mehmed'in o bankadaki parası var** would mean 'Mehmet's money, which is in that bank, exists', which is hardly worth saying.

[2] In fact **Gazi** 'warrior for the Faith' could also precede the name, as could **Hoca** and **Molla**, both titles of religious dignitaries. The proverbial expression **ha Hoca Ali ha Ali Hoca** 'both Hoja Ali and Ali Hoja' means 'it all amounts to the same thing'.

male and female: **Cem Sultan** 'Prince Jem', **Esma Sultan** 'Princess Esma'.

Apposition is not used, however, as it is in English, to show a person's occupation in such expressions as 'Ahmet Bilen, a professor of the Faculty of Letters'; Turkish idiom demands 'from the professors of the Faculty of Letters Ahmet Bilen': **Edebiyat Fakültesi Profesörlerinden Ahmet Bilen.**[1] So **genç aktrislerimizden Ayşe Güzel** 'Ayesha Güzel, one of our young actresses'; **gümrük memurlarından Orhan Soysal** 'Orhan Soysal the Customs officer'. These expressions in the ablative, like other qualifiers, precede the word they qualify; cf. **efendiden, dürüst insanlar** 'respectable, honest people' (**efendi-den** 'from the class "gentleman"').

Expressions consisting of a numeral and the name of a container or a unit of measurement are followed by the name of the commodity in apposition:

bir bardak su	a glass of water
iki şişe süt	two bottles of milk
üç kutu kibrit	three boxes of matches
dört araba odun	four cartloads of wood
beş yıl hapis	five years' imprisonment
altı avuç dolusu şeker	six handfuls of sugar
or **altı avuç şeker**	
yedi kaşık dolusu çorba	seven spoonfuls of soup
sekiz dakika gecikme	eight minutes' delay
dokuz metre kumaş	nine metres of cloth

The same construction is used with words like **takım** 'set', **grup** 'group', **cins, nevi, çeşit** 'sort': **bir takım çamaşir** 'a set of linen' (cf. V, 9), **bir grup talebeler** 'a group of students', **bir çeşit armut** 'a sort of pear'.

With **ara** 'interval' the distributive numerals are used in such sentences as **otobüsler, beşer onar dakika ara ile geçiyordu** 'the buses were passing at five- or ten-minute intervals'.

Either or both of the nouns in apposition may be separately qualified: **bir tencere pis su** 'a saucepan of dirty water', **iki büyük şişe beyaz şarap** 'two large bottles of white wine'.

---

[1] With the insertion of **Sayın** 'esteemed' before the name, this is how letters would be addressed to him.

XVII

THE NOUN CLAUSE AND THE SUBSTANTIVAL SENTENCE

1. The verbal noun in -me and the personal participles. It will be recalled that the personal participles have three functions: as adjectives, as nouns meaning 'that which I do' and as nouns meaning 'the fact that I do'. It is the third of these functions which will be discussed in this section, as it must be distinguished from the functions of **-me** in its sense of 'the act of doing'.

-me is used in indirect commands, **-diği** and **-eceği** in indirect statements and questions: **çocuklara aşağıya inip kendisini sokakta beklemelerini söyledi** 'she told the children to go downstairs and wait for her in the street'. With the substitution of **beklediklerini** for **beklemelerini**, the sentence would mean 'she told the children that they went down and waited . . .'; with **bekliyeceklerini**, '. . . that they would go and wait . . .'. **onbaşıya köşeye doğru koşmasını emretti** 'he ordered the corporal to run towards the corner'; **yarın sabah gelmeniz için telefon etti** 'he has telephoned for you to come ("for your coming") tomorrow morning'; **kapıyı kilitlemeyi unutmayınız** 'do not forget to lock ("the locking") the door'; **kapıyı kilitlemenizi tavsiye ederim** 'I recommend that you lock ("your locking") the door'; **kapıyı kilitlediğinizi unutmayınız** 'do not forget that you have locked the door'; **lokantamızda müskürat istimal olunmadığından (XI, 24) talep edilmemesi muhterem müşterilerimizden rica olunur** 'as intoxicants are not used in our restaurant, our honoured clients are requested not to ask for them' ('their-not-being-demanded is-requested from our . . . clients'); **en çok bunun yapılmasını isterdim** 'most of all I should like this to be done'.

Although the function of the personal suffixes is to indicate the agent of the **-me** verbal noun, the third-person suffix is used with the **-me** verbal noun of impersonal passives: **de Gaulle, dolar yerine altın esasına dönülmesini istedi** 'de Gaulle has demanded a return to the gold standard instead of the dollar';

kapılara dayanılmaması rica olunur 'it is requested that one should not lean against the doors' ('its-not-being-leaned to-the-doors is-requested'); **ileri gidilmesi rica olunur** 'going ("the being-gone-of-it") forward is requested'. The -si in these examples performs no visible function and is to be ascribed to analogy with sentences of the type of the two preceding examples; cf. also II, 22. For a similar use of the third-person personal participle of an impersonal passive, see XVIII, 3 (*a*).

Indirect questions are made by turning the verb of the direct question into the appropriate personal participle: **ne yapıyorum?** 'what am I doing?' **ne yapacağım?** 'what am I going to do?' **ne yaptığımı, ne yapacağımı bilmiyorum** 'I do not know what I am doing ⟨or⟩ what I am going to do'. **parayı kimden aldınız?** 'from whom did you take the money?' **parayı kimden aldığınızı soracak değilim** 'I do not intend to ask from whom you took the money'. **kızların yanına yaklaştı ve kendilerine ne şekilde yardım edebileceğini sordu** 'he approached the girls ("came near to the side of the girls") and asked in what way he could help them' (for the conditional force of **edebileceğini** see XX, 9).

In the following example (from Eyüboğlu), the **-me** verbal noun is used in the indirect question beginning with **nasıl**, to convey the idea of necessity; i.e. there is an indirect command within the question: **her yerde, her zaman okuryazarlar toplum hayatının nasıl bir düzene girmesi gibi meseleler üzerinde az çok bir fikir sahibidirler** 'everywhere and always the literates have more or less of an idea on questions such as what sort of order social life should enter'. The personal participle **girdiği, girmekte olduğu,** or **gireceği** would mean not 'should enter' but 'has entered', 'is entering', or 'would enter'.

When a personal participle is the subject of a sentence whose predicate is a noun, care must be taken not to read it as qualifying the noun: **Bu iki kaygu bir araya gelmiyorsa kabahat kimin? Orası ayrı mesele: ama gelmediği ortada** 'If these two concerns do not coincide ("do not come to one place"), whose fault is it? That is a separate problem, but that they do not coincide is manifest' ('their-not-coming is in-the-middle'; cf. XVI, 6, first paragraph). To take **gelmediği** as an adjectival qualifier of **ortada** ('but in the middle to which they do not come') is grammatically possible but fruitless.

In the next example, both subject and predicate are personal participles: **zaten bizim de merak ettiğimiz bunların kimler olduğu** 'just so; what *we* are curious about is who these people are'. The subject is **merak ettiğimiz** 'that pertaining to our exercising curiosity', the predicate **bunların kimler olduğu** 'of-these, their-being who?'

2. The substantival sentence. This term has been coined to cover a situation which often arises in Turkish: a complete sentence functions as a noun clause or adjectival clause within a longer sentence. An obvious example, not peculiar to Turkish, is in reporting speech: **güneş daha batmadı, dedi** '"the sun has not yet set", said he', where the words quoted are the direct object of **dedi** 'said he'. But in Turkish the substantival sentence is more widely used than this, especially in the colloquial and therefore in the latest literary idiom. The regular literary practice with a sentence that is to be the subject or object of a verb is to turn it into a noun clause by substituting a personal participle for the finite verb of the original sentence: **kumar meraklısı idi, bir gece bin lira kaybettiği olurmuş** 'he was a gambling enthusiast; it was said that there were occasions when he lost a thousand lira in a night'. Here the subject of **olurmuş** 'was said to happen' is **bir ... kaybettiği** 'his losing ...'.[1] **'yer yok' diye müşteri çevrildiği görülmüş şey değildir** 'for customers to be turned away with the words "no room" is a thing that has never been witnessed'. Here the subject is **çevrildiği** 'their being turned'. In the next example, however, the original finite verb has not been changed to a personal participle: **düşünüş ayrılıkları hayatımızı allak bullak eder de dostluğa dokunmaz olur mu?** (Eyüboğlu) 'is it possible that differences of ways of thinking can throw our lives into chaos and not affect friendship?' The subject of **olur mu** is the complete sentence **düşünüş ... dokunmaz**.

3. The substantival sentence as adjectival qualifier. **kuş uçmaz kervan geçmez bir yer** 'an inaccessible spot', lit. 'a bird-does-not-fly, caravan-does-not-pass place', with the sentence **kuş ...**

---

[1] The same construction is rarely found with **-mişlik**, the abstract noun of the **-miş** participle; for examples see XXIV, 29, 30.

geçmez occupying the normal position of the attributive adjective, before **bir**. **aşağı tükürse sakalı, yukarı tükürse bıyığıydı** 'he was in a quandary', lit. 'he was if he spits down, his beard; if he spits up, his moustache'; cf. XII, 10 (*d*). **seyahattan üfür uçtum, tut kaçtım, döndüm** 'I came back from the trip so skinny that a breath of wind would have blown me away', lit. 'I came back, puff! I've flown; hold me! I've gone!' The four-verb sentence **üfür . . . kaçtım** stands in place of an adjective such as 'emaciated'. In the essay which gives its name to Eyüboğlu's *Mavi ve Kara*, the 'blue' and the 'black' symbolize respectively art and money: **Hiç bir şey vermez mi olur paranın kulu olmuş sanatçı? Verir, kolayına kaçtığı için daha da bol verir; ama ne? Kirli bir mavi, olmasa da olur bir mavi** 'Is it totally unproductive that he becomes, the artist who has become the slave of money? He produces; because he has taken the easy course ("fled to-the-easy-of-it") he produces even more abundantly, but what? An impure blue, a blue we could well do without'. The adjectival qualifier of the final **bir mavi** is the complete conditional sentence **olmasa da, olur** 'even if it were non-existent, that would be all right'. It may be noted that whole sentences can similarly be used as qualifiers in colloquial English: 'a headmaster of the "this is going to hurt me more than it's going to hurt you" breed'; 'a sheriff of the "shoot first and then ask questions" school'.

4. *The substantival sentence as qualifier in izafet*. The sentence so used can be of as little as one word, or longer: **öldü haberi** 'the news that he has died', lit. 'he-has-died the-news-thereof'; **olmaz cevabı** 'the answer "it is not possible"'; **kalk borusu** 'reveille' ('the "Rise!" trumpet'); **nereye gidiyoruz sorusu** 'the question "where are we going?"'; **ben yaptım iddiasiyle** 'with his claim of "I did it"'; **ne oldum delisi** 'a parvenu who gives himself airs' ('"what I have become!" madman'). **nasıl olup da . . . tabiri fransızca'nın** 'comment se fait-il que' **sü manasiyle kullanılır** 'the expression **nasıl olup da . . .** is used with the meaning of the French "comment se fait-il que"'. The **sü** is the third-person suffix linking the French phrase to its qualifier **fransızca'nın** and it has this particular form because it takes its vocal colour from the 'que', whose vowel is more or less the Turkish **ö**.

5. The sentence with case-endings. As a whole sentence can thus take the syntactic place of a substantive, it is not surprising that we sometimes find a sentence to which a case-ending is attached: **komşu hasta, geçmiş olsun'a gidelim** 'the neighbour is ill; let's go and wish him better', lit. 'let-us-go for-the-"may-it-be past"'. The literary Turkish for 'he makes no mention of when it will be finished' is **ne zaman bitirileceğinden hiç bahsetmiyor,** with the personal participle (here in the ablative because that is the case required by **bahsetmek** 'to mention') replacing the finite verb of the simple question **ne zaman bitirilecek?** But a vivid colloquial version could be **ne zaman bitirilecek'ten hiç bahsetmiyor,** which has the feeling of 'When will it be finished? *That* he doesn't mention'.

Neden milli olsunu kabul ettiniz de, sonra tekrar yabancı sermayede karar kıldınız?[1] 'Why did you accept that it should be national (i.e. Turkish) capital and afterwards again decide on foreign capital?' The final **u** of **olsunu** is the direct object suffix, which makes **milli olsun** 'let it be national' into the object of **kabul ettiniz** 'you accepted'. See also the last example in XI, 32.

[1] Sakıp Sabancı, *İşte Hayatım* (n.p., 1985), p. 153.

XVIII

ADJECTIVAL PHRASES AND PARTICIPIAL QUALIFIERS

1. The başıbozuk construction.[1] This type of qualifier derives from a statement whose subject is an izafet group: **şu adam-ın baş-ı bozuk** 'that man's head is deranged'. The words **başı bozuk** 'his head deranged' can be used to qualify **adam** by placing them before it: **başı bozuk adam** 'the his-head-deranged man, the man whose head is deranged'. Now when a sentence whose verb is not simply the copula (as it is in **başı bozuk**) is made into a qualifier, the verb becomes the corresponding participle; thus to make a qualifier out of **Üniversitede hukuk okuyor** 'she is reading law at the University' we substitute for the present tense **okuyor** the present participle **okuyan**: **Üniversitede hukuk okuyan kız** 'the girl who is reading law . . .'.
But this construction offers further possibilities. Beginning with the statement **kızı hukuk okuyor** 'his daughter is reading law', we can turn it into a qualifier on the **başıbozuk** pattern: **kızı hukuk okuyan adam** 'the man whose daughter is reading law' ('the his-daughter-reading-law man'). **babası Çin'de doğdu** 'his father was born in China'. The participle corresponding to **doğdu** is **doğmuş**: **babası Çin'de doğmuş bir tanıdığım var** 'I have an acquaintance whose father was born in China'. **ihtiyar-ın ak sakal-ı kana boyanası** (VIII, 23) 'may the old man's beard be dyed in blood!' The future II may be used adjectivally (IX, 3): **bu ak sakalı kana boyanası ihtiyar tütün kaçakçılariyle ortaktır** (F. Celâlettin) 'this damned old scoundrel (lit. "this may-his-white-beard-be-dyed-in-blood oldman") is in league with the tobacco-smugglers'.
When the verb of the original statement is simply the copula, the use of a participle meaning 'being', i.e. **olan** or **bulunan**, is optional in the derived başıbozuk qualifier. It is perhaps a little more usual when the subject of the qualifying phrase, the **baş** of

[1] This name has been chosen because it may be familiar to the reader; the word exists in English in the form bashibazouk, 'irregular soldier'.

başıbozuk, is something or someone external to the thing or person qualified, as in the first three of these examples: **evi büyük olan bir arkadaşım** 'a friend of mine whose house is big'; **bahçesi büyük olan bir ev** 'a house which has a big garden'; **amcası bakan bulunan bir çocuk** 'a child whose uncle is a minister'; **eli açık bir dost** 'an open-handed friend'.

The construction provides a large number of phrases, some so common that, like **başıbozuk** itself, they are generally written as one word, e.g. **gelişigüzel** 'random' (lit. 'its-way-of-coming beautiful', i.e. however it comes it is all right); **karnıyarık** 'stuffed aubergine' ('its-belly split'); **dini bütün** 'pious' ('his-religion complete'); **gözü pek** 'unyielding' ('his-eye firm'); **eli boş** 'empty-handed'.

If such an adjectival phrase is used predicatively, the personal suffix may vary with the person: **sakın elin boş gelme** 'mind you don't come empty-handed'; but **sakın eli boş gelme** is also possible.

As alternatives to many such expressions, phrases with **-li** may be used: 'a garden with a tumble-down wall' may be **duvarı yıkık bir bahçe** or **yıkık duvarlı bir bahçe**; 'the keen-eyed colonel' may be **bakışları keskin albay** or **keskin bakışlı albay**; 'the open-windowed room' **pencere-si açık oda** or **açık pencere-li oda**.

2. Translation of English relative clauses. The purist may object that such a heading as this has no place in a Turkish grammar. The uses of the Turkish participles, however, are difficult to grasp through a purely descriptive treatment and the author is therefore emboldened to hope that he may be forgiven for approaching the topic from the wrong end.

The English-speaker composing in Turkish must resist the temptation to translate his relative clauses with the help of **ki** (XIII, 15); this use is regarded as alien and is increasingly rare in modern Turkish.

(a) When the relative pronoun (i.e. the English relative pronoun) is in the nominative, use **-en, -miş (olan), -ecek (olan)**: 'the man who is now speaking' **şimdi konuşan adam**; 'the letter which came yesterday' **dün gelmiş olan mektup**; 'the congress which will begin tomorrow' **yarın başlıyacak olan kongre**.

(b) Use the personal participles:
(i) When the relative pronoun is in the accusative, either as object of the verb in the relative clause—'the letter which I wrote' **yazdığım mektup**; 'the lawyer whom he chose' **seçtiği avukat**—or as object of a preposition other than 'of' or one whose Turkish equivalent is a secondary postposition (in which cases rule (d) applies): 'the ship on which they came' **geldikleri vapur**; 'the door from which we emerged' **çıktığımız kapı**; 'the beggar at whom you looked' **baktığınız dilenci**. When the relative pronoun is the object of 'with', **birlikte** or **beraber** 'together' may be inserted: 'the friends with whom he drank' **birlikte içki içtiği arkadaşlar**.

(ii) When the relative pronoun is 'whose' or object of 'of', while the noun it governs is the complement of the verb in the relative clause, the equivalent noun in Turkish takes the third-person suffix: 'the man whose servant I am' **hizmetçisi bulunduğum adam**; 'a society of which I am a member' **üyesi bulunduğum bir cemiyet**; 'the province of which you are going to become governor' **valisi olacağınız vilâyet**; 'the village of which he has been elected mayor' **muhtarı seçildiği köy**.

The rule holds good if the verb with the complement is dependent on another verb: 'the society of which I intend to become a member' **üyesi olmak niyetinde bulunduğum cemiyet** ('the society pertaining-to-my-being-found in-the-intention-of to-become its-member'); 'the village of which he wishes to be elected mayor' **muhtarı seçilmek istediği köy** ('the village pertaining-to-his-wishing to-be-elected its-mayor').

(c) When the relative pronoun is 'whose' or object of 'of', while the noun it governs is in the nominative as subject of the verb in the relative clause, use the başıbozuk construction with participle: 'the man whose father is now speaking' **babası şimdi konuşan adam**; 'the jockey whose horse came first' **atı birinci gelmiş olan cokey**; 'the society whose congress will start tomorrow' **kongresi yarın başlıyacak olan cemiyet**.

(d) When the relative pronoun is 'whose' or object of 'of', while the noun it governs is the object of the verb in the relative clause, or when the relative pronoun is object of a preposition whose Turkish equivalent is a secondary postposition, use the başıbozuk construction with personal participle: 'the artist whose pictures we are seeing' **resimlerini görmekte olduğumuz ressam**; 'the artist at whose pictures we looked' **resimlerine baktığımız ressam**; 'the society to whose congress we shall go tomorrow' **kongresine yarın gideceğimiz cemiyet**; 'the house from the inside of which we emerged' **içinden çıktığımız ev**; 'the ideal for whose sake he died' **uğrunda öldüğü ülkü**.

(*e*) Rules (*c*) and (*d*) apply also if the noun following the 'whose' is itself in the genitive. If the noun governed by that noun in the genitive is the subject of the verb in the relative clause, proceed according to rule (*c*); if it is the object of the verb or of a preposition, proceed according to rule (*d*): 'the man whose father's house is near ours' **babasının evi bizimkine yakın olan adam**; 'the man whose father's house we bought' **babasının evini aldığımız adam**; 'the man in ("with") whose father's car we came' **babasının arabasiyle geldiğimiz adam**.

3. Two variant types of participial qualifier. The previous section does not cover two situations which arise in Turkish.

(*a*) When an impersonal verb is made into a qualifier, it can be either as a participle or as a personal participle: **sağlık istatistiklerine göre mart en çok hastalanılan, hattâ en çok ölünen aydır** 'according to health statistics, March is the month in which most illness arises, indeed in which most deaths occur'. **hastalan-ıl-an** and **öl-ün-en** are the participles corresponding to **hastalan-ıl-ıyor** and **öl-ün-üyor**, impersonal passives meaning 'being-ill-is-done' and 'dying-is-done'. **yavaş yavaş normale dönüldüğü bir sırada, o adam iktidara geçti** 'at a time when things were slowly returning to normal, that man came to power'. **dönüldüğü** is the adjectival form of the statement **dön-ül-üyordu** 'returning was being done', but the function of the third-person suffix here is not obvious. The best explanation is that the use of the personal participle in expressions of time with **sıra, zaman**, etc., has become habitual.

(*b*) Sometimes in situations where rule (*b*) of the previous section would lead us to expect a personal participle, we find instead **-en, -miş,** or **-ecek**: **güneş girmiyen eve hekim girer** 'the physician enters the house which sunshine does not enter'; **yangın çıkan bir Amerikan uçak gemisinde 47 ölü sayıldı** 'on an American aircraft-carrier on which fire broke out, 47 dead have been counted'; **sırtındaki gömlek, bazı kasabalarda** ilkokul öğrencilerine göğüslük yapılan yerli gri bezdendi 'the shirt on his back was of the local grey cloth of which in some towns pinafores are made for primary-school pupils'; **elektrik getirilen Istıranca köyünde sanayileşme başladı** 'in the village of I., to which electricity is being brought (or 'has just been brought'; see IX, 1, penultimate paragraph), industrialization has begun'; **Atatürk'ün istediği ilk öğretim raporunu hazırlamak üzere çıktığımız bir köy gezisinde hiç öğretmen girmemiş bir köyde okur yazar çocuklar bulduk** 'on a village-tour, which we went on in order to prepare the report on primary education which A. wanted, we found children who could read and write in a village which no teacher had ever entered'; **kaliteyi muhafaza etmek için tütün ekilecek sahaları tesbit edeceğiz** 'in order to preserve the quality, we shall fix the areas where tobacco is to be sown'. In these examples we might have expected:

not	*but*
girmiyen	girmediği
çıkan	çıktığı
yapılan	yapıldığı
getirilen	getirildiği
girmemiş	girmediği
ekilecek	ekileceği

In fact, only in the penultimate example is the personal participle a possible alternative; in the others, it would sound too specific in conjunction with the broad and amorphous subjects sunshine, conflagration, pinafores by and large, electricity, and tobacco. In the penultimate example, **girmediği** would be possible; it may be that in the writer's mind there was an echo of the proverb which is the first example in this paragraph.

XIX

THE SUBJUNCTIVE AND Kİ

1. Clauses of purpose. Clauses of purpose containing a subjunctive may, but need not, be introduced by **ta ki** (sometimes written **tâ ki**) or by **ki** alone: **o vakit, bir sağa bir sola başvurmağa başlıyorsunuz; tâ ki daldığınız bu toprak deryası içinden kendinize bir iz bulup çıkasınız** (Yakup Kadri) 'then you begin to cast about, now to right, now to left, so that you may find a track for yourself and emerge from this ocean of earth into which you have plunged'. **oturdum ki bir dakika dinleneyim** 'I sat down so that I might rest a minute'. Note that there is no 'sequence of tenses'; the main verb in the past tense is followed by the present subjunctive. **herkesten çok koşacaksın ki paçayı kurtarasın** 'you will run more than everybody so that you may save your skin'. **şimdi uyuyun ki, sabah kalkınca dinç kafayla çalışasınız** 'now sleep, so that when you get up in the morning you may work with a sound head'.

It must be emphasized that **ki** merely introduces such clauses; it is the subjunctive (in which term is included the third-person imperative) that expresses purpose, so that the **ki** may be omitted from purpose-clauses, especially in less formal language: **pencereyi aç, oda havalansın** 'open the window, so that the room may air'. **Muhalefet ne yapılsın istiyor? Söylesinler öğrenelim** 'What does the Opposition want done? Let them say, so that we may learn', lit. 'What does the Opposition want should-be-done? Let-them-say let-us-learn'.

2. The subjunctive after a negative main verb. After a negative main verb, the subjunctive with **ki** is used to show what would have been the consequence had the main verb been positive but, as things stand, is now impossible: **ödediğim taksitlerin makbuzları yanımda değil ki çıkarıp göstereyim** (Aziz Nesin) 'the receipts of the instalments I have paid are not on me that I should produce and show them'. **Ölüm bu. Siyaset hayatı değil ki, bir o yana bir bu yana dönesin** (idem) 'It's death, this. It isn't political life, that you should swing now to that side, now to this'. **Ben gittikçe öfkelenmeğe başlıyorum: / "Nasıl arabacılık bu! diyorum; ne yol bilirsin, ne de . . . / "Yol nerede efendi? yol yok ki bileyim; diyor. / Biçarenin hakkı var. Evet yol yok ki** . . . (Yakup Kadri) 'I am gradually beginning to get annoyed. "What sort of driving is this!" I say; "you neither know the road, nor . . .". "Where is the road, Sir? There is no road for me to know," says he. The poor fellow is right. Yes, there is no road' (for the untranslated final **ki** cf. page 214, second paragraph).

The subjunctive occurs in other types of subordinate clause after a negative main verb: **gün geçmiyor ki turistik tesislerden bir şikâyet mektubu almamış olalım** 'a day does not pass without our receiving ("that we be not-having-received") a letter of complaint from the tourist establishments'. The logic behind the use of the subjunctive here is that as we in fact receive such letters daily, our not receiving one is only a concept, with no objective reality. **hiç bir gerici yoktur ki, Atatürk düşmanı olmasın** 'there is no reactionary who is not an enemy of Atatürk'. **hiç tahmin etmiyorum ki vaziyette bir değişiklik olabilsin** 'I do not reckon that there can be a change in the situation'. **zannetmem ki paşa veyahut valdesi buna razı olsunlar** 'I do not think that the Pasha or his mother will agree to this'.

3. *The subjunctive in noun clauses.* The third-person imperative without **ki** makes noun-clauses which can be the subject or object of a verb; for an instance of the latter use, see the last example in section 1, above. **böyle bir millet esir yaşamaktansa mahvolsun evlâdır** (Atatürk) 'for such a nation, to be annihilated is preferable to living ⟨as⟩ slaves'. The predicate is **evlâ-dır** 'is preferable'; the subject is **böyle . . . mahvolsun** 'that such a nation be annihilated'. **ev benim olsun da ziyanı yok tek katlı olsun** 'let the house be mine and it doesn't matter if it is single-storied'. Here the subject is **tek katlı olsun** 'that it be single-storied', the predicate is **ziyan-ı yok** 'there is no harm in it'. **düşüncelerimizin yönü bir olsun yeter** 'that the direction of our thoughts be one ⟨and the same⟩ is sufficient'. **hiç olmasın daha iyi** 'that it should not exist at all is better'. **bu kadar basit bir işi yapamasın, hayret doğrusu** 'that he should be unable to do a job as simple as this is truly surprising'. **Çalıkuşu ağlasın?**

On senedir ne muallimelerinden, ne arkadaşlarından bunu gören olmamıştı 'That the Wren should weep? For ten years none of her teachers or friends had seen this' ('neither of her teachers nor of her friends the-one-seeing this had-not-occurred'). The question-mark indicates wonder at the unusual event; the author[1] could not make a question of it in the normal way, because Çalıkuşu ağlasın mı could only mean 'Is the Wren to weep?'

An alternative way of analysing these examples is to explain them as consisting of two separate main clauses, e.g. mahvolsun evlâdır 'let it be annihilated; ⟨that⟩ is preferable'. Against this is the fact that such sentences are pronounced with no pause between the two verbs. Nor will this explanation fit the example bu kadar basit . . . doğrusu, or indeed the last one.

[1] Reşat Nuri, *Çalıkuşu* (Istanbul, 1928, in the old alphabet), p. 8.

XX

CONDITIONAL SENTENCES

1. Open conditions. These are expressed by the conditional form of the appropriate tense.

(*a*) *present*: **halkımızı gerçekten seviyor-sak, onun için çalışıyoruz derken yalan söylemiyor-sak, onu kuşkulardan korkulardan kurtarmak ilk işimiz olmalı** 'if we truly love our people; if, when saying we are working for them, we are not telling lies, our first task ought to be to deliver them from suspicions and fears'.

(*b*) *future*: **söyliyecek-sen** (or **söyliyecek olur-san**) **söyle** 'if you are going to tell, tell'.

(*c*) *aorist*: this is by far the commonest tense of open conditions: **hülâsa eder-sek şuraya varıyoruz** 'if we summarize, we arrive at this point'; **şunu bir anlar-sanız bana büyük bir iyilik etmiş olacaksınız** 'if you will only understand this, you will have done me a great kindness'; **bu tirene yetişemez-sem işi kaçıracağım** 'if I cannot catch this train I shall lose the job'.

(*d*) *past*: **o, yola çıkmış-sa biz niye oturuyoruz burada?** 'if he has started out, why are we sitting here?'; **günah mı işledik beş on para kazandı-ysak?** 'have we committed a sin if we've earned five or ten coppers?'

2. Alternatives to the conditional verb. Open conditions may also be expressed without a conditional verb, in four possible ways:

(*a*) The protasis may have a personal participle with **takdirde**; see XI, 23.

(*b*) The protasis may be a question in the di-past: **o geldi mi ben burada durmam**, lit. 'has he come? I do not stop here', which may mean either 'if he has come I'm not stopping here' or 'if ever he comes I don't stop here'. A macabre old saying ran **asıldın mı İngiliz sicimiyle asıl** 'if you are hanged, be hanged with English rope'.

(*c*) The sentence may be cast as a *reductio ad absurdum*; the protasis concedes what the speaker regards as false, the apodosis (introduced by **de**) asks for an alternative: **sen yapmadın da kim yaptı?** 'you didn't do it and who did?'; **Tanrı, Doğru'nun, İyi'nin, Güzel'in yardımcısı olmaz da neyin yardımcısı olur?** 'if God does not help the True, the Good, the Beautiful, what *does* He help?' (lit. 'God does not become the helper of ... and of what does He become the helper?').

(*d*) In the colloquial, the protasis may have an imperative instead of a conditional verb: **uzatma bırakır giderim** 'don't prolong ⟨the discussion because if you do⟩ I shall abandon ⟨it and⟩ go'.

3. Remote and unfulfilled conditions. The verb of the protasis is in the appropriate tense of the conditional mood; the verb of the apodosis is, with unfulfilled conditions, in the aorist past or, less commonly, the future past; with remote conditions or for greater vividness with unfulfilled conditions, the aorist present or future simple: **evimiz döşeli dayalı ol-sa buyurun de-r-dim ama görüyorsunuz . . .** 'if our house were properly furnished I should say "please come in" but you see . . .'; **başka bir kaynağımız ol-ma-sa-ydı bu âlimin tetkikleri bu meseleye tam bir cevap vermiş olacaktı** 'if we had no other source, this scholar's researches would have given a complete answer to this problem'; **bakmakla usta olun-sa köpekler kasaplık öğrenir** 'if one became (impersonal passive) a master craftsman by watching, the dogs would learn the butcher's trade'; **sen ol-san ne yap-ar-sın?** 'if it were you, what would you do?' Note that the idiom is 'if you were', i.e. the person responsible; cf. the next example. **medeniyetin maddisi mânevisi diye ayırmalar yapanlara ben ol-sam hiç elektrik ver-mem** 'to those who make distinctions between material and spiritual civilization ('distinctions saying "of civilization, its material, its spiritual"'), I should give no electricity, if it were up to me'. **şu adam karşımda ol-sa gırtlağına sarıl-acağım** 'if that man were facing me I should wrap myself round his throat'.

4. Apodosis to an unexpressed protasis. The aorist past is used in expressions like **ol-ur-du** 'it would be' and **iste-r-dim** 'I should like', which are apodoses of an implicit remote or unfulfilled protasis such as 'if possible, if it were so': **Baudelaire'in**

Dördüncü Mehmed'in hayatını tanımış olmasını isterdim; hakkı olan bir şöhret kazan-ır-dı 'I should have liked Baudelaire to have been acquainted with the life of Mehmet IV; he would have won a fame which is his due'. **Sen sarışınsın, ben de esmerim. Ne güzel çocuğumuz olurdu, dedi** '"You are blonde, I am dark. What a beautiful child we should have", said he'.

5. Alternative protases. Pairs of alternative protases are expressed in the remote form (because the two conditions, being mutually exclusive, are not both open), with a **de** after the verb in each protasis: **biz iste-sek te iste-me-sek te kız beğenmediği adama var-maz** 'whether we want it or whether we do not want it, the girl will not marry a man she does not like'; **şehirde iş bul-sam da bul-ma-sam da köye hiç dönmem gayrı** 'whether I find work in the city or not, I shall never go back to the village any more'.

6. Concessive clauses. A single conditional verb followed by **de** is concessive: **köye iste-sem de dönemem gayrı** 'I cannot go back to the village any more even if I should want to'; **içmem! dedi-ysem de, ısrar ettiler** 'though I said "I don't drink!", they insisted'. See also XI, 22 and 29, last paragraph.

7. 'Whatever, whenever, whoever, wherever'. Such words are expressed by **ne, ne vakit, kim, nerede,** etc. (with or without a preceding **her**; cf. the penultimate paragraph of XI, 14) followed by a conditional verb: **ne yıkılmış-sa softalar yıkmıştır bu memlekette** 'whatever has been demolished the bigots have demolished in this country'. **ne kadar yukarıdan in-er-se o kadar derine gir-er** 'the greater the height it falls from, the deeper it goes in' (of a pile-driver; lit. 'from whatever amount high it descends, to that amount deep it enters'). **ne vakit evlerine git-sek veyahut onlar bize gel-se-ler hep kavga ederler** 'whenever we go to their house or they come to us, they always quarrel'. **nasıl** with a conditional verb is usually best translated not 'however' but 'just as' or 'in just the same way that': **nasıl İstanbul mimarlığı bir günde teşekkül etmemiş-se, Boğaziçi de tek bir zamanın eseri değildir** 'just as Istanbul architecture did not take shape in one day, so too the Bosphorus is not the work of a single time'.

Sentences of the type discussed in the preceding paragraph can be put into the past tense by making the verb of the apodosis past: **ne kadar yukarıdan inerse o kadar derine gider-di** 'the greater the height it fell from, the deeper it used to go in'.

8. eğer, şayet. Both these words are Persian in origin and they mean respectively 'if' and 'if perchance'. As the sense of 'if' is conveyed in Turkish by the conditional verb, neither is syntactically essential. **eğer** is useful in a long and complicated sentence to give warning that a conditional verb is coming, in the same sort of way that Spanish uses a premonitory inverted question mark and exclamation mark. **eğer Augsburg ittifakı akdedilmemiş olsaydı bugünkü dünyanın pek başka bir dünya olacağı düşünülemezse de, eğer Newton 1687 de Principia'sını neşretmemiş olsaydı, bugünkü ilim ve medeniyet âlemini aynı vaziyette bulacağımız pek şüpheli idi** 'although it cannot be thought that if the Treaty of Augsburg had never been concluded the world of today would be a very different world, if Newton had not published his *Principia* in 1687 it is very doubtful whether we would find today's world of science and civilization in the same position'. One might have expected the last verb to be **olurdu** 'it would have been', but the conditional notion, the 'would have', is contained in the **bulacağımız**; see the following section.

The use of **şayet** has much the same effect as stressing the 'if' in an English conditional sentence:

gelse	if he should come (as he might)
şayet gelse	*if* he should come (which I doubt)
gelirse	if he comes (as he perfectly well may)
şayet gelirse	*if* he comes (which I am not guaranteeing)

9. Conditional sense of the future personal participle. The future personal participle is used to turn into a noun- or adjective-clause the apodosis of a remote or unfulfilled condition: **bugünkü Hanhay bölgesinde Rus jeologlar araştırmalar yapmışlar ve sonunda bu bölgenin 1000 yıl önce ormanlı olamıyacağını ortaya koymuşlar** 'Russian geologists have conducted researches in the present-day Hanhay region and have finally shown that a thousand years ago this region could not have been forested'. Here **olamıyacağı** (in the accusative as object of ortaya koymuşlar 'they have placed into the middle, have revealed') is the nominal form not of **olamıyacak** 'it will not be able to be' but of **olamıyacaktı** 'it could not be'. **eskiden kimbilir kaç gün, kaç gecede kona göçe gidecekleri bir yere şimdi üç beş saat içinde kuş gibi uçacaklardır** 'to a place to which in the old days they would have gone (**gidecekleri** here not "pertaining to their future going" but "pertaining to their future-in-the-past going") in who knows how many days and nights, constantly camping and moving on, they will now fly like birds in a few hours'. See also XXIV, §§ 27, 28.

10. The conditional base. Turkish grammarians call the conditional base, e.g. **gelse**, 'the wish-condition mood', **dilek-şart kipi**. When it expresses wishes it may be introduced by **keşki** or **keşke** (P), which syntactically is as redundant as **eğer**. The reader is once again reminded to distinguish between (*a*) the past tense of the conditional mood and (*b*) the conditional mood of the past tense:

(*a*) **bil-se-ydim** if I had known
(*b*) **bil-di-ysem** if I knew

(*a*) can be the protasis of an unfulfilled condition—**bilseydim buraya gelmezdim** 'had I known I should not have come here'—or can stand alone as a hopeless wish: **bilseydim!** or **keşki bilseydim!** 'if only I had known!' (*b*) is the protasis of an open condition in the past: **o zaman bildiysem şimdi unutmuşum** 'if I knew then, I have forgotten now'.

The first persons of the conditional base can express a diffident first-person imperative: **çarşıya gitsek** 'if we were to go to the market', with an implied apodosis **olmaz mı?** 'would it not be all right?' Cf. **eşyalarımı şuraya koysam olmaz mı** 'wouldn't it do if I were to put my things over there?' An intermediate stage may be seen in **çarşıya gitsek mi?** 'how about going to the market?' (lit. 'if we were to go . . .?'). But no apodosis seems to be implicit in **ne yapsam?** 'what should I do?' For an alternative explanation of these uses, see the end of § 14, below.

11. -sene, -senize. The interjection **e/a** is suffixed to the second persons of the conditional base to make an imperative: **desene!** 'do say!' **otursanıza!** 'do sit down!' This may be followed by **ya**

for greater emphasis. 'Oh if you'd say/sit' is a literal translation but has a petulant note not found in the Turkish, which can be courteous or impatient according to the tone of voice in which it is said.

12. -se beğenirsin? The second persons of the aorist present of **beğenmek** 'to like, to approve', are idiomatically used with a conditional verb: **bana ne deseler beğenirsin?** The meaning is a mixture of 'can you guess what they called me?' and 'what they called me will amuse you'. Note that the expression is a question, despite the absence of **mi. bizi görünce ne yapsa beğenirsiniz?** 'you'll never guess what he did when he saw us!' **ertesi gün ne olsa beğenirsiniz?** 'you'll never guess what happened the next day!' **şekercinin süprüntüleri arasında ne arasalar beğenirsiniz?** 'you'll never guess what they were looking for among the confectioner's sweepings!'

13. olsa olsa. This expression will be found in the dictionary with the meaning 'at the very most'. This is a special case of an idiomatic use of the conditional, namely that when repeated it conveys a sense of limitation. In this use, the subject can be (*a*) that of the main verb or (*b*) impersonal:

(*a*) **bilse bilse bunu bilir** 'if he knows anything he knows this'; **bu takım, yese yese, bir gol yer** 'if this team gives away anything, they'll give away one goal' (**gol yemek** 'to have a goal scored against one').

(*b*) **bilse bilse kardeşim bilir;** 'if anyone knows, my brother knows'; **dinsizin hakkından gelse gelse ancak imanı çok kuvvetli olan gelir** 'if anyone can get the better of the atheist, only he whose faith is very strong can'; **arada olsa olsa bir derece farkı vardır** 'if there is anything in between, there is a difference of degree'; **ben sana olsa olsa, bir okuyucu olarak bu pazarda iyiyi kötüden ayırmanın beylik şartlarını söyleyebilirim** 'at the very most I can tell you, as a reader, the conventional rules for distinguishing the good from the bad in this market'.

The paucity of examples under (*a*) reflects the infrequency of this use. Elöve gives one example each in the first and second persons (p. 1096): **bir saat içinde okusam okusam kırk elli sahife okuyabilirim** 'if I read at all in an hour I can read forty or fifty pages'; **onlara yardım için versen versen elli lira verirsin** 'if you give anything to help them you will give fifty lira'.

14. olsa gerek 'it must be'. **Sait Faik'in kaygusu yeni olmak değil, sahih olmak: gerçekten yeni olmasını sağlayan da bu olsa gerek** (Eyüboğlu) 'Sait Faik's concern is not to be original but to be authentic: this must be what in fact makes him original'. The construction is rare with other verbs than **ol-**: **bu fikir, yanlış bir dünya telakkisinden doğsa gerek** 'this idea must originate from a mistaken attitude to the world'. **gerek** means 'necessary' and the construction may be an elliptical conditional sentence: 'if it be, ⟨then it is as⟩ is necessary'. This possibility is supported by **olsa yeridir** in the next example, 'if it be, it is its place', i.e. it is appropriate: **Biz insanlar Allahı arayıp dururuz. Bulsak acaba ne diyeceğiz? İlk şikâyetimiz tıynetimizin bozukluğu hakkında olsa yeridir** (Hüseyin Rahmi) 'We humans are constantly seeking God. If we should find ⟨Him⟩, I wonder what we shall say? It is appropriate that our first complaint should be about the corruptness of our clay'. An alternative explanation is that the conditional sometimes overlaps the subjunctive and that the literal sense of **olsa gerek** is 'necessary ⟨that⟩ it be'. Thus we might explain **ne yapsam?** (§ 10) as synonymous with **ne yapayım?**

XXI

ASYNDETIC SUBORDINATION

JUST as co-ordination can be expressed asyndetically, i.e. with no conjunction (cf. XIII, 1), so subordination may be expressed with no visible subordinating link.

(*a*) With verbs of thinking and perceiving: **vagonu doldu sanarak başka yere gidecekler** (Reşat Nuri) 'thinking the compartment full, they will go elsewhere'. **sanmak** 'to think' regularly has two objects, e.g. **seni arkadaş sanıyordum** 'I used to think you ⟨a⟩ friend'. The second object here is the finite verb **doldu** 'has become full'. **zihni gayriihtiyarî bir hayal yaptı; bir lâhza yumurcağı yatağında sapsarı yatıyor gördü** (idem) 'her mind involuntarily created a vision; for an instant she saw the awful child lying deathly pale in his bed'.

The common locution **sizi gitti mi bildi** is puzzling unless one remembers that **bilmek** means not only 'to know' but also 'to consider, to guess': 'he wondered whether you had gone', lit. 'he considered you ⟨as⟩ "has he gone?"' Note that the verb of the subordinate clause is in the third person, although a second person is being addressed, as in this example: **ben seni öldü biliyordum** 'I was thinking you were dead' ('I was considering you ⟨as⟩ "he has died"'). The subordinate verb may be in the first or second person and the pronoun object can then be omitted: **memleket geri, diyoruz; ve memleketi geri bulduğumuzu açıklamakla, ileri olduğumuzu isbat ettik zannediyoruz** 'we say "the country is backward" and, by making it clear that we find the country backward, we think we have proved that we ⟨personally⟩ are advanced'. This is a neat and euphonious alternative to . . . **ileri olduğumuzu isbat ettiğimizi zannediyoruz. şimdi ona ne söylüyordur biliyor musun?** (Y. Kemal) 'do you know what he must be telling him now?' A more literary way of saying this is **şimdi ona ne söyliyeceğini biliyor musun?** 'do you know what he will be telling him now?' **bana ne oldu hiç sormayınız** 'don't ask what has happened to me'. **bir de bak-**

tım, son durağa gelmişiz 'I gave a look, we have arrived, I realized, at the last stop'. To assume an ellipsis of **ki** in this last example is natural for English-speakers, who feel that 'I saw we had arrived' is short for 'I saw that we had arrived'. But, as we have seen in XIII, 15, **ki** is not an essential element of such sentences.

Besides the verbs already mentioned, the construction is found with **saymak** and **addetmek** 'to count', **farzetmek** 'to suppose', and **duymak** and **hissetmek** 'to feel'.

(b) With expressions of time formed with **-dir** and **ol-**. To the examples given in XII, 23 may be added: **iki defa-dır muhacir olduk** 'it is twice we have been exiles'; **o gün bugün-dür devam eder** 'it has been going on like that ever since', lit. 'that day is today it continues'; **yazarın da günleri olur, kaleminden bal akar** 'the writer too has days ⟨when⟩ honey flows from his pen'. Cf. **bir gün gelecek, beni unutacaksın** 'a day will come ⟨when⟩ you will forget me'.

(c) **bir kıyamettir koptu**, etc. See VIII, 43.

(d) **söylenir durur**, etc., and **ısınamadım gitti**, etc. See XI, 35 (a) and (h).

(e) Conditional sentences with a question or an imperative as protasis. See XX, 2 (b) and (d). **insan çalıştı mı her şeyi başarır** 'if one works one accomplishes everything', lit. 'has man worked? he accomplishes everything'. **yapma, öldürürüm seni** 'stop it, I'll murder you'. In translating, we provide a subordinating link by inserting respectively 'in that case' and 'or', but the subordination is clear to a Turk without any such device.

It is debatable whether to include under this heading the use of the subjunctive past in unfulfilled conditionals, e.g. **bileydim buraya kadar gelmezdim** 'had I known, I should not have come thus far'. On the whole it seems best to call it an instance of asyndetic subordination, as this explains how the use arose, whereas the stock explanation, that in such sentences the subjunctive past is synonymous with the conditional past, explains nothing.

XXII

PUNCTUATION

THERE is no general agreement among Turkish writers or printers on how to punctuate and the reader must be prepared for anything. The advice given in the Introduction to *OTD* is not without its occasional value: 'If you are completely bewildered by some sentence, try cutting out all the punctuation marks and often you will find it quite easy to translate.' One cannot, however, afford to ignore a comma near the beginning of a sentence, which usually indicates the subject: **bu Bakanın kararıdır** 'it is this Minister's decision'; **bu, Bakanın kararıdır** 'this is the Minister's decision'. **üniversiteli kızı kaçırdı** 'he abducted the girl undergraduate'; **üniversiteli, kızı kaçırdı** 'the undergraduate abducted the girl'.

The semicolon is rare; the reader may have noticed how often the translation of an example uses a semicolon where the original has a comma.

A colon, even when it does not introduce direct speech, may be followed by a capital letter: **Karışık bir duygu var içimde: Bu yıl bana hem pek kısa, hem de pek uzun geliyor** (N. Ataç) 'I have a mixed feeling inside me: I am finding this year both very short and very long'. **Bir gün evvel kurduğu kapana baktı: Kapan nasıl bıraktıysa öyle duruyordu** 'He looked at the trap he had set a day earlier: the trap was standing just as he had left ⟨it⟩'.

Quotation marks are entirely dispensed with by some writers: **Niçin diyeceğim geliyor** (S. Kocagöz) 'I feel like saying "Why?"'. Some use a dash and quotation marks or a dash alone to mark a change of speaker in a dialogue:

—Öyleyse sorumlu kim? diye bağırdım.
Gözleri gözlerimde,
—Sorumlu düzen, bütün suç düzenin . . . dedi (A. Nesin)

'"In that case, who is responsible?" I cried. His eyes on mine, he said, "The one responsible is the system; all the fault lies with the system"'. See also the last example in the first paragraph of XIX, 2, where each speaker's words are introduced but not closed by quotation marks.

The word of saying, etc., may be written with a capital letter: **Babamın beni dinlemesi lâzım ... Deyip duruyordu.** '"My father must listen to me", he kept saying'. Cf. **Diye** in the next example.

In Ottoman printing, brackets were used round quotations and words especially emphasized. Relics of this practice may still be encountered: **İleri giden de yok. Tramvayda: [İleri gidilmesi rica olunur] Diye yazılı olmasına rağmen** (B. Felek) 'Nor is there anyone going forward, in spite of the fact that there is a notice in the tram reading "Passengers are requested to go forward"' ('in spite of its being written saying ...'). **Ziyaretçiler, heyet mensuplarından birinin deyimi ile (utanç verici) bir kabul görmüşlerdir** 'The visitors, in the words of one of the members of the group, had a "shameful" reception'.

The use of three dots, to indicate that something has been left to the imagination, is very common, especially after a final **ki**: **Niçin cevap vermiyorsun?—Birşey sormadın ki ...** 'Why don't you answer?'—'You haven't asked anything ⟨for me to answer⟩'.

The question mark is often omitted after rhetorical questions or replaced by an exclamation point: **Çocuk cop ile dövülür mü!** 'Does one beat a child with a truncheon?' It may be omitted after polite requests couched in question-form: **Bir su lûtfeder misiniz** 'Will you be so kind as to let me have a glass of water?'

Question marks and exclamation points, in or out of brackets, are used to indicate irony, presumably by those who do not trust their readers to detect it unaided. **akrabam olacak o zat-ı şerifin ? bana etmediği kalmadı** 'there is nothing which that honourable gentleman who is supposed to be my kinsman has not done to me'. **her yazdığı cümlede iki üç hata yapan büyük âlim (!) şimdi ne diyor?** 'what is the great scholar, who makes two or three mistakes in every sentence he writes, saying now?'

Names of months and days are generally written with a small letter and not a capital.

XXIII

SENTENCE-ANALYSIS

THE present chapter is devoted to a word-by-word demonstration of how one sets about translating a complicated sentence. The great difficulty is not so much that the Turkish verb usually comes at the end. It is that as the Turkish qualifier precedes whereas the English qualifier generally follows, a native English-speaker has virtually to read the Turkish sentence backwards. One important reservation must, however, be made. When the author has used side by side several words or clauses of the same grammatical function, whether or not they are joined by a conjunction they must be translated in the order in which he wrote them and not backwards, i.e. not in the order in which one meets them as one works back from the verb.

The two specimen sentences are both somewhat longer than the modern norm. The first, which is from İ. H. Danişmend's *İzahlı Osmanlı Tarihi Kronolojisi* (Istanbul, 1947–1955), ii. 183, also exceeds the modern norm in its proportion of Arabic words.

Muazzam bir devletin fakir, zayıf ve muhtaç bir millete bir takım siyasî ve askerî mülâhazalarla sadaka şeklinde vermiş olduğu bu müsâadelerin istiklâl mefhumunu ihlâl eden birer siyasî imtiyaz mahiyetini alması, Osmanlı imparatorluğunun inhitat devrinde kuvvetlenmiş olan Avrupa devletlerinin bu eski müsâadeleri gittikçe sûiistimâl etmelerinden mütevellit ve bilhassa son devirlere münhasır bir vaziyettir.

In a sentence of some length one can generally rely on the author's putting a comma after the subject. The first comma is after **fakir** 'poor', which is indeed in the absolute case. But if this is the subject the preceding words must qualify it; being in the genitive case, however, they can only qualify in izafet, i.e. they can only qualify a word with the third-person suffix. Moreover, as the three words after the comma mean 'weak and needy', they clearly belong with **fakir** as qualifiers of **bir millete**. The next comma is after **alması**, which is also in the absolute case, so we shall assume that this is the subject and that everything preceding qualifies it. **alma-sı** 'its taking' is the second element of an izafet, the first element of which cannot be **mahiyetini** as that is in the accusative; the izafet must be a definite one and we have to find its defining genitive. The nearest preceding genitive is **bu müsâadeler-in** 'these concessions'. So far we have 'these concessions' taking', i.e. 'the fact that these concessions take', and the object of 'take' is **mahiyet-i-ni** immediately preceding it: 'its nature'. This is the second element of an izafet of which **imtiyaz** 'privilege' is the qualifier. It is itself qualified by **birer siyasî** 'each-one-a political ':'the nature of so many political privileges'. The four preceding words must also qualify **imtiyaz: ihlâl ed-en** 'which violate' / **istiklâl mefhum-u-nu** 'the concept of independence'. Now we fit together the izafet **muazzam bir devlet-in** 'a mighty State's / **olduğ-u** 'pertaining to its being' / **vermiş** 'having-given'; i.e. 'which a mighty State had given' / **sadaka şekl-i-nde** 'in the form of charity' / **-la** 'with' / **bir takım . . . mülâhazalar** 'a number of political and military considerations' / **fakir . . . bir millet-e** 'to a poor, weak, and needy nation'.

So much for the subject. Now we look at the end of the sentence for the verb: **-tir** 'is'. Its complement is **bir vaziyet** 'a situation'. The remainder of the sentence qualifies this word. **münhasır** 'confined' / **son devirler-e** 'to the latest periods' / **bilhassa** 'especially'—but now we come to **ve** 'and', so we should first have translated the clause before it. **mütevellit** 'originating' / **etme-leri-nden** 'from their doing' / **sûiistimâl** 'abuse';[1] i.e. 'from their abusing' / whose abusing? **Avrupa devletler-i-nin** 'the European States'' / **kuvvetlenmiş olan** 'which had grown strong' / **Osmanlı . . . devr-i-nde** 'in the Ottoman Empire's period of decline'. All we need now is an object for 'abusing': **bu eski müsâadeler-i** 'these ancient concessions' / **gittikçe** 'gradually'.

'The fact that these concessions, granted by a mighty State as charity to a poor, weak, and needy nation in view of a number of political and military considerations, assumed the character of political privileges which violated the concept of independence is a situation born of the gradual abuse of these ancient concessions

---

[1] This is a Persian izafet group, composed of two Arabic words: *sū'* 'evil' and *isti'māl* 'use'.

by the European powers, which had grown strong during the Ottoman Empire's period of decline; a situation peculiar to the latest times.'

The second specimen is from Niyazi Berkes's *200 Yıldır Neden Bocalıyoruz* (Istanbul, 1965), pp. 129–30.

Türkiyede devletçilik programının uygulanılışına girişilirken, plânlamanın yalnız sanayi alanına teksif edilmesi, toprak hukuku reformunun önlenmesi, sanayileşme ilerledikçe bunun tarımsal makineleşmeye hem teknik hem ekonomik sebeple tesir edememesi tarım alanının plânlama dışında ayrı bakanlıkların sürekli olmayan, çok defa birbirini tutmayan gelişi güzel tedbirlerine bırakılması, özellikle eğitim alanı ile tarım alanı arasında hiçbir plânlı ilişiklik kurulmaması, okuma-yazma öğretmekle köylünün kalkınacağına inanılması, ve en sonunda da sanki çok kahramanca bir iş imiş gibi köylüye mükâfat tevzi eder gibi, toprak dağıtma gibi sözde - reformlara gidilmesi devletçiliğin başarısızlığa uğratılmasında başlıca rolleri oynamıştır.

We may leave aside for the moment the first clause, ending in **-ken** 'while', as it is obviously an adverbial clause of time, and concentrate on the main sentence.

A rapid glance shows seven **-me** verbal nouns with the third-person suffix but no case-ending, five of them followed by a comma (the omission of commas after **edememesi** and **gidilmesi** must be accidental). These verbal nouns we may take to be the subjects of the final verb **oynamıştır** 'has played', or rather, as there are several subjects, 'have played' / **başlıca rolleri** 'the principal roles' / **devletçiliğ-in uğratılma-sı-nda** 'in étatism's being brought' / **başarısızlığ-a** 'to successlessness'. Now we deal with the seven noun-clauses. **plânlama-nın teksif edil-me-si** 'planning's being condensed' / **yalnız sanayi alan-ı-na** 'only to the sphere of industry'. Then comes a four-word izafet chain: **toprak hukuk-u reform-u-nun önlenme-si** 'land-law-reform's being-prevented'. **sanayileşme ilerledikçe** 'as industrialization advanced' / **bu-nun tesir ed-eme-me-si** 'its inability to affect' / **tarımsal makineleşme-ye** 'agricultural mechanization' (**tesir etmek** 'to affect' is construed with a dative) / **sebep-le** 'by reason' / **hem teknik hem ekonomik** 'both technical and economic'. **tarım alan-ı-nın bırakılma-sı** 'the agricultural sphere's being-left' / **plânlama dışında** 'on the outside of planning' / **ayrı bakanlıklar-ın gelişi güzel tedbirler-i-ne** 'to the haphazard measures of different ministries'. **sürekli olmayan** and **çok defa birbirini tutmayan** both end in present participles and amount to relative clauses qualifying **tedbirlerine**: 'which were not continuous and which often did not hold each other', i.e. which were often mutually conflicting. **hiçbir plânlı ilişiklik kurul-ma-ma-sı** 'no planned relationship's being-established'; the izafet is indefinite as the qualifier, being negated, cannot be definite. **özellikle ... arasında** 'especially between the sphere of education and the sphere of agriculture'. **inan-ıl-ma-sı** 'its being believed'; the passive is impersonal, as **inanmak** 'to believe' is construed with a dative and cannot be made into a true passive. The **-sı** has no visible antecedent and its presence is clearly due, at least in part, to analogy with the five preceding verbal nouns which are in izafet with qualifiers. The dative required by **inan-** is **köylü-nün kalkınacağ-ı-na** 'the villager's future progressing': 'the belief that the villager would progress' / **okuma-yazma öğretmek-le** 'by teaching reading–writing'. **ve en sonunda da** 'and at the most last of it' / **gid-il-me-si** 'going's being done', another impersonal passive, which we may paraphrase by 'having recourse' / **sözde-reformlar-a** 'to reforms in word', i.e. 'to so-called reforms' (the hyphen is not strictly necessary but is probably intended to make **sözde** a sort of prefix, corresponding to 'pseudo-'). **toprak dağıtma gibi** 'such as distributing land' / **sanki ... imiş gibi** 'as if it were a very heroic action' / **köylüye ... eder gibi** 'as if distributing largess to the villager'. Now the first clause. The verb **girişmek** 'to enter upon' is another of those which take a dative, so the passive **giriş-il-** is impersonal. 'In Turkey, while one was entering upon' / **devletçilik program-ı-nın uygulan-ıl-ış-ı-na** 'the étatism-programme's being-applied'. It is probably best to break up this huge structure when translating into English.

'The following factors, which existed when the programme of étatism was being put into effect in Turkey, were largely responsible for the failure of étatism. (*a*) Planning was confined to the sphere of industry. (*b*) The reform of the land-law was prevented. (*c*) As industrialization advanced, the less was it able, for technical as well as economic reasons, to affect agricultural mechanization. (*d*) The sphere of agriculture was left outside the scope of planning and abandoned to the haphazard measures taken by several ministries; measures which were discontinuous and often in conflict with each other. (*e*) No planned relationship was set up, in particular between education and agriculture. (*f*) There was a belief that the advancement of the peasant could be achieved by teaching him to read and write. (*g*) Recourse was had to such pseudo-reforms as land-distribution, as though this were a piece of great magnanimity; as if bestowing largess on the peasantry.'

XXIV

FURTHER EXAMPLES

The sentences below are intended to provide supplementary illustrations of various points of grammar.

1. Baraj'da toplanacağı hesaplanan 15 milyar metreküp su ile, kurak bölge rahatça sulanabilecektir 'with the 15 thousand million cubic metres of water which, it is calculated, will be collected at the dam, it will be possible for the dry zone to be irrigated quite easily'. **toplanacağı hesaplanan su** 'the water whose future collecting is being calculated'. See XVIII, 2 (c).

2. Üçü yumurtlıyan on bir tavuğu var 'he has eleven hens, three of them laying'.

3. Solcu piyes sağcı piyes diye bir ayırım yapmak, ömürlerinde kaç defa tiyatroya gittikleri meraka değer bu sayın üyelerin haddi değildir 'to make a distinction between "leftist plays" and "rightist plays" is not the place of these honourable members, concerning whom one may well wonder how many times in their lives they have ever been to the theatre'. **kaç defa . . . gittikleri merak-a değer** 'their having gone how many times ⟨being⟩ worthy of curiosity'.

4. Karısının evlerine sığındığı iki kadını öldüren adam tevkif edilmiştir 'the man who killed the two women in whose house his wife had taken refuge has been arrested'; lit. 'the two women pertaining to his wife's taking refuge in their house'.

5. İki hemşeri olduğu sözlerinden anlaşılan iki hamlacı (Sait Faik) 'two chief rowers, from whose speech it was evident that they were two fellow-townsmen'; **. . . olduğu sözlerinden anlaşılan** 'their-being . . . being-understood from their words'.

6. Şayet görürsem elini öpeceğim bir okuyucu, bir mektup yazdı bana 'a reader, whose hand I shall kiss if I see him, has written me a letter'.

7. **Sen de ticaret mi yapıyorsun, delikanlı?—Neden yapmıyacakmışım?** 'Do you engage in commerce too, young man?'—'Why shouldn't I?' **yap-mı-y-acak-mış-ım** is the inferential form of **yapmıyacaktım** 'I should not', and the literal meaning of the question is 'why is it inferred ⟨by you⟩ that I should not?'

8. **Kız, şansın varmış . . . Ya bu herifle evlenseymişsin . . . —Allah korumuş** (Aziz Nesin). 'Daughter, it is to be inferred ⟨from what we now see⟩ that you were lucky (**şans-ın** "your luck" **var** "existent" **-mış** "I gather that it was"). And what if—I see it all now—you had married this scoundrel!' (the words in dashes in this last sentence represent the difference between the conditional inferential **evlen-se-y-miş-sin** and the conditional past **evlen-se-y-din**)—'God protected me, I now realize' (the last three words of the translation would have been unnecessary had she said **koru-du** instead of **koru-muş**).

9. **Geçmiş olsun, evinize hırsız girmiş.—Girdi, dedim** (idem). ' "I hope you soon get over it; I hear your house has been burgled."—"It has indeed", I said.' Lit. ' "May it be past; thieves, I gather, have entered your house."—"They have in fact entered", I said.'

10. **Hanım şoförlerimiz iftihar etsinler, yalnız Türk erkeklerinden değil, Belçikalı şoförlerinden de ihtiyatlı vasıta kullanıyorlar. Daha doğrusu kullanıyorlarmış.** 'Let our lady drivers take pride; they drive ("use vehicles") not only more carefully than Turkish men, but also more carefully than Belgian drivers. To be more accurate, they are said to do so.'

11. **Eski bir nahiye müdürü (ki orada kalsa imiş şimdi vali olabilirmiş) kendini sinemaya vermiş** (Doğan Nadi). 'A former regional director (who, if he had stayed there, it is said (**imiş**), could now have been a provincial governor, it is said, (**-miş**)) is said to have dedicated himself to the cinema.'

12. **Güya** (*sic*, for **gûya**), **sen, çıkardığın Kadro mecmuasında iktisadî siyasetimizi baltalayan ve hattâ Parti Umumî Kâtibinin iddiasına göre Rejim'in temellerini sarsan (!)**

neşriyatta bulunuyormuşsun. Bu, böyle giderseymiş Ticaret Vekili tuttuğu yolda emniyetle ilerleyemezmiş. Öte yandan Cumhuriyet Halk Partisi de hizipleşmek tehlikelerine maruz kalırmış (Yakup Kadri). 'Allegedly, you, in the magazine *Kadro* which you put out, have been engaging in publication which sabotages our economic policy and even, according to the assertion of the Party General Secretary, shakes the foundations of the régime. If this were to go on like this, they say, the Minister of Commerce would not be able to proceed safely on the road he has taken. Furthermore, the RPP would, they say, be left exposed to risks of breaking up into factions.'

13. Dokunmuşlar; dedi ve gittikçe ağırlaşan bir sesle ilâve etti: Dokunmuşlar değil dokundular; benim gözümün önünde ... (idem). '"I heard they assaulted ⟨her⟩", he said and added, in a voice which grew gradually heavier, "No, I didn't hear it; they actually did it, in front of my eyes"'.

14. Ama bütün bu gerçeklere inat, sanatı paranın, maviyi karanın üstüne çıkaranlar var ya? Binde bir de olsun var ya? İşte onlar sanatçı: üst tarafı manatçı! Çok mu sert oldu bu yargı? Yumuşatalım biraz: bütün manatçıların sanatçı olduğu zamanlar vardır (Eyüboğlu). 'But despite all these facts, there are, are there not, those who put art above money, the blue above the black (XVII, 3)? They exist, even though they be one in a thousand, don't they? It is precisely these who are artists; the remainder are etceteras! Has it been very stern, this judgement? Let's soften it a little: there are times when all the etceteras become artists.' Here **manatçı** has been extracted from the m-doublet **sanatçı manatçı** 'artists and so on' and given an independent life of its own.

15. Ama devletliler bir kuşkulanmıya görsün (XI, 35 (*d*)), en merhametliler en zalimler bir anda birleşiyor, din kardeşi min kardeşi dinlemiyorlardı (idem). 'But let the Establishment not grow suspicious; the most compassionate and the most tyrannical would unite in an instant and would not heed any considerations of common religion or common anything else' ('they would not listen to "religion-brother" or "anything-else-of-the-sort-brother"'; **din min**).

16. **Bizde eleştiri olup olmadığı yıllardır tartışılıp durur. Varılan sonuç, bizde eleştiri olmadığıdır.** 'Whether or not we have any literary criticism has been constantly debated (XI, 35 (*a*)) for years. The conclusion that has been reached is that we have no literary criticism.' **varmak** 'to arrive' is intransitive; **varılan** is therefore the impersonal passive participle; see XVIII, 3 (*a*).

17. **Bazı günler, Café Soufflot'nun mutad toplantılarına Sami Paşazade Sezai Bey ve Prens Mehmet Ali gibi— Ragıp Beyin tabirince—aristokratların da iştirak ettiği olurdu.** 'On some days it would happen that—to use Ragıp Bey's expression—aristocrats too like Sezai Bey, the son of Sami Pasha, and Prince Mehmet Ali, would join in the usual meetings at the Café Soufflot.' **iştirak ettiği** 'their participating' / **olurdu** 'used to occur'.

18. **Hacı olan bir Müslüman sosyal hayatını nasıl tanzim etmelidir?—İslâmiyet "hüsn-ü ahlâk" olduğuna göre hacı olup olmamaklığın buna bir tesiri yoktur.** This is a reader's question and a newspaper columnist's answer (Refik Ulunay). Its grammatical interest lies in the use of the somewhat rare verbal noun in **-meklik** (see X, 6). There could have been no possibility of ambiguity, as the verbal noun is negative and in the genitive, if the writer had chosen to use the **-me** verbal noun instead: **olmamanın**. But clearly he wished to be both precise and impressive, hence **olmamaklığın** 'of the fact of not being'. 'How ought a Muslim who is a pilgrim to arrange his social life?'— 'As Islam is moral excellence (II, 26), the fact of being or not being a pilgrim has no effect on this.'

19. **Ey Allahım, bütün insanlara, onların senin çocukların ve birbirlerinin kardeşi olduklarını öğretmen zamanı gelmedi mi?** (Halide Edip). 'O God, has not the time come for You to teach all mankind that they are Your children and each other's brothers?' The termination of **çocuklar-ın** is not the genitive suffix but the second-singular possessive. Probably a writer of a later generation would have chosen an alternative to **öğretme-n** 'your teaching', because of the far commoner **öğretmen** 'teacher'.

20. **İsmet Paşanın şimdiye kadar olmaz dediğinin, olduğunu bilen varsa parmak kaldırsın.** 'Hands up anyone who knows of anything ever yet happening which Ismet Pasha said wouldn't happen!' Lit. 'if there is anyone who knows the happening of that pertaining to Ismet Pasha's saying "It won't happen" (**olmaz**) till now, let him raise finger'.

21. **Bayramlaşamadıklarımız.** 'Those of our number with whom we cannot exchange the season's greetings.' This splendid word headed the obituary column of a newspaper at the **Bayram**, the festival which ends the month of fasting, in 1960. **Bayramlaş-** 'to exchange **Bayram** greetings'; -**ama**- VIII, 55 (*b*); -**dıklarımız** IX, 7.

22. **Daha sonra Hazreti Muhammedin Medineye ilk ayak bastığı gün devesinin kapısında diz çöktüğü Eyüp Sultanın türbesi ziyaret edildi.** 'Later a visit was paid to the mausoleum of Eyüp Sultan, at whose door the Prophet Muhammad's camel knelt the first day he set foot in Medina.' This is an involved example of the rule given in XVIII, 2 (*d*): 'Eyüp Sultan, characterized-by-its-bending knee' (**diz çöktüğü**) 'at-his-door' (**kapısında**). The 'its' refers to **Hazreti Muhammedin devesinin** 'His Excellency Muhammad's camel's'. **Hazret** (A) means 'Presence' and is used in Persian izafet as an honorific.

23. **Bu konuda yayımlanan değerli bir makalede, Şinasi'nin piyes yazması, "Fransız tiyatrosunu yakından görüp tanıdıktan sonra Garp tiyatrosunun edebî değerini iyice kavramış olması" ile yorumlanmış, ve, eserini "oynatma ümidi olmaksızın yazdı"ğı söylenmiştir** (Cevdet Kudret). 'In a valuable article published on this theme, Shinasi's writing of plays has been interpreted by "his having thoroughly grasped the literary value of the Western theatre after seeing and becoming acquainted with the French theatre from close at hand" and it has been said that "he wrote with no hope of producing" his play.' This sentence is remarkable for a grammatical oddity not unlike the one discussed on p. 49, n. 3. The quotation **oynatma ümidi olmaksızın yazdı** 'he wrote without there being hope of getting-performed' has been turned into a noun clause, subject of **söylenmiştir**, by adding **ğı** to the finite verb **yazdı**, instead

of closing the quotation at **olmaksızın** and then putting **yazdığı**. This is not a unique example of this use.

24. **Görülüyor ki, verdiğimiz örneklerde ve veremediğimiz binlercesinde, dil mantığı kolaylık ve sürati sağlamak için, kendi kanunlarına aykırı da olsa bazı tasarruflara gitmektedir.** 'It is seen, in the examples we have given and the thousands that we have not been able to give, that the logic of language, in order to ensure ease and speed, resorts to certain economies, even though they be contrary to its own laws.' Noteworthy here is the substantivizing effect of the third-person suffix on the adverb **binlerce** 'by thousands'.

25. **Dün Köprüden geçerken Fatih camiinin minaresine bayrak çekildiğini gördüm. Yarın öbürgün bu âdet de yerleşirse Demokrat Parti zamanında yerleşmiştir diye tarih kitapları yazsın için, ben de buraya yazıyorum.** (B. Felek). 'Yesterday while crossing the Bridge I saw that flags had been hoisted on the minaret of the Fatih mosque. I am writing ⟨this⟩ here so that if this custom too takes root, tomorrow or the next day, the history-books may write that it took root in the Democrat Party era.' The rare use of **için** with a third-person imperative to express purpose (VIII, 41) is doubtless to avoid repeating **diye**; **yazsın diye** would have been the natural way of saying 'so that they may write'.

26. **Kırık Ali—Müsaaden olursa kalkalım hoca efendi, bize izin ver! dedi. Vara demiyeydi. Bekir Hocadır bu, yakasına bir yapışmasiyle Kırık Ali'yi sandalyeye çökertti** (Aziz Nesin). 'Broken Ali said, "With your permission (lit. 'if your permission exists'), Hoja Efendi, let's go; give us leave!" He might as well not have said it (VIII, 37 (*a*), at end). It was Bekir Hoja that he had to deal with (p. 71, foot); with one grab at his collar he sent Broken Ali flopping back into the chair.'

27. **Eğer öğrenci bir sorunun beş şıkkından birini işaretliyecekken iki şıkkı doğru diye işaretlemişse, elektronik beyin oyuna gelmeyip hemen bunu farketmekte ve öğrencinin kurnazlık yaptığını ortaya koymaktadır.** 'If the student, while supposed to mark one of the five alternatives (i.e.

alternative answers) of a question, has marked two alternatives as correct, the electronic brain is not taken in but notices this at once and reveals that the student is guilty of sharp practice.' Note the necessitative sense of the future participle in **işaretli-y-ecek-ken**, not 'while he will mark' but 'while he should mark'. See the first sentence of VIII, 21, and cf. the next example.

28. **Tavuklar en yumurtalayacakları zamanda yumurtalamayıverdiler.** 'The hens, at the time when they should most have laid, suddenly stopped laying.' See XI, 35 (*f*).

29. **Benim de gülmüşlüğüm vardır Atatürk'ün kullandığı bir çok terimlere** (Eyüboğlu). 'There have been times when I too have laughed at a number of terms which Atatürk used.' For the construction, see XVII, 2, and cf. the next example.

30. **Bu seçimin bütün problemleri çözmüşlüğü ileri sürülemez.** 'It cannot be suggested that this election has solved all the problems.' Lit. 'this election's state-of-having-solved all the problems cannot-be-pushed forward'.

31. **Proje gerçekleşseydi hakikaten Türklere düşecek pay ancak bu ruhanî pay olacaktı; maddî payı varsın maddeye tapan gâvurlara kalsındı** (Niyazi Berkes). 'If the project were to materialize, in fact the share that would fall to the Turks would be only this spiritual share; the material share of it might as well be left to the infidels, who worshipped the material.'

32. **Âdeta kalın bir yağmur bulutu içine girdik ve etrafı değil, gittiğimiz yolu göremez olduk** (R. N. Güntekin). 'We entered what was virtually a dense rain-cloud and ceased to be able to see not ⟨just⟩ our surroundings but the ⟨very⟩ road on which we were going.' For **göremez olduk** 'we-became unable-to-see', see IX, 11.

INDEX

In arranging the entries in alphabetical order, no regard has been paid to divisions between words or suffixes; thus **-mesine** precedes **-mesin mi**.

Suffixes which may begin with **c** or **ç**, **d** or **t**, **ġ** or **k**, will be found in the forms beginning with **c**, **d**, and **ġ** respectively. See p. 12. Suffixes subject to vowel harmony which may occur with **e** or **a** will be found in the forms with **e**; similarly, those which may occur with **i**, **ı**, **ü**, or **u** will be found in the forms with **i**. See p. 18. Thus for **-tük** see **-dik**; for **-ça** see **-ce**. The index includes not only individual suffixes but also such combinations of suffixes as may be difficult for the novice to unravel; for example, **-emememe**.

A hyphen following a suffix indicates the omission of **-mek**; i.e., it shows that the suffix makes verb stems. The few exceptions to this rule are noted; see, for example, the first of the two entries **-n-**.

Bold figures are used under some entries which have several page references, to indicate the principal reference if it does not happen to be the first.

The following abbreviations have been used in addition to those shown on p. xxiv:

adj.	adjective	neg.	negative
adv.	adverb	subj.	subjunctive
interj.	interjection	vbl.	verbal

-a, -â (adv.), 196.
abbreviated nouns, 232.
ablative, 28–29, 37–38, 195, 201.
  followed by adjective, 54.
  followed by adverb, 196–7.
  partitive use of, 47, 253.
  postpositions with, 89, 95.
absolute case, 28, 35, 36, 85–86, 195.
  postpositions with, 85–87.
abstract nouns, 62, 173, 256[1].
acaba, 190, 196.
accentuation, 20–24, 47, 74, 76.
  of adverbs, 22–23.
  of compound words, 23, 74.
  of diminutives, 23.
  of foreign words, 22.
  of gerunds, 176, 177, 179, 182.
  of izafet groups, 47.
  of place-names, 21–22, 47.
  of verbs, 23, 109, 120–1, 122, 132, 143, 170.
accusative, 28–29, 34, 35–36, 53, 248.
adaş, 64.
âdeta, 198, 289.
adjectives, 43, 44, 51, **53–56**, 57–66.
adverbial forms of the verb, 174.
adverbs, 35, 43[1], 92, 138, 177, 182, **193–205**.
  of place, 198–9.
  of time, 200–5.
ağabey, 40, 232.
ait, 89.
akşamüstü, 92, 201.
alafranga, 95.
aleyh, 94–95.
alphabet, xxi, 1–2.
alt, 90, 92, 93.
ama, 210.

INDEX

aman, amanın(ız), 137.
Amerikalı, Amerikan, 44–45.
amma (da), 210.
-an (adv.), 196
-an (Persian pl.), 28.
ana (dat. of ol), 67.
anbean, 196.
ancak, 211.
anda, andan, 67.
anı, anın, 67.
-ane, see -hane.
-ane (adj. and adv.), 53, 66.
anlam, 225.
anlaşılmak, 150.
anne, 13, 22.
aorist tense, 107, **115–22**, 136.
  participle, 161–2.
apodosis, 268–9.
apostrophe, 2, 11, 34, 74, 80, 215².
apposition, 80, 86, 252–3.
ara, 90, 92, 233, 253.
Arabic borrowings, xx–xxi, 2, 3–4, 5,
  6–10, 11–12, 14–15, 17, 19–20,
  25, 55, 82–83, 86, 94–95, 154,
  156, 196, 239.
  dual, 28.
  ordinal numerals, 82–83.
  plurals, 26–28.
aracılığıyle, 147.
arasında, 90.
arka, 90, 92.
Armenianisms, 109, 119.
art, 90, 92.
artık, 203.
assimilation:
  of consonants, 12.
  of vowels, 15–16.
  labial, 16.
  palatal, 15–16.
  regressive, 16.
asyndetic subordination, 274–5.
aşağı, 198–9.
aşırı, 199.
-at (Arabic pl.), 27, 51.
-*at*-, 8¹, 65.
atak, 223.
atmasyon, 172.
auxiliary verbs, 154–6, 176.
aykırı, 89.
ʿayn, 8, 14, 15, 32–33.
aynı, 51, 76–77.
aynısı, 77.
ayni, 76.

az, 54, 75.
azar azar, 194.
az buçuk, 81.
Azeri, Azerbaijani, xix, 179¹.

bakımından, 93, 196.
bakındı, 138.
bases, tense- and mood-, 107–8, 136.
baş (classifier), 80.
  (secondary postposition), 90, 92.
başa baş, baş başa, 195.
başıbozuk construction, 234, 259–62.
başka, 76, 89, 184.
başkası, 76.
baştan başa, 195.
bazı, 51, 74.
-be- (Persian preposition), 196.
ben, 67–69.
bencileyin, 205.
bendeniz, 68.
beraber, 187, 261.
beri, 89, 181, 191, 198–9.
beriki, 72.
bıldır, 204.
bile, 166, **197–8**.
bilir bilmez, 182.
bir, 36, 53–54, 79, 83, 197, 248.
bir an evvel, 195.
biraz, 75.
bir bir, 193–4.
birbir, 76.
birçoğu, birçok, 75.
bir daha, 204.
bir de, 197.
birdenbire, 195.
bir de ne göreyim, 197.
birer, 83, 194.
biri, 48, **74**.
biribir, 76.
birisi, 49, 74.
birkaç, 76.
bir -ler, 54.
birlikte, 187, 261.
bir takım, 23, 75, 253.
birtakım, 23, 75.
bir türlü, 197.
biz, 67–69.
bizler, 68.
böyle, boylesi, 72, 74, 75.
böylece, böylecene, 195.
boyunca, 94.
brackets, 277.
bu, 24, 71–72.

buçuk, 81, 83.
buffer-letter, 29.
bugün, 17, 201.
bugünden, 201.
bulunmak, 149, 159.
bundan böyle, 203.
bununla beraber, birlikte, 211.
bura, 199-200.
burası, 48, 199-200.
burda, burdan, 199.
bu yana, 89.
buyurmak, 155.
bütün, 76.

caba, 196.
canım, 41.
capital letters, 1-2, 276-7.
cardinal numerals, 79-80.
cases:
  endings after figures, 80.
  endings after sentences, 258.
  summary of endings, 34.
  uses, 28-29, 35-38.
causative verb, 144-8, 155.
-ce (adv.), 23, 58, 183, 185-6, **194-5**, 196, 200, 237.
-ce (diminutive), 57-58, 161.
-ceğiz, 57.
-cek, 57.
-cene, 195.
-cesine, 23, **188-9**, 195.
characteristics, tense- and mood-, 107.
CHP, 46[1].
ci (after figures), 82.
-ci, 12, 19, 50, **59-60**, 220.
-cik, 23, **57**, 200, 203.
-cil, 66.
-cileyin, 205.
circumflex accent, 2, 3-4, 7.
classifiers, 80-81.
collective numerals, 84.
colloquialisms, xxii, 40, 44, 49[2], 85-86, 87, 97-98, 102, 112, 121-2, 133, 138, 161, 165, 168, 172, 175, 180, 182, 189, 191, 192, 195, 196, 203, 204, 212, 214, 218-19, 233, 235-6, 237-8, 242-4, 256, 258, 268, 274.
comparison of adjectives, 38, 54-55.
  of adverbs, 196-7.
complements to the verb, 36, 184, 240, 241.
compound adjectives, 23, 231-6, 260.
compound nouns, 23, 43, 231-8.

compound verbs, 191-2.
concessive clauses, 186, 187, 269.
conditional, 99-103, 106, 107, 109-10, 113-14, 119-20, 124, 125, 127, 128-9, **130-2**, 136.
conditional sentences, 24, 112, 257, **267-73**, 275.
consonants, 2-12.
  alternation of, 10-12, 19, 31, 96, 99, 108, 113.
  assimilation of, 12.
  clusters, 9-10.
  doubled, 8-9, 11-12, 33.
  non-final, 3.
  non-initial, 2-3, 20.
  unvoiced, 12.
  voicing of, 12.
converbs, 174.
co-operative verb, 143-4.
culinary terms, 48.
cursing, 115.

-ç, 222.
çala, 174-5.
çevre, 90, 91.
çeyrek, 82, 202.
çıkagelmek, 191.
-çin, 87.
çoğu, 75.
çok, 13, 75, 197.
çoktan, 201.
-çün, 17, 87.
çünki, çünkü, 215.

daha, 54, 203-4.
dahi, 207.
dahil, 90, 91.
dair, 87-88.
dara dar, 195.
-daş, 64.
dates, 201.
dative, 28-29, 36-37, 87-89, 147, 157, 189, 195, 198-9, 248-9.
de, 24, 178-9, 197, **206-7**, 235, 268, 269.
-de (loc.), 12, 28-29.
-de- (denominal verb), 231.
decimals, 79.
değil, 103-5, 106, 166.
değil mi (ki), 211.
değin, 87-88, 180.
dek, 88[1].
-dek, 87-88, 180.

demek, 204, 215-16.
demek ki, 216.
demonstratives, 35, 43, 71-72.
-den (abl.), 28-29.
-dense, 188.
denominal verbs, 227-31.
derken, 204-5.
destur, desturun(uz), 137.
determinatives, 74-77.
deverbal substantives, 220-6.
devrik cümle, 159, 240, **241-5**.
devşirme, 171[1].
dış, 90, 91, 92, 198.
dışarda, dışardan, 199.
dışarı, 198-9.
di, 138, 139.
-di, 99, 127, 169, 181.
dice numbers, 84.
-di gitti, 192.
diğer, 5, 76.
diğer taraftan, 216.
-diği, 163-6, 184-7, 254-6.
-diği gibi, 187.
-diği halde, 186.
-diği için, 186, 187.
-diği kadar, 186-7.
-diği müddetçe, 185-6.
-diğince, 183.
-diğinden, 186.
-diğinden başka, 184.
-diğine göre, 165.
-diği nispette, 186.
-diği sürece, 185-6.
-diği takdirde, 186, 267.
-diği vakit, zaman, 185.
-dik (first pl.), 99, 127.
 (participle), 162-3, 183.
-dikçe, 183, 186.
-dikleri, 163-4, 246-7.
-dikse, 101, 129.
-dikte, 183.
-dikten başka, 184.
-dikten sonra, 183-4.
-dikti, 128.
dilek-şart kipi, 271.
dilenci, 220.
-diler, 99, 127.
-dilerdi, 128.
-dim, 99, 127.
-dimdi, 128.
-di -medi, 182.
-di mi (conditional) 267.
diminutives, 23, 57-58, 200, 203.

-dimse, 100-1, 129.
-din, 99, 127.
-dindi, 128.
-diniz, 99, 127.
-dinizdi, 128.
-dinizse, 101, 129.
-dinse, 101, 129.
di- past tense, 107, 108, 122, **127-30**.
diphthongs, 10.
-dir, 96-98.
 after verb, 109, 139-40.
 with following verb, 141, 192.
-dir- (causative), 144.
-dir gider, 141, 192.
-dirler, 96.
-dirt-, 146.
distributive numerals, 83-84, 193-4, 253.
-diydi, 128.
diye, 139, 174, **175**.
-diyse, 101, 129, 131.
doğru, 87-88, 193.
doğrudan doğruya, 195.
dolayı, 89.
doublets, 236-8.
dönme, 171[2].
dual, 10, 28.
durmak, 191.
durmamacasına, durmamasına, 189.
durmamasıya, 181, 189.
durundu, 138.
durup dururken, 179.
düzeltme işareti, 2.

e (interj.), 271.
-e (dat.), 28-29.
-e- (denominal verb), 227.
-e (gerund), 174-6, 177, 191, 192.
-e (subj.), 107, 132.
-ebil-, 151-2.
-eceği, 163-6, 254-5.
-eceği gel-, 165.
-eceği gibi, 187.
-eceği kadar, 186-7.
-eceği tut-, 166.
-eceğim, 113, 163-6.
-eceğimiz, eceğin(iz), 163-6.
-eceğine, 187, 189.
-eceğiz, 113.
-ecek, 112, 158-61.
-ecek kadar, 160.
-ecekleri, 163-6.

-ecekti, 113.
-ecek yerde, 187.
ede, 176.
-e dur-, 191.
-egel-, 151, 191, 192.
egemen, 221.
-egör-, 191.
-eğit-, 192.
-eğen, 223.
eğer, 5, 270.
-eh, 133.
-ek (deverbal), 223.
-ek (subj.), 133.
-ekal-, 191.
-ekle-, 148.
ekmek, 170.
-e ko(y)-, 191.
el, 14, 81.
-el (adj.), 53, 65.
-el- (denominal verb), 229.
-ele-, 148.
-eli, -eliden, 181.
-elim, 107, 132.
-elt-, 229.
-em, 133.
-eme-, 151, 189.
-ememe, -emememe, 171.
-emez, 151.
-emi-, *see* -eme-.
-emiyebil-, 151-2.
-emiyeceğine, 189.
emphatic pronouns, 70-71.
emzirmek, 146.
en, 55, 197.
-en (adv.), 196.
-en (participle), 158, 180.
enclitics, 23-24.
-ene değin, -dek, kadar, 180.
-en, -ene, 248-9.
English borrowings, xxi-xxii, 9, 10-11, 17, 22, 154, 221, 232.
epenthetic vowels, 9-10, 19, 58.
epey, 56.
-er (aorist), 116.
-er- (causative), 145.
-er- (denominal verb), 229-30.
-er (distributive), 83-84, 193-4.
erat, 28.
-erek, 24, 174, 176-7.
-erekten, 177¹.
erenler, 26.
ermek, 96, 165, 190.
-ert-, 146.

INDEX

ertesi, 202.
-esi, 115, 161, 165, 180, 259.
-esice, 161.
-esi gel-, 165.
-esin, -esiniz, 107, 132.
-esiye, 180-1.
-esiye kadar, 181.
eskiden, 201.
esnasında, 94.
etmek, 151, 154-6.
etraf, 90, 91.
-ev, 226.
-evüz, 133.
evvel, 89, 181-2, 195, 203.
evvelâ, 196.
evveline kadar, 203.
evvelki, evvelsi, 202-3.
extended stem, 143, 153.
-ey, 226.
-eyaz-, 191.
-eydi, 134.
-eyem, 132.
-eyim, 107, 132.
-eyin, 132.
eylemek, 154, 155.
-eymiş, 135.
-eyn, 28.
-eyor, 108.

factitive verb, 36.
fakat, 210.
falan, falanca, 77, 238.
falan festekiz, feşmekân, 77.
fevk, 20, 90, 92.
filân, filânca, 77, 238.
filânıncı, 82.
fractions, 81-82.
French borrowings, xxi-xxii, 6-7, 9, 11-12, 17, 19-20, 25, 61, 65, 154, 221.
future I tense, 112-15, 136.
  participle, 158-61.
  conditional use of, 255, 270-1.
  depreciatory use of, 160-1, 190.
  quasi-passive use of, 159-60.
future II tense, 115.
  participle, 161, 165, 180, 259.

galiba, 4.
gayrı, 203.
gayri, 89.
-geç, 225.
geçe, 174, 202.
geçmiş ola, olsun, 134.

INDEX

gelince, 180.
-gen, 12, **223**.
gender, 25, 51.
gene, 204.
geniş zaman, 115–16.
genitive, 18, 28–29, 36, 85–86.
  logical subject in, 142, 250–2.
  predicative, 36, 68–69.
gerçi, 216.
gerek, 127, 273.
gerek . . . gerek(se) . . ., 209.
gerekmek, 127, 227.
geri, 198–9.
German borrowings, 9, 44.
gerund-equivalents, 184–7.
gerunds, 174–92.
gh, 5.
getirmek, 146.
-gi, 222.
gibi, 24, **85–86**, 187, 189.
-giç, 225.
-gil (imperative), 137.
-gil (noun), 65.
-gin, 223–4.
gittikçe, 183.
glottal stop, 7–8.
göre, 87–88, 165, 174.
göstermek, 146.
göya, 216.
Greek borrowings, 17, 22, 28, 221.
gûya, 216, 284.
güle güle, 176.
günbegün, 196.
gündem, 225.
günden güne, 195.
günü gününe, 195.
güzelim, 41.

ḥ, 5².
ha, 218.
ha . . . ha . . ., 209, 252².
hakkında, 93.
hâlâ, 196, **203**.
hâlâ daha, 203.
halbuki, 215.
halde, 186, 194.
hamza, 8, 14, 15, 33–34.
-hane, 6, **66**.
hangi, 72, **74**.
hanım, 6, 25, 232, 252.
hanım hanımcık, 193.
hani (ya), 216.
hanidir, 217.

hani yok mu, 216.
hariç, 90, 91.
hattâ, 197–8.
haydi, 23, 137, 139.
haydiniz, 137.
hele, 217.
hele hele, 217.
hem(de), 209.
hemen, hemencecik, hemencek,
  hemen hemen, 204.
hem . . . hem (de) . . ., 209.
henüz, 203.
hep, 75.
hepsi, 75.
her, 75, 164–5, 183, 269.
herhalde, 217.
hiç, 77–78.
hiçbir, 78.
husus, 93.
hyphen, 12¹, 52, 235, 281.

ıy, 4, 15.

-i, *see* third-person suffix.
-i (acc.), 28–29.
-i, -î (adj.), 2, 53, 65.
-i (deverbal substantive), 221.
-i (diminutive of personal names), 58.
icabetmek, 127.
-icek, 180.
-ici, 220.
iç, 90, 91, 92, 198.
-içe, 25.
içerde, içerden, 199.
içeri, 89, 198–9.
için, 24, 85–87, 139, 167.
için için, 193.
içre, 85.
içün, 87.
idi, idik, idiler, idim, 99.
idimse, 100.
idin, idiniz, 99.
idiyse, 100–1, 114, 124, 129.
idüğü, 165.
-ih, -ik (for -iz), 118.
-ik (deverbal substantive), 221.
iken, 142, **190–1**.
-il-, 149–51.
ilâ, 95.
ile, 85–87, 90, 95, 206.
ilen, 87.
ilerde, ilerden, 199.
ileri, 198–9.

INDEX

-im ('I am'), 96.
-im ('my'), 39.
-im (deverbal noun), 193, 224-5.
imdi, 138.
-imdir, 139-40.
'imek', 96.
imiş, 101, 102, 122, 123, 124, 284.
imişin, imişiniz, 102.
-imişse, 102-3.
-imiz, 39.
imperative, 107-8, 132-3, 134, **137-9**, 271-2.
diffident first-person, 271.
in protasis, 268, 275.
impotential verb, 151-2.
-imse-, 230-1.
-imsi, 53, 58, 231.
-imtrak, 53, 58.
-in (acc. with third-person suffix), 41.
-in (Arabic pl.), 26-27.
-in (deverbal noun), 225.
-in (gen.), 18, 22, 28-29.
-in (imperative), 107, 137-8.
-in, *see* instrumental.
-in ('your'), 39.
-in- (reflexive), 149.
(passive), 149-50.
inat, 88.
-ince (adv.), 183, 194.
-ince (gerund), 23, 179-80.
-inceye değin, -dek, kadar, 180, 181.
-inci, 82.
-inde, -inden, 40-41.
indefinite article, 53-54, 75, 248.
indefinite pronouns, 74-78.
indirect command, 170-1, 254-5.
indirect question, 254-5.
indirect statement, 254.
-ine, 40-41.
inen, 87.
inferential, 101-3, 110, 114, 120, 122-3, 124-5, 126, 128, 131, 135, 140, 211, 218, 284-5.
infinitive, 85, 87, 93, 96, 111, 167-9, 218.
with subject, 169, 187.
-ini, -inin, 40-41.
-inil-, 150.
-iniz (imperative), 107, **137-8**.
-iniz ('your'), 39.
İnönü, 233.
insan, 77.

instrumental, 22, 23, 173, 195, 201.
intensive adjectives, 55-56.
interjections, 8, 24, 137, 138, 218, 271.
interrogative particle, 24, **105**.
interrogatives, 72-74.
-inti, 222.
intonation, 24.
inverted sentences, 241-5.
-ip, 177-9, 191, 206.
-ip de, 178-9.
-ip dur-, 191.
-ir (aorist), 116.
-ir- (causative), 145.
-irt-, 146.
ise, 99-100, 217, 251.
istemezük, 121.
ister istemez, 182.
ister . . . ister . . ., 209.
-iş (diminutive of personal names), 58.
-iş (verbal noun), 167, 172-3.
-iş- (reciprocal), 143-4.
işbu, 71.
işde, 12.
-işli, 173.
işte, 12, 217.
-iştir-, 148, 149.
-it (deverbal noun), 224.
-it- (causative), 145.
Italian borrowings, 9, 17, 22, 156, 196.
İtalyalı, İtalyan, 45.
itibaren, 89.
-ittir-, 146.
-iver-, 151, **191-2**, 289.
iy, 15.
-iyor, 23, **108-9**, 136.
-iyordur, 140.
-iz (collective), 84.
-iz ('we are'), 96.
izafet, 8[1], 24, **41-48**, 85, 86, 90, 91, 93, 167, 232-3, 242-3, 246-7, 257.
frozen, 49, 233.
of material, 42.
Persian, 50-52, 68, 74, 82, 188, 190, 279.
-izdir, 139-40.
-izli, 84.

Janus construction, 48-49.
-k (deverbal substantive), 223.
-k (Type II ending), 107.
kaba Türkçe, xxi.
kaç, 72, 74.
kaçar, 83, 84.

INDEX

kaçıncı, 82.
kadar, 85–86, 87–88, 160, 180, 186–7, 191, 203.
kala, 174, 176, 202.
kaldırmak, 146.
kardeş, 17, 64.
kare, 53.
karşı, 87–88, 90, 91, 198–9.
karşın, 88.
kazaen, kazara, 196.
-ken (adj.), 12, **223**.
-ken (gerund), 23, 142, 143, **190–1**, 204–5, 288.
-kendenberi, 191.
kendi, 42, 68, **70–71**, 150.
-kenki, 251.
kesme işareti, 2.
keşke, keşki, 134, 271.
kılmak, 154–5.
kırk, 79, 80.
kıyısıra, 92.
ki, 24, **211–14**, 260, 264–5, 275.
-ki, 68, **69–70**, 72, 248¹, 251, 252¹.
kim, 72, 85–86.
kimi, 74, 75.
kimse, kimsecik, 78.
-kle-, 148.
kolay gele, gelsin, 134.
kuruş, 9.
-kü, 17, 69.
kültürel, 65¹.
küp, 53.

-l (adj.), 65.
-l- (denominal verb), 229.
lâkin, 210.
language reform, xx–xxi, 64, 220–32 *passim*, 242. *See also* neologisms.
lâzım, 127.
-le ('with'), 23, 86–87.
-le- (denominal verb), 227–8, 231.
leh, 94–95.
-len-, 228.
-ler (pl.), 18, 25–26, 29, 85, 107, 248, (also aorist of -le-).
-leri (acc. pl.), 29, 39.
-leri ('their'), 39.
-leş-, 228–9.
-leştir-, 229.
-let-, 228.
-leyin, 23, 201, 205.
-li, 44, 45, 50, **60–61**, 84, 125, 169, 172, 173, 200, 234, 260.

-lik, 62–64, 170, 241, 256¹.
. . . -li . . . -li, 61.
locative, 12, 28–29, 36, **37**, 64, 81, 94, 95, 111.
-lt-, 229.

m-, 237–8, 285.
-m ('my'), 39.
-m (Type II ending), 106.
-m (deverbal noun), 225.
maada, 89.
ma(a)mafih, 211.
madem(ki), 211.
-man, 220.
-masyon, 172.
m-doublets, 237–8, 285.
-me (vbl. noun), 57–58, 112, 125, 127, 155, 167, 168, **170–2**, 249, 254–6.
-me- (neg.), 23, **110**, 114, 115, 125, 126–7, 129, 131, 135, 138, 143, 158, 162–3, 170, 171, 177, 178, 180, 181.
-mece (noun), 57–8.
-mecesine, 189.
-mede, 112.
-meden, 170¹, 181–2, 183–4, 188.
-medense, 188.
-medeyim, 112, 168.
-medik, 129, 162–3.
-meğe, 168.
meğer, 5, 211.
meğerki, 214.
meğerleyim, meğerse, 211.
-meği, 168.
-meğin, 173.
Mehemmed, Mehmet, 6, 83.
-mek, 85, 93, 96, 111, 167–70, 187–8, 218.
-mek için, 167.
-mekle, 187.
-mekle beraber, birlikte, 187.
-mekli, 169.
-meklik, 170, 286.
-meksizin, 188.
-mekte, 111–12, 168.
-mektedir, 111–12.
-mekten, 168, 182.
-mekten ise, 188.
-mektense, 168, 188.
-mek üzere, 167–8.
-meli, 125–7, 172.
-mem, 121, 170.
-mememe, 171.
-memezlik, 173.

INDEX

-men, 220-1, 286.
merhum, 53.
meselâ, 196.
-mesine, 249.
-mesin mi, 138.
-meti(n), 182¹.
-meye (neg. subj.), 135.
-meye (dat. of verbal noun), 168.
-meyi, 168.
-meyiz, 121.
-mez, 120-1.
-mezden, 181-2.
-mezlik, 173.
-mezük, 121.
mi, 3, 24, 105, 182, 235-6, 267.
midir, 105.
mimli, 237.
'mink, schmink', 237¹.
-misin, misiniz, 105.
-miş, see inferential, miş-past.
-mişçesine, 188-9.
-mişin, -mişiniz, 102, 123.
-mişlik, 256¹, 289.
-mişmiş, 124-5.
miş-past, 107, 122-5, 136, 189.
  participle, 162.
-miştir, 122-3, 169.
-miydi, 105-6.
-miye, see -meye.
-miyebil-, 151-2.
-miyecek, 114-15.
-miyogör-, 191, 285.
-miyeli, 181.
-miyesi, 115.
miyim, miyiz, 105.
miymiş, 106.
-miyor, 110-11.
-miz, 39.
-mse-, 230-1.
-msi, 53, 58.
-mtrak, 53, 58.
Muhammed, 6.
mukabil, 89.
Mustafa Bey, 3.
müdür, 17, 105.
mütaakıp, 85.

-n (Type II ending), 106.
-n ('your'), 39.
-n- (before case-endings), 29, 40-41.
  See also pronominal n.
-n- (passive), 149-50.
  (reflexive), 149.

namına, 93, 175.
nasıl, 74, 255.
  with conditional, 269.
nasıl ki, 215.
nasıl olup da, 257.
nazaran, 87-88.
-nce, 194.
-nci, 82.
ne, 3, 59, 72-74, 195.
nece, 74, 195.
necessitative, 107, 125-7, 136.
neci, 59, 74.
negative pronouns, 74, 77-78.
negative suffix, see -me-.
ne idüğü belirsiz, 165.
-nen, 87.
ne . . . ne (de) . . ., 207-9.
neologisms, 64, 65, 66, 147, 148, 152, 195, 220-32 *passim*. *See also* language reform.
nerde, nerden, 199.
nerdeyse, neredeyse, 200.
nere, 199-200.
. . . nerede, . . . nerede, 200.
netekim, 214-15.
ne var ki, 210.
ne yapıp yapıp, 179.
nice, 74.
niçin, 74.
-nil-, 150.
nitekim, 212, 214-15.
-nız (Type II ending), 107.
-niz ('your'), 39.
nominal sentences, 239.
noun clauses, 97, 212-13, 254-6.
  subjunctive in, 265-6.
nouns, 25-52.
  of nationality, 44-45.
  of place, 22, 199-200.
  of unity, 224-5.
number, 25-28, 44, 97, 246-8.
numerals, 26, 54, 63-64, 79-84, 95, 253.

o (pronoun), 25, 67-69, 71-72, 97.
-o (diminutive of personal names), 58.
o bir, 16, 76.
object omitted, 68.
-oğlu, 43, 233.
oğul, 10, 32.
ol (imperative), 142.
ol (pronoun), 67.
olacak, ironic use of, 160, 190.

INDEX

olarak, 177, 194.
oldukça, 183.
oldum olası(ya), 181.
oldu olmadı, 181.
olmak, 23, 117-18, **141-2**, 143, 156, 166.
olmak üzere, 167-8.
olmasına, 249¹.
olmaz, 118.
olsa gerek, 273.
olsa olsa, 272-3.
olsun, 142, 210.
... olsun ... olsun, 209.
olunmak, 150-1, 155-6.
olup bitmek, 179.
olur, 23, 117-18.
olur olmaz, 182.
onomatopoeia, 193, 228, 231, 235.
optative, 132.
ora, 199-200.
orda, ordan, 199.
ordinal numerals, 82-83.
   Arabic, 82-83.
orta, 90, 91.
ortaklaşa, 174, 175.
o saat, 8, 201.
-oş (diminutive of personal names), 58.
otomatikman, 196.
Ottomanisms, 17, 26-28, 41, 50-52, 66, 67, 68, 81, 82-83, 84, 85, 87, 154-5, 165, 169, 173, 179, 180, 181, 187, 188, 191, 205, 206, 241-2, 244-5, 252-3, 277.
oxytone, 21.
oysa(ki), 215, 250.

öbür, 46, 76.
öğrenci, 220.
öğretmen, 221, 286.
ön, 90, 91, 92, 232.
önce, 89, 181-2, 201.
önceden, 201.
örneğin, 195.
öte, 198-9.
öteki, 72.
öte yandan, 216.
ötürü, 89.
öyle, öylesi, 72, 195, 215.
öylece, 195.
öyle gibime geliyor ki, 86.
özbeöz, 196.
öztürkçe, xxi.

palatalization, 3-4, 7.
parenthetic remarks, 213.
participles, 36, 87, 142, **158-66**.
passive, 149-51, 155.
   used impersonally, 150-1, 159-60, 175, 254-5, 262.
pek, 197.
percentages, 81.
periphrastic tenses and moods, 166.
Persian borrowings, xx-xxi, 3-4, 5, 6, 11, 14, 19, 25¹, 55, 66, 77, 84, 196, 211, 215, 270, 271.
   izafet, 50-52, 68, 74, 82, 188, 190, 279.
   plurals, 28.
personal names, 6, 26, 35, 43, 65, 233.
   diminutives of, 57, 58.
personal participles, 163-6, 183, 184, 185, 254-6, 261-3.
personal pronouns, 25, 35, 38-41, **67-69**, 71, 85-86.
personal suffixes, 38-41, 69, 74.
peş, 90, 91, 92.
place-names, 21-22, 35, 47, 233.
playing-card numbers, 84.
pluperfect, 123-4, 128, 136.
plural, 25-28, 54.
   Arabic, 26-27.
   honorific, 26, 27¹, 247.
   idiomatic uses of, 247-8.
   subject in, with sing. verb, 97, 246.
   Persian, 28.
popular pronunciation, 4, 5-6, 7, 11, 14, 32-33, 211.
possession, grammatical, 42².
postpositional expressions, 85.
postpositions, 24, 36, **85-95**, 97.
   primary, 85-89.
      with abl., 38, 89, 95.
      with abs., 85.
      with abs. or gen., 85-87.
      with dat., 37, 87-89.
   secondary, 85, 89-95.
potential verb, 151-2.
prefixation, 55. *See also* m-doublets, 'pseudo-'.
preposition, 95.
present I tense, 108-11, 112, 118, 136.
   participle, 158.
present II tense, 111-12.
   participle, 158.
pronominal n, 40-41, 67-68, 70, 72, 194.

INDEX

pronouns, 67–78.
protasis, 267–9.
provincialisms, 57, 65, 115, 133, 173, 177¹, 183, 203, 211, 242.
'pseudo-', 281.
pseudo-causative, -passive, 152.
punctuation, 79, 210, 250–1, 276–7.
purpose, 139, 167, 168, 175, 264.
-r, 116.
rağmen, 88.
rasgele, 134, 174.
raşidinler, 27¹.
-re, 199–200.
reciprocal verb, 143–4, 247–8.
reflexive pronouns, 70–71.
reflexive verb, 149.
-rek, 55, 57.
relative clauses, translation of, 260–3.
relative participle, 163¹.
repetitions, 235–6.
repetitive verb, 148.
-rih, -rik, 118.
-r -mez, 182.
Roman numerals, 82.
q, 4, 19.
saatbesaat, 196.
sahi, 6, 22.
sakın (ha), 208, 218.
sanki, 189, 218.
sapa, 174, 175.
sayesinde, 94.
-se, see conditional.
-se- (denominal verb), 230.
-se beğenirsin(iz), 272.
-se gerek, 273.
-sek, 100, 130.
-sel, 53, 65.
-seler, 100, 130.
-sem, 100, 130.
-se mi, 132, 271.
sen, 67–69.
-sen, 100, 130.
-sene (ya), 271–2.
seni (with terms of abuse), 68.
-seniz, 100, 130.
-senize (ya), 271–2.
sentence-analysis, 278–82.
sentence-plus, 244–5.
Serbo-Croat borrowings, 17, 25.
set, 12.
sevda, 222¹.
-seydi, 131.

sıra, 92, 185, 201.
-si, see third-person suffix.
-si (adj.), 53, 58.
-sin ('you are'), 96, 106, 137.
-sin (imperative), 107, **137–9**.
-sindi, 138–9, 289.
singular, 25–26, 44.
 verb in, with pl. subject, 97, 246.
-sin için, 139, 288.
-siniz, 96.
-sinizdir, 139–40.
-siyle, 87.
siz, 67–69, 182.
-siz, 62, 169, 182.
-sizin, 169, 188.
-sizler, 68.
solecisms, 47, 49³, 51, 60, 62, 77, 86, 95, 196, 287–8.
sonra, 48, 89, 201, 203, 233.
sonradan, sonraları, 201.
spelling, fashions in, 12, 15, 76, 168, 170, 203, 206, 211.
 rules of, 7, 20, 88, 135, 215.
su, 29, 34, 40, 230.
subjunctive, 107, 132–5, 136.
 after negative main verb, 264–5.
 in noun clauses, 265–6.
 in unfulfilled conditions, 275.
 with derken, 204–5.
 with ki, 264–5.
 with meğerki, 214.
substantival sentence, 256–8.
substantive, xxii.
suffixes, order of, 29, 40, 125, 152–3.
 consonant assimilation in, 12.
 vowel harmony in, 17–20, 29–34, 87.
Sultan, 252–3.
surette, 194.
suspended affixation, 35, 41, 43, 108, 134–5, 189.
synopsis of verb, 135–7.

-ş (diminutive of personal names), 58.
-ş- (reciprocal), 143–4.
şayet, 270, 283.
şehir, 9–10.
-şer, 83, 193–4.
şey, 27, 34, 77, 243–4.
şeyisi, şeysi, 49, 77.
şimcik, şimden(geri), şimdi, şimdicik, şimdik, 203.
şimdiden, 201, 203.

INDEX

-şin, -şiniz, 102, 123.
şol, 71.
şöyle, 72, 195.
şöylece, 195.
şöyle dursun, 218.
şöylesi, 72.
şu, 71–72.
şura, 199–200.
şurda, şurdan, 199.

-t (deverbal noun), 224.
-t- (causative), 145.
takdirde, 186, 267.
tā marbūṭa, 8, 65.
tane, 80.
tarafından, 93, 194.
tavla, 84.
-tay, 226.
teker teker, 193–4.
third-person suffix, 39, 246–7, 254–5, 262.
    doubled, 49, 74, 77, 202.
    substantivizing force of, 48, 72, 74, 288.
    vocative use of, 50.
tıpkı, tıpkısı, 77.
-ti (deverbal noun), 222.
-ti (verb), *see* -di.
time, expressions of, 184–6, 204, 275.
time, telling the, 174, **202.**
titles, 35, 65, 252–3.
-ttir-, 146.
turur, 96.
tümleç, 240.
Türk Dil Kurumu, xxi, 46, 226.
Türk iş, 44.
Türkmen, 220[1].
Types I–IV (personal suffixes of verb), 106–7.

u ('and'), 206.
uğruna, uğrunda, 94.
umak, 151.
uydurmasyon, 172.

ü ('and'), 169, 206.
üç beş, 79–80.
üsde, 12.
üst, 90, 91–92, 201, 232, 233.
üzer-, 90, 91–92, 201.
üzere, üzre, 85, 167–8.
üzerine, 88.

-v, 226.
vakit, 19, 33–34, 185.
var, 142–3, 241, 251–2.
-varî, 53, **66.**
varsin . . . (me)sin, 134, 288, 289.
ve, 186, **206.**
verb, 96–157.
    synopsis, 135–7.
verbal nouns, 142, 167–73.
    Arabic, 154–6, 176.
    passive, 171, 172.
verbal sentences, 239.
verb-stem, 11, **96,** 227.
verbs with dative, 37, 157, 165, 189.
verb 'to be', 23, **96–106,** 108, 109, 113, 118, 239.
    in temporal expressions, 204, 275.
    summary, 141–2.
veresi, veresiye, 181.
veya, veyahut, 210.
vocative, 24, 35, 50.
vowel harmony, 15–20.
    exceptions, 17, 69, 87.
    in foreign borrowings, 17, 19–20, 31–32.
    of suffixes, 17–20, 29–34, 87.
vowels, 12–20.
    alternation of, 20, 108, 110.
    classification of, 13.
    doubled, 4, 12, 15.
    epenthetic, 9–10, 19, 58.
    length of, 12, 14–15.
vu, vü, 206.

word-formation, 220–38.
word-order, 51, 54, 71, 158–9, 239–45.

-y (substantive), 226.
-y- (buffer-letter), 20, 29, 96, 112.
    narrowing effect of, 20, 112, *et passim.*
ya, 210, 216, 218–19, 271–2.
yada, 210.
yahut, 210.
yalnız, 211.
yan, 90, 92.
yana, 89.
yapiim, yapim, 133.
yapmak, 133, 155.
yâr, 19.
yarı, 81.
yarı buçuk, 81.

INDEX

yarım, 81, 83.
yarımşar, 83.
ya . . . ya . . . ya da . . ., 210.
ya . . . ya . . . veya . . ., 210.
ya . . . ya . . . yahut . . ., 210.
-yçin, yçün, 87.
-ydi, *see* idi.
yeğ, yeğrek, 55, 169, 188.
yer ('place'), 94.
(also aorist of ye-).
yerde, 187.
yerine, 94.
yılbeyıl, 196.
yine, 204.
-yken, 190.
-yle, *see* ile.
-ymiş, *see* imiş.
yok, 13, 65, **142–3**, 219, 241, 251–2.

yoksa, 219.
-yor, 23, **108–9**, 136.
yorımak, 108.
-yse, *see* ise.
yukarda, yukardan, 199.
yukarı, 198–9.
yumuşak ğe, 1, 4–5, 10–11, 113.
yüz, 79, 80, 92–93.
yüzbaşı, 49, 233.
yüzünden, 93.

-z, 84.
zaman, 185, 200.
zaten, 112.
Zat-ı aliniz, alileri, 68.
zira, 215.
ziyade, 197.
-zli, 84.